Y0-DSK-531

JOURNEY THROUGH THE UNIVERSE

THOMAS L. SWIHART
UNIVERSITY OF ARIZONA

JOURNEY THROUGH THE UNIVERSE

AN INTRODUCTION TO ASTRONOMY

HOUGHTON MIFFLIN COMPANY BOSTON

DALLAS GENEVA, ILLINOIS HOPEWELL, NEW JERSEY
PALO ALTO LONDON

To the women in my life: Merna, Louise, and Chim

Cover photograph: Comet Ikeya-Seki, photographed by J. B. Irwin, from Cerro Morado, Chile, October 1965.

PART AND CHAPTER OPENING ACKNOWLEDGMENTS

Part One: Lick Observatory photograph; **Chapter 1:** Lick Observatory photograph; **Chapter 2:** Lockheed Solar Observatory, Lockheed Palo Alto Research Laboratory; **Chapter 3:** The Kitt Peak National Observatory; **Chapter 4:** The Hale Observatories;

Part Two: Lunar and Planetary Laboratory, University of Arizona (copyright pending); **Chapter 5:** From Arthur Berry, *A Short History of Astronomy* © 1961, Dover Publications, Inc. 1961; **Chapter 6:** Robert G. Strom; **Chapter 7:** Lick Observatory photograph; **Chapter 8:** NASA; **Chapter 9:** Lick Observatory photograph; **Chapter 10:** Lick Observatory photograph;

Part Three: The Cerro Tololo Inter-American Observatory; **Chapter 11:** Sacramento Peak Observatory, Association of Universities for Research in Astronomy, Inc.; **Chapter 12:** Lick Observatory photograph; **Chapter 13:** Steward Observatory; **Chapter 14:** The Cerro Tololo Inter-American Observatory; **Chapter 15:** Lick Observatory photograph; **Chapter 16:** Lick Observatory photograph;

Part Four: Lick Observatory photograph; **Chapter 17:** Lick Observatory photograph; **Chapter 18:** The Kitt Peak National Observatory; **Chapter 19:** The Hale Observatories; **Chapter 20:** Lick Observatory photograph.

Printed in the U.S.A.
Library of Congress Catalog Card Number: 77-76343
ISBN: 0-395-25518-X

CONTENTS

PREFACE

This book is intended for anyone who is interested in astronomy. As a textbook, it is intended for a one-semester or one-term course in astronomy for the nonscience major. The purpose is to present an interesting and fairly complete story of astronomy without burying the reader under an avalanche of detail, jargon, or erudition.

Math is used only sparingly, and then it is of the grocery store variety: If eggs cost 10¢ each, how much for six of them? Although it is true that most students get algebra and geometry in high school, many do not acquire a working knowledge of them; I believe that an astronomy course should teach astronomy, not algebra and geometry.

Many scientists and teachers seem to think that science without math is necessarily limited to lists of facts and figures, and that astronomy without equations is only a thin shadow of its real self. I hope this book is at least a partial refutation of these beliefs. My goal is to get the reader to understand the physical principles behind phenomena, and at the desired level. The ability to go through a formal calculation that is poorly understood is not a part of the desired end.

In this book we study objects such as planets, stars, galaxies, and the Universe. By study I do not mean we simply list their main properties and see how these properties are determined. I mean we try to see how the objects are put together and how they may be related to each other; then we try to understand how they could have gotten that way. Does our knowledge of the laws of nature give us any clues on how the objects might have formed? How did they evolve to their present condition? What does the future hold? We cannot answer all of these questions, but we can see how far astronomers have come in trying to answer them and what the main frontiers of research are today.

Units are a problem. The metric system is well on its way in the United States, and there is no doubt that it is simpler than the hodge podge now in common use; however, I use numbers only infrequently, and when I do it is usually to impress and interest readers. Under these conditions, familiar units are called for. When kilograms, meters, liters, and degrees Celsius become as commonly used as pounds, inches, teaspoonfuls, and degrees Fahrenheit, then I will gladly make the change.

This book is not intended to be an encyclopedia of astronomy that answers all the questions that anyone might think of. There are many areas where the coverage could have been more complete, but completeness for its own sake is not my object. This is naturally a matter of judgment, and only those topics have been included which I believe deserve to be here because of their general interest and their importance.

To a scientist, precision is a very desirable attribute. In works intended for the layman, however, it is often the enemy of clarity. Must our definitions always be rigorous? Must every statement be accompanied by a declaration of its precise limits of validity? I believe there are many areas where precise knowledge is not a necessary part of understanding, and in such cases I do not hesitate to avoid it.

A number of student aids are given at the end of each chapter. There are word lists containing important terms that might be new to the reader. Simple review questions can test the reader's recollection of important points from the chapter. The much more difficult discussion questions are intended to stimulate thought beyond the regular coverage of the chapter. These could be profitably brought up for class discussion. Finally, references are given which the reader might find useful for looking further into the topics of the chapter.

Many persons helped with many parts of the book. I wish to particularly thank H. A. Beebe, B. J. Bok, W. J. Cocke, E. R. Craine, R. H. Cromwell, F. K. Edmondson, R. L. Faller, J. W. Fountain, W. K. Hartmann, J. W. Harvey, A. A. Hoag, W. B. Hubbard, S. M. Larson, R. C. Leonard, W. C. Livingston, M. S. Matthews, C. C. McCarthy, D. W. McCarthy, D. McDonald, E. W. Miller, R. B. Minton, D. L. Moore, O. R. Norton, A. G. Pacholczyk, W. L. Reitmeyer, C. S. Rodriguez, E. Roemer, E. H. Rogers, R. Schiff, N. R. Sheeley, M. K. Stein, J. W. Strebeck, P. A. Strittmatter, R. G. Strom, E. A. Whitaker, R. E. White, and R. E. Williams. T. R. McDonough and the *Sky Interpretation Bulletin*, Vol. 1, 1976, provided the idea of the cartoon on page 10.

I am grateful to J. S. Hill, Michigan State University; Morris Davis, University of North Carolina, Chapel Hill; Michael Zeilik II, University of New Mexico; John D. Eggert, Daytona Beach Junior College; Frederick R. Hickok, Catonsville Community College; Darrell Hoff, University of Northern Iowa; C. T. Daub, San Diego State University; and Howard E. Bond, Louisiana State University, for their review of the manuscript.

Gail, David, and Jennifer gave me much encouragement. Finally, I wish to thank my wife Merna for assisting in so many ways, including typing the manuscript.

T.L.S.

JOURNEY THROUGH THE UNIVERSE

PART ONE
INTRODUCTION TO ASTRONOMY

The Introduction contains background material useful for reading subsequent parts of the book. Chapter 1 presents science and gives a brief summary of the Universe and its contents. Chapters 2 and 3 describe some of the more important properties of matter and radiation, the ultimate building blocks of the Universe. Instruments used by astronomers to collect data about the Universe are covered in Chapter 4.

ONE
THE UNIVERSE

Everyone knows that astronomy is the scientific study of the Universe, but what does the term "scientific" really mean? How might the main properties of the Universe be summarized, and how has man's understanding of the nature of the Universe changed over recorded history?

1.1 Science and the Universe

THE UNIVERSE AND SCIENTIFIC PRINCIPLES

Astronomy is concerned with the Universe, so it should be made clear just what the Universe is. There are a number of ways in which the term is often used, but to astronomers the Universe is the collection of all things that we can get our hands on, scientifically speaking. The Universe contains all the space, matter, radiation, and energy whose existence we can detect, reach, feel, see, or otherwise measure using scientific principles. It also contains all of the things that we *could* detect, and so on, if we were smart enough, had powerful enough measuring instruments, and had good enough rocket ships with enough time to use them. In other words, a star is not going to be left out of the Universe just because it hasn't been found yet or because it will never be found. The only question is whether or not there is a scientific reason for believing that it could not be detected in any way, even in principle. The important qualification in this is that detection must be made according to scientific principles. This requires further discussion.

When someone confidently tells us that such and such has been scientifically proved, he is trying to convince us that there is no nonsense about it; it has been accurately and impartially determined. If we care to check it down at our neighborhood scientific laboratory, we would find that he was correct in all particulars. Thus the word "scientific" has taken on a somewhat emotional overtone in which the listener is encouraged to drop his suspicions and accept the results with confidence. This popular view of what is meant by "scientific" is a bit misleading. It is true that a scientific investigation should be impartial and as accurate as circumstances will permit and that the results should be reproducible by others. On the other hand, there is always an error or inaccuracy that clouds the proceedings and leaves a certain degree of doubt about the result. Furthermore, no matter how well we think we have explained something, we can never be sure that an even better explanation won't be thought of later. For these reasons, scientists are rather hesitant to claim that anything has really been scientifically proven.

Freddy by Rupe. Reproduced through the courtesy of Field Newspaper Syndicate.

FACTS AND THEIR INTERPRETATION

There are certain things about which essentially everyone agrees: It is brighter outside when the Sun is up than when it isn't; if water is made cool enough, it becomes ice; if you jump out of a tenth story window, you will hit the ground harder than if you jump out of a first story window; and so forth. Any such statement which our experience tells us is correct and for which we all agree there is practically no room for doubt, we will call a "fact." Over a period of time we can build up a long list of observed facts. As long as we simply note the separate facts, they have little to do with science. When we start to examine the different facts with a view to seeing if there are any patterns or relations among them, then we have the makings of science.

If we want to make a scientific study of some observed facts, the one assumption we must make is that these facts are capable of being understood through rational thought: Things do not happen at the whim of some supernatural power or in an otherwise haphazard manner but whenever physical circumstances are just right. Without this assumption one could easily "explain" all things as simply the will of God. Note that this basic assumption of science is just that, an assumption, and not a fact. Things around us may indeed be governed by the whims of supernatural powers; science only says, "Let's pretend that this isn't so and see what happens."

Science does not deny the supernatural as some people, including some scientists, seem to believe. It does not oppose religions or philosophies that require belief in things that are "beyond science." It is an error to suppose, however, that there can be scientific evidence to support opinions that are clearly nonscientific in nature. For example, it is absurd to claim that science can prove that God does not exist.

Science makes no moral judgments. It does not try to answer any questions about the purpose of things. Its job is to establish a set of ideas that will make it possible for us to know just what circumstances lead to what results. Science has given us a tremendous understanding of and control over the events and forces around us.

Science has had many successes in getting its job done, but it also has its limitations. It is not the answer to all important questions, and it does not claim to be. It is so much in our lives, however, that for our own good we ought to know what it can and can't do.

FROM FACTS TO THEORIES

In a scientific study a person considers some observed facts and relates them to some ideas.

Freddy by Rupe. Reproduced through the courtesy of Field Newspaper Syndicate.

Then he or she tries to make the ideas broad enough to cover many more facts, and often mistakes are made in the process. Eventually some rather general ideas will emerge, the fewer the better, which will tie together and explain a large number of facts, the larger the better. Meanwhile other people have come up with other general ideas in order to understand the facts. Someone will examine the different ideas to see how well they fit together in a still more general scientific principle or theory. Contradictions may be found, and then it is back to the drawing board to see how they can be resolved without ignoring any of the facts. Finally, perhaps after a long time and many trials, a very broad, self-consistent theory will come into general acceptance as the basis for understanding all of the relevant facts. This theory is, of course, subject to further change as more data come in and as comparisons are made with theories designed for facts in other areas. There are situations in which no completely adequate theory is known; this is always hoped to be only temporary.

A general scientific theory will usually have a large body of facts backing it up. This does not mean that the theory has been proved correct, as another theory might come along and explain the same facts even better. The greater the number of independent observed facts or data backing up a theory, the more confidence a person feels in the theory. If a theory has been well thought out, some adverse facts will not necessarily prove it totally wrong. Perhaps it is still basically correct but valid under somewhat narrower limits than previously believed. In this type of situation, a theory may later be found to be a special case of a still more general theory.

Let's illustrate these ideas with an example. If we drop a stone, it will fall to the ground. The same will happen to a pencil and to a loaf of bread. After thinking over these facts, we might come to the conclusion that all things will fall to the ground if they are dropped. As we test this idea with more and more examples, our confidence in our theory will rise higher and higher. But when we try a balloon filled with helium, we find that it falls up, not down. Is our idea wrong? Apparently so, but still it was right so often that perhaps we should examine the helium balloon in greater detail to see if we can understand why it is different. Perhaps the original idea needs only a small modification rather than total rejection. When we find that even a helium balloon will fall down if it is put into a vacuum with no air around it, we might change our idea to the following: Something (shall we call it gravity?) tries to make all things fall down, but if an object is light enough, it can float in the air just like other objects can float in water. Here we have made a connection between the ideas of objects falling due to gravity and objects floating or being buoyed up in a fluid, be it air or water. These ideas could be further refined by considering more facts, and we would eventually come to a limit on how far the ideas could be applied. For example, clouds are indeed held up against gravity by the air they are floating in, but the same is not true of the Sun, Moon, and stars. In finding the limits to a scientific theory, we improve our understanding of that theory and of the facts it is supposed to help explain.

While there are many facts about which essentially all scientists agree, there are other observations about which they strongly disagree. One example is extrasensory perception (ESP): Can a person read another person's mind? Some scientists claim that experiments have proved ESP is possible, while others claim just as strongly that these experiments have not been controlled well enough or are not sufficiently conclusive to be considered a proof. This is not a question of who is telling the truth and who is lying; it is an honest difference of opinion. The evidence is not strong enough in either direction for the experts in the field to have reached agreement.

This is precisely the type of area in which scientists want to spend their time and effort. They don't want to sit around confirming what is already known; they want to study things that are not understood. It is the lure of the unknown that keeps their interest. A person who deliberately spends his time among questions, uncertainties, and controversies may seem a bit odd, but scientists find it exciting.

The definition of the Universe led us to consider the nature of science. Astronomy is a science, and it is a good idea to understand from the beginning something about its methods and limitations.

1.2 The nature of the Universe

The Universe is the sum of all things that can be scientifically measured, at least in principle, so it is not possible to make a complete list of its contents. It is possible to describe a few of the major ingredients, however, and the background this gives will help to show how the details to be studied later fit into the whole.

GALAXIES

Matter is not spread evenly throughout the Universe, but is concentrated into large bunches called *galaxies*. A galaxy is an extremely large collection of stars that is held together by the force of gravity. A galaxy may contain tens or hundreds of billions of stars, each star being

FIGURE 1.1
The Andromeda galaxy, a huge system of stars, which is a twin to our own Milky Way. Two small satellite galaxies can also be seen. (Lick Observatory photograph)

about as large and bright and hot as the Sun. Billions of galaxies are known to exist. There may also be large amounts of matter in space between the galaxies, although the evidence to date indicates that this isn't the case.

The size of galaxies and the distances between them are so large as to be almost beyond comprehension. Distances in astronomy tend to be so big that the speed of light is often used to measure them. Light travels at a speed of about 186,000 miles or 300,000 kilometers each second—that's nearly *eight* times around the Earth in 1 second. At this speed, light reaches us from the Moon in a little more than 1 second, while it takes 8 minutes to arrive here from the Sun. According to scientific ideas held today, it is not possible to move faster than light: It is the ultimate in speed. Yet some galaxies are so big that it takes 100,000 years or more for light just to cross from one side to the other.

In one year light can travel a distance of about 6 trillion miles.[1] This distance is called a *light year* (not a length of time), and it is convenient to measure large distances in light years. We live in a large galaxy called the *Milky Way*, and it has a diameter of about 100,000 light years. The Milky Way is one of the larger galaxies, although it is by no means the largest.

A certain area of the sky is known as the constellation of Andromeda (an-DROM-e-da). On a very dark, clear night a person with good eyes might barely be able to see a faint hazy spot in Andromeda, although it is easily seen with binoculars. The object is called the *Andromeda galaxy* (see Figure 1.1), and it is over 2 million light years away. This means that when you look at the Andromeda galaxy, you are seeing it as it appeared over 2 million years ago; what you are seeing is the light that was emitted by the stars in that galaxy before modern man appeared on the Earth. Thus looking out at great distances is also looking back at times far in the past. In spite of its great distance, the Andromeda galaxy is one of the closest ones to us.

Most of the material in galaxies is in the form of stars, although there are also clouds of gas and dust within the galaxies in the space between the stars. The stars in the galaxies are similar to the Sun in that they are so hot that they shine brightly. Any object that is hot enough will shine brightly and give off a large amount of radiant energy. The radiation emitted by a hot body, such as a star, is called "heat" or *thermal radiation*. We can see a distant galaxy because it contains a very large number of stars, and each star emits huge

[1]A trillion is written 1,000,000,000,000 or 10^{12}. (See Appendix 2 for a description of very large or very small numbers.)

FIGURE 1.2
The radio galaxy NGC 5128. What we see in this photograph is the light from many stars plus some obscuring material. Radio telescopes reveal that this galaxy also emits very large amounts of energy in the form of radio waves. (The Kitt Peak National Observatory)

amounts of thermal radiation. Most galaxies are so far away that we cannot see the individual stars in them; only the total radiation from large numbers of stars renders the galaxies visible.

Galaxies are known out to distances of several billion light years. The light we can detect from these remote objects started its long journey toward us back when the Earth was young. Perhaps some of the radiation we can measure from the faintest and most distant galaxies is even older than the Earth or the Sun, both of which were formed around 5 billion years ago.

RADIO GALAXIES AND QUASARS

Until recently astronomers thought that stars were the main ingredients of all galaxies, but today they are not so sure. Many galaxies have been found to emit large amounts of radiation in a form that is quite different from the thermal radiation of ordinary stars. Among these galaxies are the strange *radio galaxies* (see Figure 1.2) that look more or less like ordinary galaxies when we examine them in visible light; when we point radio telescopes toward them to measure the radio waves they emit, however, these galaxies appear far larger and brighter than can be accounted for by their stars. In addition, there are the enigmatic *quasars* (see Figure 1.3), which have kept astronomers puzzled since their discovery in the early 1960s. They are probably related to ordinary galaxies in some way, but the tremendous amounts of energy they radiate into space are definitely unlike the thermal radiation of stars.

Where can all of this nonthermal radiation be coming from? What can radio galaxies and quasars contain that could outshine a whole galaxy full of stars? The nonthermal radiant energy emitted by these objects appears to be a kind known as *synchrotron radiation*. Synchrotron radiation is emitted when electrified particles, such as electrons, are made to travel past a magnet. The Milky Way and other galaxies have long been known to be large magnets, but it is not known why. It is also not known where the high-speed electrons come from, but the observed fact is that synchrotron radiation is a very important part of the Universe that was totally unsuspected, even in 1960.

Is there a limit to how many galaxies exist and how far away they extend? Is there a limit to how far into the past they have existed and how long they will last? Or does the Universe have an infinite size and age? So far, astronomers have been able to give only very incomplete answers to these intriguing questions, and I do not believe that we will ever be able to answer them with much confidence. Past experience also reveals that when answers are found to "ultimate" questions, other ultimate questions, unimaginable earlier, arise to take their place. This continuous search for answers does give astronomers a very exciting profession.

LIFE

One of the billions of known galaxies has the shape of a round, flat disk. It has between 100 and 200 billion stars, modest amounts of gas and dust among the stars, and various other odds and ends. In this galaxy there is a star, slightly above average in size as stars go, which is surrounded by some tiny, dark, and cold bodies called "planets." It is a scientific fact that on the third planet of this star, life exists.

Is life common throughout that galaxy and in the Universe? Is it very rare, occurring only in certain very special times and places? Or is it unique to that one tiny planet, all alone in the vast, unthinking Universe? These are certainly legitimate questions, but scientists pursue them with a fervor that goes well beyond ordinary scientific curiosity. Scientists and non-scientists alike, we all want to know what is really "out there."

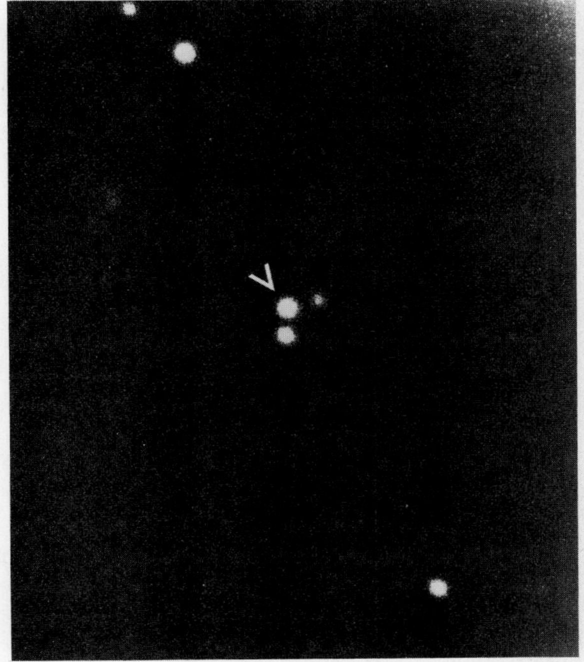

FIGURE 1.3
The quasar 3C 147. The other objects on this photograph are stars within our Milky Way. The nature of the quasars is still unknown.(Hale Observatories)

Ancient and modern views of the Universe.

1.3 Our changing view of the Universe

THE BEGINNINGS OF SCIENCE

Have you ever looked at the stars on a dark night, away from lights and smog? Most people rarely have the chance, and when they do they are usually surprised at how many stars there are and how bright they seem. The view is certainly grand; it appears that the whole Universe is open before us and that we can see to the ends of space itself. Yet all of the stars we can see make up only a very tiny part of only one galaxy.

People have been viewing the Universe for thousands of years. What they think they see has changed throughout history, especially with the rise of scientific understanding. What we know today about the Universe is the result of the efforts of many people working for many years, using a lot of complicated scientific equipment. Considering how much we do not know even now, it is no wonder that in the days of long ago people could not possibly have had the same picture of the Universe that we have today.

The earliest attempts to understand the physical Universe were based on religion and superstition. Each culture had stories of the sky that reflected its own myths and legends. Kings, priests, and fortunetellers also used the skies to try to increase their own power and wealth. A person who could convince others that he was able to read their destinies in the stars would naturally gain a lot of influence. Astrology has had many enthusiastic supporters throughout history and still does today, several centuries after astrology and astronomy have had a complete parting of the ways.

Many ancient cultures are well known for having made quite accurate astronomical observations, including the Chinese, Egyptians, Chaldeans, and Mayans. It remained for the Greeks, however, to bring into play the first signs of scientific inquiry. The Greeks were the first to combine a belief in the importance of natural causes with an atmosphere of open-mindedness that encouraged a person to make up his own mind after hearing the different sides of an argument.

Historians often give credit to Thales as being the first true scientist, although this is at least partly an oversimplification due to our ignorance of the true details. Thales lived around 600 B.C. in the city of Miletus on the west coast of what is now Turkey. He believed that water is somehow the cause of all things. While this idea seemed rather peculiar, Thales did not claim it because the gods had revealed it to him or because his king had decreed it and he thought that he'd better agree if he wanted to keep his head; he said it because of certain thoughts and experiences that he believed made it reasonable. He was glad to explain his reasons to any who were interested, and the listeners were free to agree or not. In other words, important truths did not rest on the authority of whoever had power but on what appeared to be the merits of their supporting arguments. Thales' belief, crude as it seems, is the earliest known attempt to understand the Universe in natural terms: It is the first cosmological theory.

Of course there were limits on how far a person could go in disagreeing with the generally accepted beliefs of the Greeks. In 432 B.C. Anaxagoras (an-aks-AG-o-rus) was banished from Athens for his heretical beliefs, including the idea that the Sun is a red-hot stone larger than southern Greece. Thirty-three years later Socrates (SOK-ra-teez) was forced to drink the poison hemlock for his unconventional teachings, but the world had been exposed to the atmosphere of free inquiry. While this atmosphere was destined to become dormant and remain so for many years, the passage of time has not diminished our debt to the Greeks.

THE DOME OF THE SKY

The sky gives the appearance of a large spherical dome or vault around us. It is natural to picture the Earth as the center of the Universe, the hub around which all else revolves. The stars look like tiny spots of light that are somehow fastened on the dome of the sky. They rise in the east, pass across the sky, and set in the west, all the time keeping the same fixed patterns in their positions. This daily motion of the stars gives the strong illusion that the sky is a solid dome that rotates around the Earth each day and carries the stars along with it. It is actually the Earth rather than the sky that is doing the rotating, but this is not at all obvious from first appearances.

Not all of the stars appear to be stuck on the dome of the sky. There are seven "wandering stars" that seem to move quite noticeably among the fixed stars over a period of time. These seven objects were named *planets*, from the Greek word meaning "to wander."[2]

It was commonly believed through the end of the Middle Ages that the physical Universe is contained within a very large sphere on which are attached the stars. Beyond this sphere of the stars was variously thought to be the machinery that ran things, heaven, or simply a void. The Earth was placed at the center of this system, fixed and unmoving. (See Figure 1.4, shown on the next page, for an imaginative illustration of this Universe.)

[2]The system of planets is the subject of Part II and will not be considered further here.

FIGURE 1.4

A medieval picture of the Universe. It has been suggested that this woodcut was made by the French astronomer C. Flammarion in the late nineteenth century. (The Bettman Archive)

THE SCIENTIFIC RENAISSANCE

Giordano Bruno, who lived in the late 1500s, appears to be one of the first persons to have considered that the stars are similar in nature to the Sun. To account for how faint the stars appear, Bruno placed them at enormous distances. He also imagined that they were scattered over infinite space instead of all at the same distance from us. He argued that the stars were not fixed at all but that each had its own motion; the great distances to the stars made these motions imperceptible to us except over long periods of time.

These views were a tremendous advancement in accuracy, but they were based only on speculation without evidence, and Bruno was not able to convince many others. Furthermore, these ideas were strongly heretical, since they attacked beliefs about the structure of the Universe which had been accepted by the Church. Bruno apparently had a forceful personality and a rather meager tact. That combination proved too much, and after a number of stormy confrontations with authorities, Bruno was burned at the stake in 1600, refusing to the last to repent his heresies.

Times were changing. In 1609 Galileo became the first person to look at the stars through a telescope. His telescope was very small and of poor quality, yet what he saw revolutionized the scientific thought of the day. Before the end of the seventeenth century, it was apparently common knowledge among astronomers, although certainly not among the general public, that the Earth is only one of a number of planets that move in orbits around the much larger Sun and that the stars are of the same general nature as the Sun and, perhaps, might have planets around them, too. The scientific revolution was at hand. According to Bagley and Rowley,[3]

> The new scientists of seventeenth-century Europe readily dismissed . . . [earlier views] as outdated legend, but the poets and romantics—to say nothing of the astrologers and Christian fundamentalists—still found . . . [them] far more acceptable than the difficult mathematical concepts of the new astronomy.

THE DISTANCE OF SIRIUS

Let us consider some of the not so difficult mathematical concepts of this "new astronomy." The light given off by an object spreads out as it moves away from the object, becoming thinner and weaker as it travels. That is why anything looks fainter if we see it from farther away. If we go twice as far away from a

[3] J. J. Bagley and P. B. Rowley, *A Documentory History of England*, Penguin, Harmondsworth, England, 1966, Vol. I, pp. 136–137.

glowing body, it doesn't seem twice as faint as before; it seems four times fainter. Likewise if a lamp is moved 10 times farther away, it appears 10 × 10 or 100 times fainter.

How far away are the stars? In early times, people had no way to tell; they could only guess, and the guesses might be 10 miles, 1000 miles, a trillion miles, or any other number. By assuming that the stars were more or less like the Sun, astronomers were finally able to get a handle on stellar distances.

The brightest star in the night sky is named *Sirius*. It is in the area of the sky known as the constellation Canis Major, the Big Dog, and it is best seen toward the south during winter. As bright as Sirius appears to us, the Sun looks 10 billion times brighter. It follows that if the Sun and Sirius are really the same type of object, as Bruno suggested, we can find the distance to Sirius. We must conclude that Sirius looks fainter because it is farther away. We stated before that the effect of distance can be found by multiplying distance by itself. The distance to Sirius multiplied by itself must equal 10 billion. Now, the number 100,000 satisfies this since 100,000 × 100,000 = 10 billion. It therefore follows that Sirius must be 100,000 times farther away than the Sun if the Sun and it are identical.

Sirius appears about 1000 times brighter than the faintest stars that can be seen without a telescope. Since 30 × 30 is about 1000 (we are not trying to be exact here; only a rough idea is enough), we might expect these faint stars to be about 30 times farther away than Sirius. Far fainter stars yet can be detected with a telescope, so by the end of the 1600s astronomers had enough data to begin to appreciate the immense distances covered by the system of the stars.

It is not so important that the assumptions just made are not exactly correct. The stars are indeed similar to the Sun but not identical. Sirius actually gives off more light than the Sun and is therefore even farther away than we calculated. What is important is that the calculation was roughly correct, and that man's view of the Universe took a large step forward when people began to realize that the Sun is not unique but is only one of many stars.

NEBULAE AND GALAXIES

With the passage of time, telescopes became bigger and better, and astronomers discovered many new objects that attracted their attention. A large number of faint, fuzzy objects were found, and they were called *nebulae* (NEB-u-lee) (singular nebula), from a Greek work meaning "cloud." While the nebulae look like clouds of glowing gas, some astronomers thought that they might be large groups of stars that are too densely packed to be seen individually. In the mid-eighteenth century the famed philosopher Immanuel Kant proposed that the nebulae are huge systems of stars similar to our own Milky Way galaxy. Kant called these systems "island universes." While Kant's ideas were basically correct, they were mainly guesses and had little influence on the scientists of his day.

By the late nineteenth century, astronomers realized that some nebulae were indeed clouds of glowing gas within our own galaxy of stars, but the nature of the other nebulae remained in doubt. This doubt led to many lively disputes. The situation remained this way until 1925 when data found by J. C. Duncan and E. P. Hubble proved that most of the nebulae of uncertain status are actually distant galaxies, huge collections of stars far beyond the boundaries of the Milky Way.

Are any more major changes in our picture of the Universe in store for the future? It is true that what we know today is known with greater accuracy than what was known or thought to be known in the past. Still it would be presumptuous for us to take the point of view that after all of the mistakes of the past, today we know the Universe as it really is. Before the invention of the telescope, astronomers had no way to know about other galaxies. Perhaps discoveries yet to be made are destined to open up much larger and broader views of the Universe—views that will make today's opinions small and pale in comparison. To claim that this will happen is pure speculation. To claim that it cannot happen is to have missed an important lesson from the past.

IMPORTANT WORDS

Universe
Fact
Theory
Galaxy
Light year
Milky Way
Thermal radiation
Radio galaxy
Quasar
Synchrotron radiation
Thales
Planet
Giordano Bruno
Galileo
Sirius
Nebula

REVIEW QUESTIONS

1 What is the Universe?

2 What is a fact?

3 Is science simply a collection of facts?

4 Does science deny the supernatural?

5 Why can't science say whether a person or a star is more important?

6 Why will some adverse facts not necessarily prove a scientific theory to be totally wrong?

7 Why are scientists always studying about areas that are highly uncertain?

8 What are galaxies?

9 Approximately how many stars may a galaxy contain?

10 What is a light year?

11 What is thermal radiation?

12 In what ways do radio galaxies and quasars differ from ordinary galaxies?

13 Who among ancient peoples were the first to introduce scientific principles?

14 Who is often given credit for being the first scientist?

15 In what way do the motions of the planets appear to differ from those of the stars?

16 What idea did Bruno introduce in the late 1500s?

17 Who first looked at the stars through a telescope?

18 When was the true nature of the galaxies determined?

QUESTIONS FOR DISCUSSION

1 If a person accepts new ideas too readily, he is gullible, yet if he resists them too much, he is stubborn. Discuss how a scientist might try to find a balance between these extremes.

2 Suppose that it were proven "scientifically" that supernatural powers, such as ghosts, exist. Would this be a blow to the authority of science?

3 In discussing factual matter, such as what the Moon is made out of, the opinion of a scientific expert in that field should carry a lot of weight. After all, who knows more about it than he or she does, unless it's another expert? On the other hand, should the expert's opinion be given the same great weight when considering questions of judgment, such as how much money the country should spend trying to find the answer?

4 According to one idea, all galaxies were formed at the same time about 15 billion bears ago. Explain how measurements made of galaxies at different distances might be used to find out how galaxies change with age.

5 How might you try to convince a skeptical person that the Earth is only a very tiny part of the whole Universe? Can you do more than simply say, "That's what the scientists say, so it must be true"?

6 The guess that stars are more or less like the Sun turned out to be correct. Yet even if it hadn't, most scientists would agree that it represented a great advance in thinking over earlier views. Why?

7 In 1611 the Florentine astronomer Francesco Sizzi stated his reaction to Galileo's claim that he (Galileo) with his telescope found four satellites that move around Jupiter:

There are seven windows in the head, two nostrils, two ears, two eyes, and a mouth; so in the heavens there are two favorable stars, two unpropitious, two lumaries, and Mercury alone undecided and indifferent. From which and many other similar phenomena of nature such as the seven metals, etc.,

. . . we gather that the number of the planets is necessarily seven. . . . Besides, the Jews and other ancient nations, as well as modern Europeans, have adopted the division of the week into seven days, and have named them from the seven planets: now if we increase the number of planets, this whole system falls to the ground. . . . Moreover, the satellites are invisible to the naked eye and therefore can have no influence on the earth and therefore would be useless and therefore do not exist.[4]

How would you try to convince Mr. Sizzi that his reasoning isn't quite up to modern scientific standards?

REFERENCES

There are many good references on science and its historical development. Two examples, both by G. Holton, are *Introduction to Concepts and Theories in Physical Science* (Addison-Wesley, Cambridge, 1952) and *Thematic Origins of Scientific Thought* (Harvard University Press, Cambridge, 1973). A delightfully written book on how the methods of science are misapplied, by M. Gardner, is *Fads and Fallacies in the Name of Science*, 2nd ed. (Dover, New York, 1957). The first two parts of A. Koestler's *The Sleepwalkers* (Grosset & Dunlap, New York, 1963) give a good account of Greek and medieval science, although much of the development has to do with the planetary system. *Exploring the Universe* (Oxford, New York, 1971), edited by L. B. Young, contains essays and excerpts from many scientists on the nature and methods of science with particular emphasis on astronomy.

Two important histories of astronomy are J. L. E. Dreyer's *A History of Astronomy from Thales to Kepler*, 2nd ed. (Dover, New York, 1953) and A. Berry's *A Short History of Astronomy* (Dover, New York, 1961). The Berry book covers through the end of the nineteenth century. An excellent book on more modern developments by O. Struve and V. Zeburgs is *Astronomy of the Twentieth Century* (Macmillan, New York, 1962). Extracts from the works of astronomers are contained in *Source Book in Astronomy*, edited by H. Shapley and H. E. Howarth (McGraw-Hill, New York, 1929) and in *Source Book in Astronomy 1900–1950*, edited by H. Shapley (Harvard University Press, Cambridge, 1960).

[4]G. Holton, *Introduction to Concepts and Theories in Physical Science*, Addison-Wesley, Cambridge, 1952, pp. 164–165.

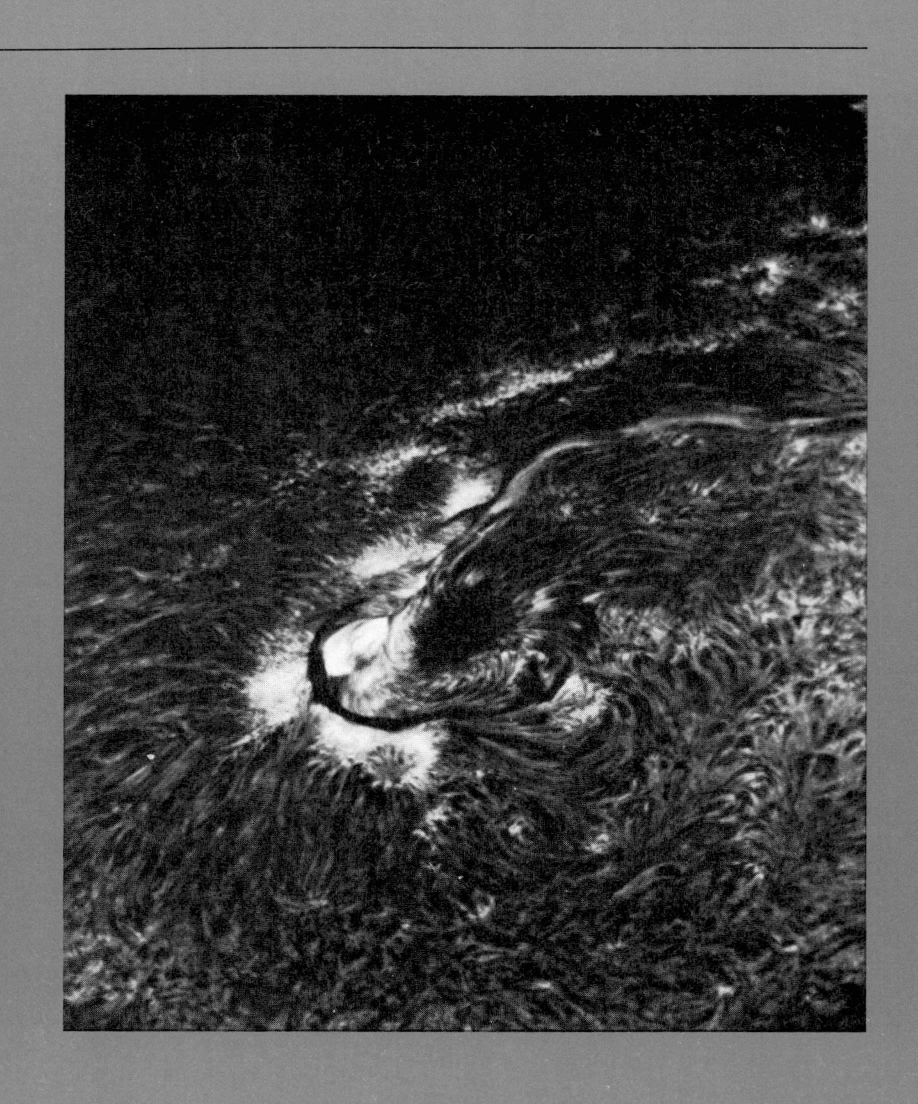

TWO
MATTER AND ENERGY

The Universe contains material objects, such as planets, stars, and galaxies. These objects are made of matter and possess an important quantity known as energy. The study of some of the basic properties of matter and energy is an important first step in trying to understand our Universe.

2.1 Matter

MASS, WEIGHT, AND DENSITY

The total amount of matter in an object is known as its *mass*. Mass is often confused with *weight*, but the two are quite distinct. Weight is the force of gravity pulling on an object.[1] A 10-pound weight is an object for which the Earth exerts a gravitational force of 10 pounds, if the object is on the surface of the Earth. It is true that a 20-pound weight has twice the mass of a 10-pound weight, and this correspondence has led to weight often being used as a measure of mass. Suppose, however, that we take a 10-pound weight out into space, very far from the Earth or other bodies. The object now weighs almost nothing because the force of gravity on it has been reduced to almost zero, but its mass is still the same. The object can still be called a 10-pound mass, even though it no longer weighs 10 pounds.

We might take the same 10-pound mass to the Moon. Gravity on the surface of the Moon is six times weaker than it is on the surface of the Earth. On the Moon the object would weigh a little under 2 pounds, although it would have the same mass as before. On the Moon a 10-pound mass weighs only 1⅔ pounds. Only where the pull of gravity is the same as on the surface of the Earth does a 10-pound mass weigh 10 pounds.

The *volume* of an object tells us how much space it takes up. The larger the object, the greater is its volume. Volume is measured in cubic inches, gallons, liters, or some equivalent unit.

A large object is one that has a large volume, but it does not necessarily have a large mass. A large pile of feathers may have less mass than a small lump of lead. The difference is in *density*, or how concentrated in a small volume the matter is. Lead has a lot of matter squeezed into a small volume, which means that it has a large density. The feathers have matter spread rather thinly over a large volume, which means that they have a low density. The density of an object is its mass divided

[1] Sometimes weight is defined as the force an object actually feels. The point of this definition is that there are conditions in which the effects of gravity can be modified, such as in an accelerating car, a diving airplane, or an orbiting satellite. We will avoid these complications here.

FIGURE 2.1
Forces in nature. (a) The gravitational force. (b) The electric force. (c) The nuclear force.

by its volume. An object may have some regions of high density and other regions of low density, in which case one finds the average density by dividing the total mass by the total volume.

So far we have been thinking of pieces of matter that are large enough to see and measure easily. What can we find out about the smallest fragments of matter, those on the microscopic level?

FUNDAMENTAL PARTICLES

A large piece of matter can be broken into pieces, and these in turn can be broken into smaller pieces. Is there any limit on how far this process can be continued? Is there an ultimate size or an ultimate type of particle beyond which matter cannot be further divided?

Philosophers and scientists have long talked about the possible existence of *fundamental particles* as the basic things from which all material objects are built. These were envisioned as permanent particles which could not be created, destroyed, or changed in any way. In particular they could not be broken into smaller pieces. Different forms of matter were supposed to result from different arrangements of the fundamental particles among themselves. The atomic theory of Leucippus (loo-KIP-pus) and Democritus (de-MOK-ri-tus) of 400 B.C. in Greece is one of the earliest forms of this idea. The word *atom* comes from the Greek, meaning something that cannot be cut or divided.

A related idea concerns the so-called "elements," basic substances from which all other substances are supposed to be made. As we have noted, Thales considered water the single element from which all materials are made. Empedocles (em-PED-o-kleez), a Greek who lived around 450 B.C., originated the idea of the four elements, earth, water, air, and fire. All materials were supposed to be mixtures of these four elements. Each substance was supposed to have its own special proportions of the different elements, and each substance derived its properties primarily from the element or elements that predominated in its make-up. This idea was accepted by many of the later Greek philosophers and scientists, and it was passed on to the Middle Ages where it was an important part of medieval science.

By the eighteenth century, the elements earth, water, air, and fire had been replaced by the *chemical elements*, such as hydrogen, tin, iron, and so on, as the ultimate substances that were believed to make up all matter. In the early 1800s the Englishman John Dalton combined the concepts of "element" and "atom" to form the basis of modern atomic theory and of chemistry. Dalton thought of atoms as funda-

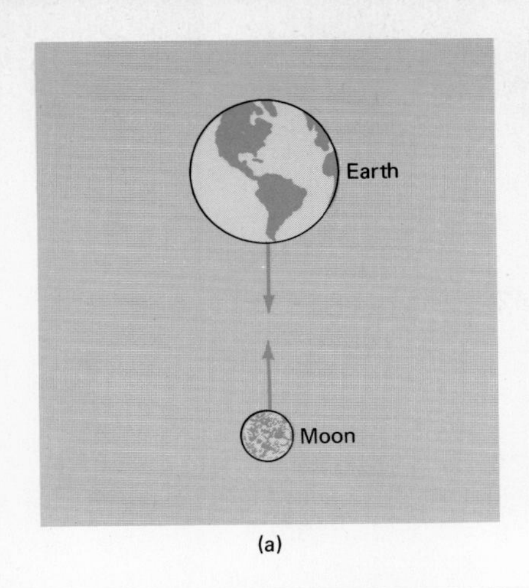

(a)

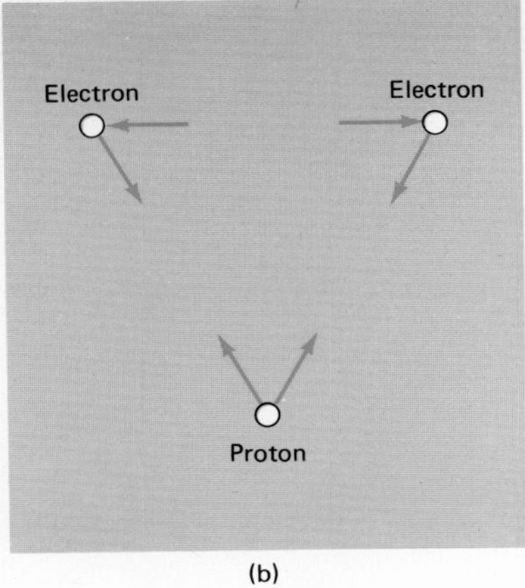

(b)

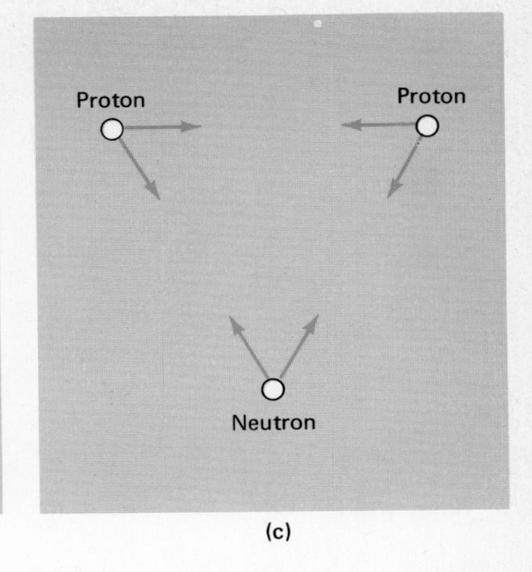

(c)

mental particles, that is, the basic particles of all matter. Each element has its own special type of atom, and complex substances are formed by combinations of different types of atoms.

At the beginning of the twentieth century, scientists realized that atoms can be separated into still smaller and more basic particles. They are not the indivisible particles that their name implies. We still consider atoms as the basic units of the chemical elements, but they are no longer believed to be fundamental particles.

Today it is suspected that there are no fundamental particles in the original meaning of the term.[2] Under the right conditions, it appears that any particle can be changed into other types of particles, and particles can even be created or destroyed as material particles. The term "fundamental particle" is now applied to those particles for which it is not too obvious that they are built out of smaller and simpler parts. There are dozens of such fundamental particles known today. Most of them are very unstable and change into something else after a very short existence.

There are three types of fundamental particles that are especially important for understanding the structure of matter: *electrons*, *protons*, and *neutrons*. These particles are extremely tiny in size and in mass. It takes 30 billion billion billion or 3×10^{28} electrons to add up to an ounce of mass. Protons and neutrons are both about 2000 times as heavy as electrons, but it still takes 100 billion billion of them to equal the mass of the left hind leg of a housefly.

FORCES BETWEEN PARTICLES

To see how electrons, protons, and neutrons can combine with each other to form atoms and larger pieces of matter, it is necessary to consider the forces that hold the particles together. A force is a push or a pull that can be exerted on matter. The fundamental particles exert various forces on each other, and these forces are of four different kinds: gravitational, electric, strong nuclear, and weak nuclear (see Figure 2.1). Magnetic forces are closely related to electric forces and are not considered a separate kind.[3]

The gravitational force is quite familiar to us as the force that holds us onto the surface of the Earth. All particles that have mass, including electrons, protons, and neutrons, exert a pull on each other. This pull is the gravitational

[2]Hypothetical particles called "quarks" are believed by some persons to be the next candidates for the role of fundamental particles.

[3]The weak nuclear force is not important for our purposes, and we will not be concerned with it here.

force. This force is very weak, so it takes an enormous number of fundamental particles, as many as the number of particles in a small planet, before the gravitational force becomes important. The gravitational force between two fundamental particles is completely negligible because the particles have such small masses.

The electric force is a force that is exerted between certain fundamental particles and not between others. Those particles that do exert electric forces and respond to them are said to have electric charges, and the other particles that do not are said to be neutral. The electric charges are of two kinds, called positive and negative. The force between like charges (both positive or both negative) is repulsive, which means that the particles try to push each other farther apart. Opposite charges (one positive, the other negative) attract each other. Protons have a positive electric charge, electrons have a negative electric charge, and, as the name indicates, neutrons are electrically neutral. Through the electric force, protons repel each other, electrons repel each other, and a proton and an electron attract each other. Neutrons are not affected by the electric force.

The strong nuclear force or simply *nuclear force*, as we will call it, is similar to the electric force in that some fundamental particles have it and some do not. The proton and neutron do, the electron does not. The nuclear force is attractive, and it is of no importance unless the particles get extremely close together. If protons and neutrons do get close enough together for the nuclear force to act, the nuclear force easily overpowers the electric force which would try to push the protons apart. How close to each other do protons and neutrons have to get before the nuclear force can act? Less than about one ten-trillionth (10^{-13}) of an inch.

One interesting fact about the neutron is that it needs company. If left alone it changes spontaneously into an electron and a proton. The neutron is unstable by itself, but if one or more protons or neutrons are brought close enough to a neutron to exert the nuclear force on it, then the neutron will be stable.

ATOMS AND MOLECULES

Complicated structures of matter are formed as follows: One or more protons come together with or without neutrons and form a *nucleus*. The smallest nucleus has just one proton and no neutrons. The nucleus is strongly held together by the nuclear force, although if the protons and neutrons are not in the proper proportion or if there are too many protons, the nucleus will be unstable and will not be able to hold together. Such an unstable nucleus is said to be *radioactive*.

Because of the protons, the nucleus always has a positive electric charge. Electrons with their negative electric charges are attracted to a nucleus, and enough electrons

FIGURE 2.2
Atoms of the chemical elements hydrogen, helium, and lithium. The letters e, p, and n stand for electron, proton, and neutron. (a) Hydrogen. (b) Helium. (c) Lithium.

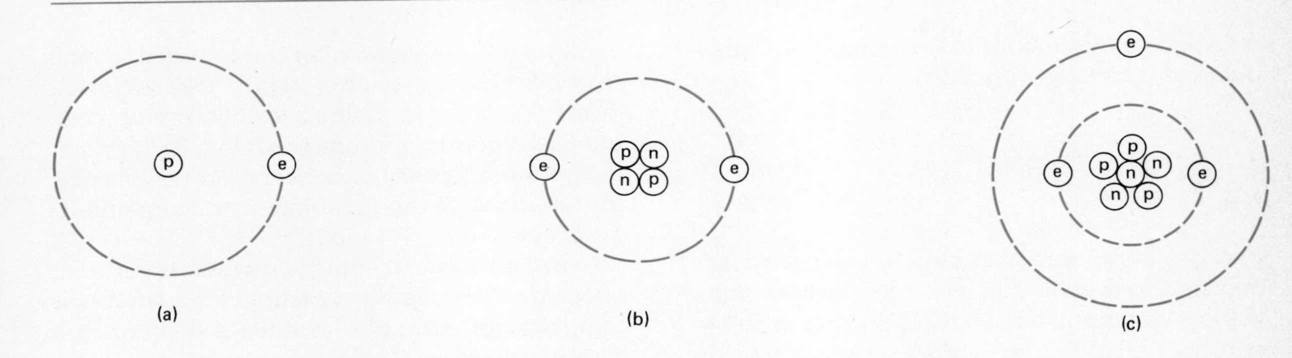

congregate around a nucleus to exactly balance the electric charge. The electrons do not come into the nucleus but move around it in paths, much like the planets move around the Sun. The strengths of the electric charges on protons and on electrons are the same, so normally there are the same number of electrons around a nucleus as there are protons in the nucleus. The resulting structure is an atom. As we have already seen, atoms are the basic particles of the chemical elements.

Atoms which have the same number of protons in the nucleus (and, therefore, which normally have the same number of electrons around the nucleus) are of the same chemical element. The element hydrogen has one proton, helium has two, oxygen has eight, iron has 26, uranium has 92, and so forth. (Figure 2.2 illustrates several atoms.)

Atoms having the same number of protons but different numbers of neutrons are called *isotopes*. There are three isotopes of the element hydrogen that have zero, one, and two neutrons along with the single proton. All three are hydrogen atoms and have very similar properties, except that the heaviest isotope of hydrogen is radioactive. All elements have a number of isotopes, some of which are radioactive.

Different atoms often stick together through the electric force and form larger particles called *molecules*. Molecules are the basic building blocks of the chemical compounds. Sodium atoms and chlorine atoms have a strong affinity for each other, so an atom of each will come together and form a molecule of the compound we know as table salt. Two hydrogen atoms and one oxygen atom combine to form a molecule of water. All of the material objects that we are familiar with are aggregates of huge numbers of atoms or molecules, or mixtures of both.

2.2 Energy

Energy allows things to happen. An energetic person is one who can get a lot of work done without getting tired easily. Energy is whatever makes all of this hard work possible. Energy is usually defined in science as a measure of the ability of something to do work, that is, its ability to exert forces on objects and move them.

In popular usage, energy is a somewhat vague concept, but to a scientist it is quite specific. Scientists can even measure the amount of energy and assign a number to it. For example, it takes a certain amount of energy to lift a 10-pound mass up 6 feet, against the force of gravity on the surface of the Earth. It takes twice as much energy to lift the same mass twice as far; it only takes half as much energy to lift a 5-pound mass 6 feet.

CONSERVATION OF ENERGY

Energy can take many different forms. When objects interact with each other and with radiation, exchanges of energy take place. What makes energy so important is that in any interaction the total amount of energy does not change; it is the same after the interaction as it was before. The energy might take different forms and it might belong to different objects, but the total amount remains fixed. This principle is widely known as *conservation of energy*, and it is one of the most important statements in all science.

Appealing to energy conservation can often greatly simplify a complicated situation. Suppose, for example, that we see a car that has been wrecked after being run into a brick wall. Knowing that the total energy of the car and its surroundings was the same before the accident as after, we know that the car must have had at least enough energy to do the observed damage. This fixes the approximate speed at which the car was traveling.

If we want to use a large amount of energy, we must provide that energy. Energy is very definitely one item for which we cannot get something for nothing. This is true for society, as we are being made aware by the various energy problems we face, and it is no less true for the Sun itself.

The Sun gives off a tremendous amount of energy each second in the form of light and other radiation. We know that it has been doing this for a very long time because life needs the Sun's warmth to survive, and fossil evidence shows that life has been in existence on the Earth for 2 or 3 billion years. Where can all of this energy come from? The Sun must have had all of this energy stored up inside somehow. In what form has it been stored? How is it being changed into radiation and released a little at a time? In trying to answer these and related questions, astronomers have developed theories on how stars are built and how they age. We will discuss these theories in some detail in later parts of the book, but first we will look at the many different kinds of energy.

It is convenient to divide energy into four general types: (1) energy of motion, usually called *kinetic energy;* (2) *heat energy;* (3) energy of position, usually called *potential energy;* and (4) *radiation energy.* The first three types will be discussed in the remainder of this section. Radiation is special enough to deserve a chapter of its own.

KINETIC ENERGY

Kinetic energy is possessed by any material object that is moving. The combination of mass and speed (or velocity) produces kinetic energy. The larger the mass, the larger the kinetic energy. The same is true for speed, but we must multiply the speed by itself to find its effect on the kinetic energy. If an object is accelerated to twice its original speed, the kinetic energy is $2 \times 2 = 4$ times larger than before.

Many of the properties of kinetic energy are familiar from everyday experience. We

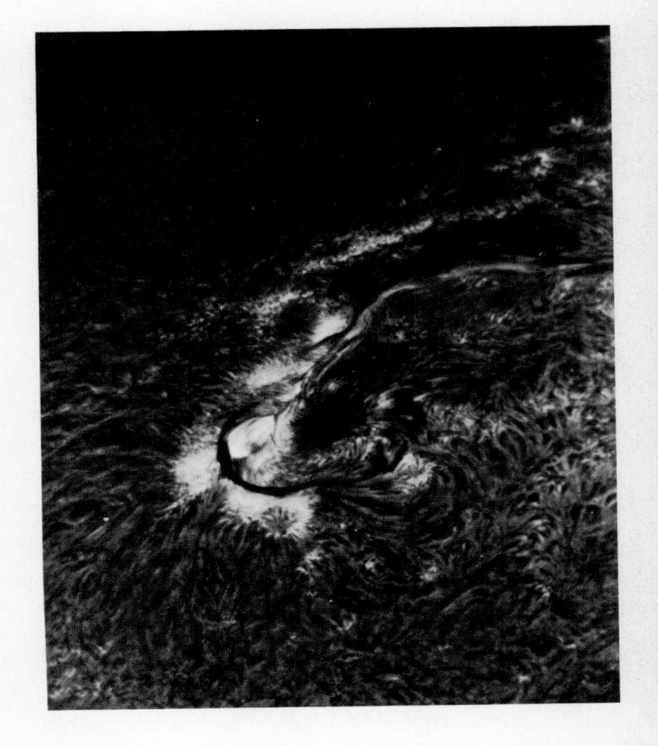

FIGURE 2.3
The Sun is the Earth's main source of energy. (Lockheed Solar Observatory, Lockheed Palo Alto Research Laboratory)

must throw a small stone faster than a large one if the two are to do the same amount of damage to the target, for it is the kinetic energy of the stone that does the damage. A car moving at 50 miles per hour has 4 times as much kinetic energy as it would have at 25 miles per hour; at the faster speed, the brakes must get rid of (that is, change into other forms of energy) that much more kinetic energy before they can bring the car to a halt.

HEAT

Heat is another form of energy that is common to our everyday experiences. If we want an object to become hot, we must see to it that some form of energy is added to it. Any object that is hot will, if left to itself, get rid of some energy by radiation or by other means, and this will usually cool the object down.

The atoms and molecules that make up matter are constantly moving around, colliding with each other and bouncing off in random directions. When an object is made hotter, these molecular motions become more rapid, and the collisions are more violent. The motions of the molecules and atoms are directly responsible for heat energy. As each tiny particle moves, it has a very small amount of kinetic energy. The sum of the kinetic energies of all the molecules and atoms that make up an object is the total heat energy of the object. Heat energy is a special form of kinetic energy. If an object is cooled off, the molecular motions slow down and their kinetic energies become less, so the total heat content of the object is decreased. Likewise when an object heats up, its molecules move faster and have more kinetic energy.

We commonly use temperature to indicate how hot something is. In fact, temperature is a direct measure of how fast the atoms and molecules are moving. If the random molecular motions in a body increase or decrease, the temperature must be going up or down.

We should be careful to distinguish between the temperature of an object and its total heat energy. A single match may burn with the same temperature as a large forest fire, but it obviously does not have the same amount of heat. The difference, of course, is that there is a lot more material burning in the forest fire than in the match. Temperature tells us how much kinetic energy each atom or molecule has. One could say that temperature tells how concentrated the heat energy is. The total heat energy of an object depends both on its temperature and on how many molecules it is composed of. A burning match is much hotter than an iceberg; due to its much larger mass, however, the iceberg has more total heat energy than the burning match.

How fast do the particles actually move?

That depends on the temperature and also on how heavy the atom or molecule is. The fastest atoms are those of hydrogen, since they are the lightest. At a temperature of 72°F, a hydrogen atom will average about 6000 miles per hour. The much heavier iron atom will have the slower speed of about 800 miles per hour. It is hard to believe that the objects we see every day, including our own bodies, are composed of tiny particles that are darting around with such high speeds. At the center of the Sun the temperature is about 30,000,000°F. There the hydrogen atoms average nearly 1½ million miles per hour.

POTENTIAL ENERGY

When a push or pull is applied to an object in a certain direction, that force tries to move the object in that direction, giving it some kinetic energy. Any object in a position in which a force is exerted on it has a certain amount of energy that, at least potentially, can be converted into kinetic energy. This energy is called *potential energy*.

When an outside agent moves an object against the prevailing force, potential energy is stored in the object. When the object is free to be pushed along by the force, potential energy is released and changed into kinetic or another form of energy. Winding a clock, setting a mousetrap, and stretching a rubber band all involve storing up potential energy by making something move against its prevailing force. When the object is released, it springs back to its normal position—quickly, for the mousetrap and the rubber band; in a slow, controlled way, for the spring of the clock. The potential energy is changed back into kinetic energy, and work is done. Just as much energy is needed to set these contraptions as is regained when their springs are released, so we fail to create energy out of nothing. We merely convert the energy into what we hope is a more useful form.

It takes energy to lift a mass up against the force of gravity, and this gives potential energy to the mass. The higher we lift the mass, the more potential energy it gets. When we release it, the mass will fall to the ground, changing its potential energy into kinetic energy as it goes and in this way picking up speed. The kinetic energy is then responsible for the damage it causes when it hits the ground.

Suppose we take the mass off into space, far away from the Earth. According to our discussion, the farther from the Earth we take it, the more potential energy it possesses. But the gravitational force becomes weaker when we get farther from the Earth, so we have the rather strange result that the weaker the force of gravity on the mass, the more potential energy the mass possesses. The most potential energy it can have is when it is moved far enough away that the force of gravity is essentially zero. Potential energy does not manifest itself except when it is in the process of changing into other forms of energy or they into it.

In the last section we saw that nuclear and electric forces hold electrons, protons, and neutrons together to form atoms and moleclues. These forces have their own potential energies, and under the proper conditions these potential energies can make themselves known by being converted into heat, radiation, and kinetic energy. In a burning match the molecules in the wood are combining with oxygen in the air to produce water, carbon dioxide, and other types of molecules. During these processes the electric forces holding the molecules together release a large amount of potential energy, and this potential energy is changed into the heat and light energies that we associate with fire. At the center of the Sun, violent collisions sometimes cause certain nuclei to combine with others, producing different types of nuclei. Extremely large amounts of nuclear potential energy are released in the process, and this supplies the energy the Sun needs to keep shining for so long. The food we eat stores potential energy of the digested food inside us; with an effort on our part, some of this energy can be converted into kinetic energy to do work or exercise.

TEMPERATURE AND THE STATES OF MATTER

A solid object is made out of molecules (or atoms) which exert electric forces on each other that are strong enough to hold themselves together with some degree of rigidity. Each molecule moves back and forth around a fixed position, bumping into its neighbors, but it is not free to wander around to other parts of the object. If the solid is heated to a higher temperature, the molecules increase their back-and-forth motions, and the collisions become more violent. If the temperature continues to rise, a point will be reached at which the molecules have enough kinetic energy to break the bonds that tend to hold them in place. Then the molecules are free to wander far and wide, and the solid has melted to become a liquid or, in the more extreme case, has vaporized to become a gas.

Some molecules exert stronger forces than others, and these remain solid at higher temperatures. All matter becomes solid (or at least liquid) at a low enough temperature, and all matter becomes gaseous if the temperature is made sufficiently high. A liquid is intermediate between a solid and a gas in that the electric forces between molecules are important but not strong enough to hold the molecules in place. Practically all of the matter in the Universe is in the gaseous state; the solids and liquids that are common on the Earth are quite rare in the Universe at large.

While there is no limit to how much kinetic energy a particle can have, there is a limit to how little. After all, you cannot have less motion than none at all. When an object is made so cold that all of its atoms and molecules are essentially standing still, all of the heat energy has been removed, and the object is as cold as it is possible to get. The temperature at which this occurs is called *absolute zero*. Absolute zero is equivalent to about −460°F or −273°C.

If heat energy is added to a gas, the molecules gain in kinetic energy and, at a certain temperature, the molecules will be torn apart by their impacts on each other. In this way, large molecules are broken into smaller ones, and at sufficiently high temperature, all molecules are broken down into their component atoms.

The breaking-down process does not stop with atoms. An atom consists of one or more electrons held near a nucleus by the electric force. If the temperature is high enough, collisions will tear the electrons away from the nuclei, and the gas particles will then have electric charges: negative charge on each electron and positive charge on each nucleus. In this case, the gas is said to be "ionized." The nuclei themselves will finally break down into various kinds of fundamental particles at extremely high temperatures, above about 100,000,000,000°F. Radiation will also play a part in the processes described here.

Matter itself is also a form of energy, as there are conditions under which matter disappears, and radiation or other forms of energy appear in its place. The reverse also occurs: Energy can change into matter. Einstein's famous equation $E = mc^2$ shows that the mass m is equivalent to the amount of energy E; c is the numerical value of the speed of light. In most cases of interest, the creation and destruction of matter are not of great importance, and we can assume with little error that the amount of matter present remains constant.

IMPORTANT WORDS

- Mass
- Weight
- Volume
- Density
- Electron
- Proton
- Neutron
- Fundamental particle
- Atom
- Chemical element
- Energy
- Conservation of energy

Gravitational force
Electric force
Nuclear force
Nucleus
Radioactivity
Molecule
Kinetic energy
Heat
Temperature
Potential energy
Absolute zero

REVIEW QUESTIONS

1 What is the difference between mass and weight?

2 How can you change the weight of a 10-pound mass?

3 What is density?

4 What did the word "atom" originally mean?

5 What are the three main types of elementary particles?

6 What are the four basic types of forces in nature?

7 What do we mean when we say that a particle has an electric charge?

8 Why do neutrons not exist by themselves in nature?

9 What holds the nucleus of an atom together?

10 What holds the electrons onto an atom?

11 What is the distinction between chemical elements and isotopes?

12 How are molecules related to atoms?

13 What is meant by conservation of energy?

14 What are the four main types of energy?

15 What is kinetic energy?

16 What does temperature tell us about the atoms or molecules in an object?

17 Why does an iceberg have more heat than a burning match?

18 What kind of energy do we give to an object when we lift it up against the force of gravity?

19 Why will a solid object melt or vaporize at a certain temperature?

20 What does Einstein's famous equation $E = mc^2$ tell?

QUESTIONS FOR DISCUSSION

1 A balloon will float in the air if its average density is less than the density of air. By putting helium gas in the balloon, we actually give it a little more mass. How can this help the balloon to float in the air?

2 Suppose the electrical charges on electrons and protons were made weaker so that the forces between charges would become correspondingly less. Would this change be obvious to us, or would it be noticed only by scientists working with their high-power instruments?

3 Most of the energy we use comes directly or indirectly from the Sun. When we burn wood, coal, or oil products, for example, we release potential energy that was stored up by plant life with the aid of sunlight. How does water power depend on the

Sun? Can you think of any energy supply that is independent of the Sun?

4 Inventors sometimes claim that they have made a perpetual motion machine, that is, a gadget with moving parts that is supposed to be able to keep moving forever without any outside energy sources. Do you think a machine like this is possible? Why?

5 A rubber ball is bouncing on a hard surface. What are the three specific forms of energy of the ball that play an important part in the action? Tell how the energy of the ball is exchanged back and forth between these forms as the ball goes up and down and collides with the ground. Why does the ball eventually stop bouncing? What happens to all of the bouncing energy it used to have?

6 Suppose the ball in the above question is dropped onto a surface of soft sand. Now a new form of energy becomes important. What is it, and how does it help explain what happens?

REFERENCES

The material in this chapter is covered in greater detail in almost any elementary text on physical science, physics, or chemistry.

There are also two recent articles in *Scientific American* concerning fundamental particles. The first is "Electron–Positron Annihilation and the New Particles" by S. D. Drell [232 (June 1975), 50] and the second, by Y. Nambu, is "The Confinement of Quarks" [235 (November 1976), 48].

THREE
RADIATION

The main source of information we have about astronomical objects is the radiation that they send our way. This chapter introduces radiation and describes those properties that are most important in astronomy.

3.1 Waves

Radiation is most familiar to us in the form of the visible light that we see by, but there are many other kinds of radiation which we cannot see. All kinds of radiation are of interest to us because they all contain energy and can use that energy to affect matter.

MAKING WAVES

If we take a particle with an electric charge, such as an electron or a proton, and make it move back and forth, it will be found to emit radiation. The type of radiation it emits will depend on how fast the particle is moving back and forth. If we try the same thing with an electrically neutral particle, such as a neutron, no radiation appears. It is apparent that the radiation is somehow related to the electric charge. Just what does the charge have to do with radiation?

Two electric charges exert a force on each other, and the force becomes stronger as the particles get closer together. How can the charges reach out and affect each other over the empty space between them? Apparently empty space is modified by the presence of the charge just as it seems to be modified by the presence of a mass that can exert a gravitational force on another mass. Scientists say that the space around a charge contains an *electric field.*[1] An electric field is not made out of matter, so what is it? All we can do is describe it in terms of its properties: An electric field can exist in space, and it exerts a certain force on particles that have electric charges. We cannot see, feel, hear, or taste an electric field by itself, but we can note its effects on charged particles.

When we move a charge back and forth, its electric field waves back and forth with it. If we suspected that radiation consists of this waving electric field, we would be on the right track. The situation is more complicated than this, however. A moving electric charge[2] also brings into existence what is called a *magnetic*

[1]Likewise a gravitational field exists in the space around a mass.

[2]A moving electric charge is called an electric current. The electricity that we use in our homes consists of a stream of electrons moving along the wires and into the light bulbs and appliances.

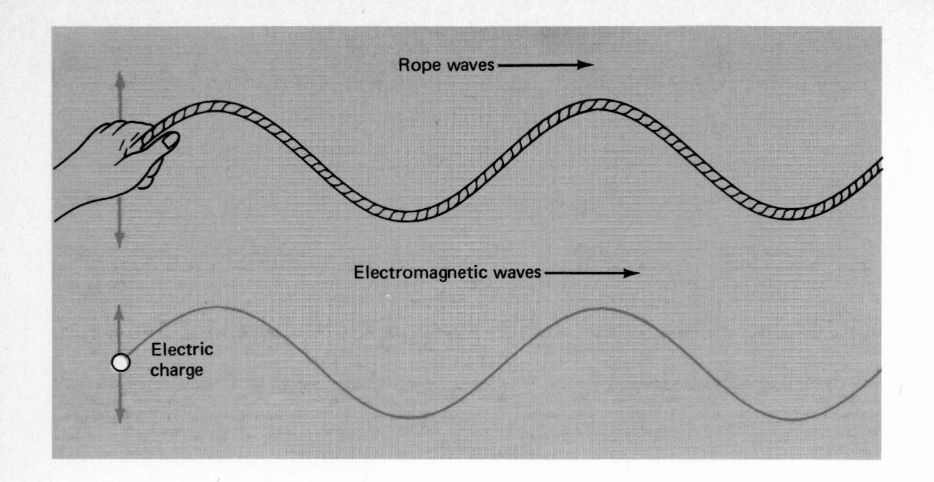

FIGURE 3.1
Waves in a rope and the electromagnetic waves produced by a vibrating electric charge.

field. As you might guess, a magnetic field also exists in the space around a magnet, and it exerts a certain force on another magnet. As the electric charge oscillates back and forth, it sends out an electric field and a magnetic field in all directions. These two fields wave back and forth in step with each other and with the charge that caused them. Together the two fields make up what is called an *electromagnetic wave*, and this wave is radiation. We can cause a wave to travel along a rope by waving the end of the rope up and down; in the same way, we can send electromagnetic waves or radiation out into space by waving a charged particle up and down (see Figure 3.1).

FREQUENCY AND WAVELENGTH

A wave consists of a series of hills, or crests, and valleys, as illustrated in Figures 3.1 and 3.2. The distance between the crests is called the *wavelength* of the wave. The crests and valleys do not hold still but move in some direction. A person will see a certain number of crests go by each second of time; that number is known as the *frequency* of the wave. In the case of the rope in Figure 3.1, we see that a new crest is produced in the rope every time the person's hand reaches the top of its up-and-down motion. The new crest travels out along the rope and makes way for the next one

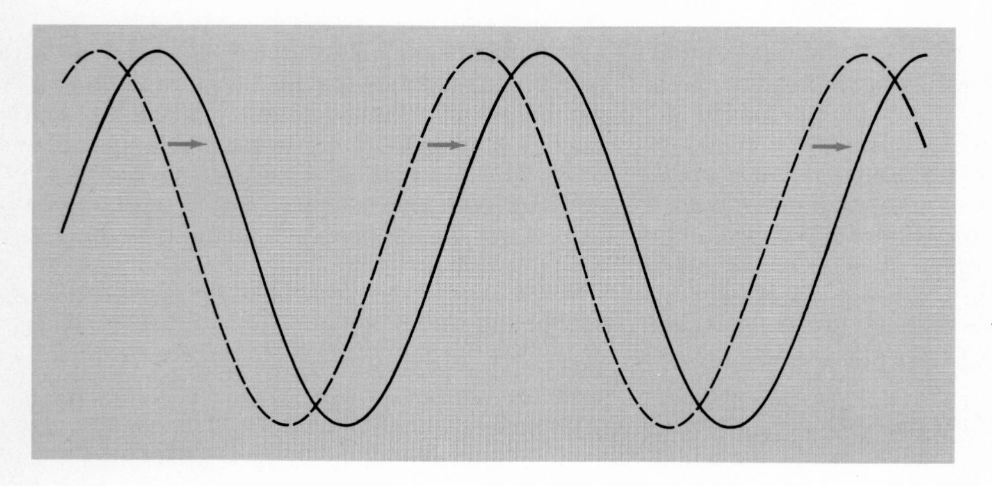

FIGURE 3.2
A moving wave. The full curve represents the position of the wave at a slightly later time than the dotted curve. The wave is moving toward the right.

Type	Frequency (cycles per second)	Wavelength (inches)
Radio waves	lower than 3×10^{11}	longer than 0.04
Infrared	3×10^{11} to 4×10^{14}	3×10^{-5} to 0.04
Visible light	4×10^{14} to 8×10^{14}	1.5×10^{-5} to 3×10^{-5}
Ultraviolet	8×10^{14} to 1×10^{16}	1×10^{-6} to 1.5×10^{-5}
X rays	1×10^{16} to 3×10^{19}	4×10^{-10} to 1×10^{-6}
Gamma rays	higher than 3×10^{19}	shorter than 4×10^{-10}
Red light	4×10^{14}	3×10^{-5}
Yellow light	5×10^{14}	2×10^{-5}
Violet light	8×10^{14}	1.5×10^{-5}

[See Appendix 2 for a description of exponential notation.]

TABLE 3.1
Frequencies and wavelengths of radiation.

which is produced one cycle later. The frequency of the wave is fixed by how rapidly the person moves his hand up and down. In the same way the frequency of the electromagnetic wave is determined by how rapidly the charged particle that produces the radiation is oscillating up and down. Frequency is also known as "vibration rate," since it tells how fast the wave is vibrating.

TYPES OF RADIATION

The different types of radiation[3] consist of electromagnetic waves with different wavelengths and frequencies. Radiation is usually divided into six types according to the frequency or wavelength: *radio waves*, *infrared waves*, *visible light*, *ultraviolet rays*, *x rays*, and *gamma rays*. The approximate frequencies and wavelengths of these types of radiation are given in Table 3.1.

Radio waves have the lowest frequencies of all, while gamma rays possess the highest frequencies. Wavelength goes in the opposite way: Radio waves are the longest and gamma rays the shortest. In all cases, the higher the frequency, the shorter the wavelength. Figure 3.3 shows the different types of radiation on a frequency scale. Note the narrow range of frequencies connected with visible light.

COLOR

Our eyes can not only detect visible light, they can also distinguish the specific frequencies or wavelengths within the visible light band. This is done by color. The lowest frequencies and longest wavelengths of all visible light appear red to our eyes. Light of slightly higher frequency (and shorter wavelength) looks orange, and this progresses through the complete rainbow of colors: yellow, green, blue, and finally violet. Violet light has the highest frequencies and shortest wavelengths of all visible light. Other colors, such as tan, white, purple, or pink, are obtained by the proper mixture of light of different frequencies. The terms "ultraviolet" and "infrared" refer to radiations whose frequencies and wavelengths place them next to the violet and to the red, respectively.

Note, for example, that in Table 3.1 red light has a frequency of about 4×10^{14} cycles per second. This is a very large number whose name is four hundred trillion; written out it is 400,000,000,000,000. This is how many times the electric and magnetic fields of red light are

[3]Just as oscillating electric and magnetic fields can produce an electromagnetic wave, so it is believed that an oscillating gravitational field can produce a gravitational wave. Evidence for such "gravitational radiation" is currently being sought. Unless explicitly stated to the contrary, however, the term "radiation" will be used to mean only waves of the electromagnetic variety.

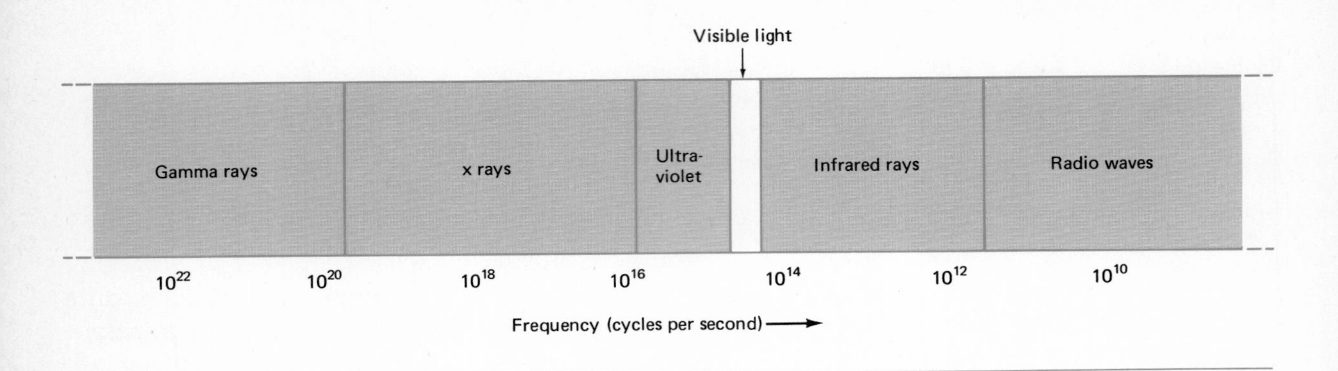

FIGURE 3.3
The frequencies of the different types of electromagnetic waves.

wiggling back and forth *each second.* This is how many wave crests of red light strike our eyes each second. If we wish to create red light by causing a charged particle to oscillate back and forth, the particle will have to do it this many times per second. Such an extremely large rate of vibration seems incredible, yet blue light, ultraviolet rays, x rays, and gamma rays have even higher frequencies. This is only one of many examples we can find in the Universe of quantities that seem outrageously extreme in size or magnitude.

While the frequencies listed in Table 3.1 and Figure 3.3 are very large numbers, the wavelengths of the radiations are generally quite small by our everyday standards. Again using red light as an example, we see from Table 3.1 that its wavelength is about 3×10^{-5} inch = 0.00003 inch. This means that 30,000 wave crests of red light stretched out one after the other, as they appear in Figure 3.1, cover only 1 inch.

THE SPEED OF LIGHT

There is a very interesting association between the frequency and the wavelength of a wave. Suppose we see a wave (not radiation) coming toward us with a frequency of, say, 6 cycles per second. Then six wave crests will pass us each second. Let us also suppose for this example that the wavelength is 4 inches. Then the wave crests are 4 inches apart, and this 4 inches is measured along the direction that the wave is moving. Now if six objects pass us in single file each second, and if these objects are 4 inches apart, then the objects must be moving 6×4 or 24 inches per second (see Figure 3.4). In other words, the speed at which the wave is moving equals the product of its wavelength multiplied by its frequency. This is true not only of radiation but of other waves, too.

It is an observed fact that when matter is not around to interfere with it, all radiation travels at the same speed. This is commonly referred to as the "speed of light," although this speed is the same speed as that of gamma rays, radio waves, and so on. This speed is about 186,000 miles per second or nearly 670 million miles per hour. We can now understand why higher frequency radiation has a shorter wavelength. The product of frequency and wavelength is equal to the speed of light,

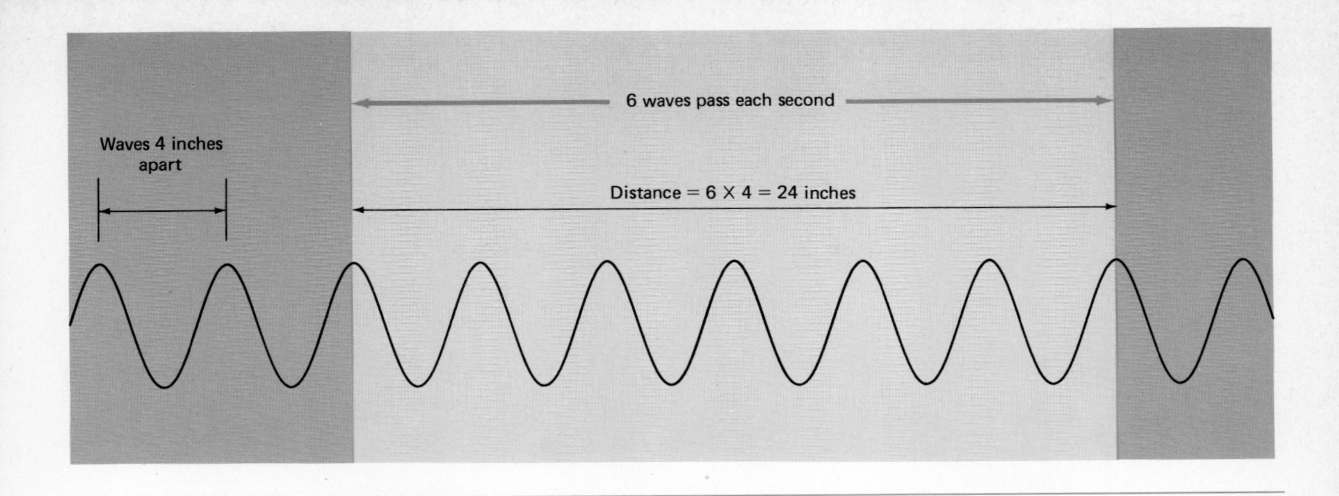

FIGURE 3.4
The speed of a wave is equal to its frequency multiplied by its wavelength.

and this is the same for all radiation. If the product of two numbers always has the same value, then as one of the numbers is made larger, the other must become smaller.

3.2 Photons

There is an interesting and rather mysterious way in which radiation differs from other types of waves: Radiation comes in packages or pieces, called *photons*. Each photon has its own frequency and wavelength and moves with the speed of light. It is difficult to understand how waves can come in packages. Figure 3.5 shows a representation of a photon. The wave is appreciable only in a small region, and it rapidly goes to zero within the distance of a few wavelengths. Photons are the "fundamental particles" of radiation.

PHOTONS AND ENERGY

Each photon has a certain amount of energy, and that energy is proportional to the photon's frequency. Gamma rays have the highest frequency of all radiation, and they also have the greatest energy of all photons. A radio-wave photon has the least energy of all, and visible-light photons are intermediate in energy.

Photons can be created and destroyed by matter; these processes are called *emission* and *absorption*. We have already seen that a

FIGURE 3.5
A schematic representation of a photon.

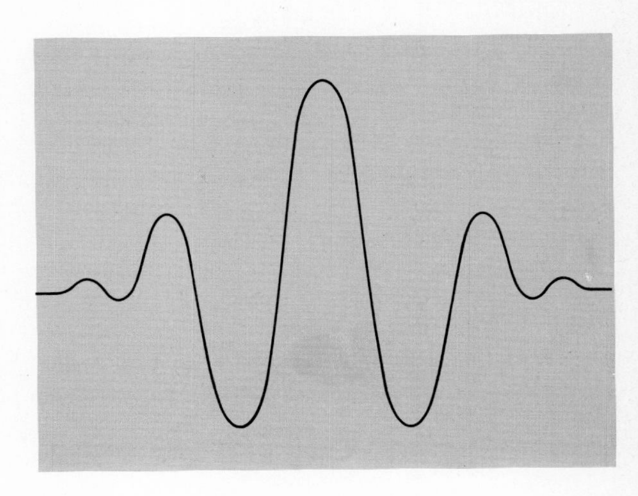

vibrating charge causes radiation to be emitted. Such a charge creates photons and sends them off in all directions. The photons have a frequency that depends on the vibration rate of the charge. The photons carry off a certain amount of energy, and that energy must be provided by the charge. Thus the charge will slowly lose its energy of vibration to the photons it creates, and its oscillations will slowly decrease as time goes on unless another energy souce keeps it going. Whenever a piece of matter emits radiation, it loses the energy that is carried off by that radiation. Likewise when matter absorbs radiation, it gains the amount of energy that is brought in by the absorbed photons.

RADIATION IN OUR LIVES

The higher frequency photons—ultraviolet rays, x rays, and gamma rays—have such large energies that they can do considerable damage to objects that absorb them, and they are highly injurious if absorbed in large amounts. The Sun radiates dangerous amounts of high-energy photons, but the air around us absorbs practically all photons of frequencies greater than about 10^{15} cycles per second. Thus on the surface of the Earth we are protected from most of the Sun's harmful radiation although enough ultraviolet photons do get past the Earth's atmosphere to be able to cause a good sunburn. Ultraviolet rays, x rays, and gamma rays all have well-known medical applications, but care must always be taken that a person does not receive too much exposure to them.

Infrared photons have a rather low energy content, but a large enough number of them together can make up for their low energy. Logs burning in the fireplace emit very large numbers of infrared photons, and these produce a warming effect as they are absorbed by the skin. Ovens operate by concentrating infrared rays or, in the case of microwave ovens, high-frequency radio waves on food and heating it by the absorption of the photons.

What we hear coming from radios are not radio waves but sound waves. Sound consists of vibrations of the atoms and molecules in matter and is a totally different kind of wave from radiation. A radio station sends out or broadcasts radio waves, and a radio is able to absorb the radio photons and convert their energy to sound. It is the sound we hear, not the radio waves. Similarly, television waves, which are rather high frequency radio waves, carry energy that a TV set is able to change into visible light.

Radar is another example of the practical application of radio waves. Radar uses radio waves which have frequencies of about 10^{10} cycles per second. A transmitter sends out radar photons, and if a solid object is nearby, photons will reflect off of it. Some of the reflected photons will come back to a receiver, which accurately times their arrival. Since the photons are known to travel with the speed of light, measuring the time interval between the transmitted and reflected signals indicates how far away the reflecting object is. Radar is commonly used to locate airplanes, ships, and speeding cars. Astronomers use radar to find very accurate distances to the Moon and the nearer planets and to study their surfaces.

HEAT AND RADIATION

Causing an electric charge to vibrate is not the only way to create photons. Consider the atoms and molecules that make up a substance. They are constantly moving around as a result of the heat content of the substance; the higher the temperature, the more violent are their motions. The particles are not usually disturbed by the collisions they have with each other—they bounce off like billiard balls. Sometimes, however, the electric forces that hold together an atom or molecule are strained by a collision. Part of the energy of the collision is stored as potential energy, and the atom or molecule is said to be excited to a higher potential energy. It is as though the electric forces between the electrons and the nuclei were springs, and the collision stretched the springs. The particle wants to relieve the strain by snapping the springs back to their original positions. This action releases the stored up potential energy. Where does it go? It may go back to the heat energy of the substance, or it may appear as a photon, created on the spot. In this way, part of the heat energy of any substance is constantly being changed into radiation.

What happens to a photon that is emitted in the way just described? It travels a certain distance, and then it probably will be absorbed by another atom or molecule that it happens to meet. The photon disappears, and its energy is taken up by the particle, which strains it and excites it to a higher potential energy. Once again the particle wants to relieve the strain by getting rid of this stored-up potential energy; once again it can do so by adding it to the heat energy of the substance or by creating new photons and sending them on their way. In this way there is a continual exchange between the heat energy of a substance and radiation.

An object that has any heat energy at all will be constantly emitting and absorbing photons in the way just described. Photons that are emitted by an atom or molecule somewhere in the middle of the object probably will be absorbed by other particles before they travel very far. Photons emitted at the edge of an object, however, can escape from it without being absorbed. These escaping photons carry energy away from the object, draining it of its heat. Any object, if left completely to itself, will eventually radiate away all of its heat energy and cool down to a temperature of absolute zero—the temperature at which essentially all molecular motion ceases. This occurs at about −460°F. In practice it is not possible to completely isolate an object, so it always receives at least a little energy from its surroundings. This is what keeps an object's temperature above absolute zero.

The rate at which radiative energy escapes from an object obviously depends on how hot it is. A high temperature means that more heat energy is available to be converted into radiation. Temperature also affects the type of photons being emitted. At low temperatures, the collisions between the particles are mild, and only slight strains can result. This leaves only small amounts of energy available for photon emission; hence the photons will be of low energy and of low frequency. At high temperatures, large collision energies are available, the particles can be excited to very high potential energies, and high-frequency photons can be emitted. At room temperature, the molecules in an object do not have enough energy to emit significant numbers of visible-light photons; the emission from such an object is almost completely in the low-energy infrared and radio regions. If a hot object emits lots of x rays, on the other hand, it must have an extremely high temperature. A cool star emits more red photons than blue, while a hot star radiates more blue than red. In these examples it should be clearly understood that we are talking about photons that are created or emitted in the object itself. We can see a table because it reflects light that was emitted by a lamp or by the Sun. The heat energy of the table has nothing to do with the radiation it reflects.

3.3 Spectra

The radiation emitted by any object is a mixture of photons of different frequencies. We can get some important information about the object by adding up the energies of all the photons it emits each second. Even more important is the information about the quality of the radiation; in other words, an indication of how many photons there are at the different frequencies. Any kind of display, such as a listing or graph, that shows how the radiation

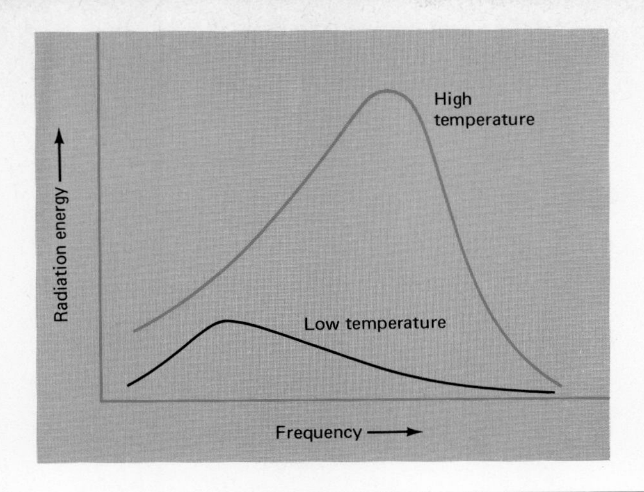

FIGURE 3.6
Continuous spectra of a high-temperature object (top curve) and of a low-temperature object (lower curve).

given off by an object is distributed over different frequencies is called a *spectrum* of the object. Most of the things astronomers are interested in depend on the measuring and the understanding of spectra of all kinds of astronomical objects.

RAINBOWS AND COLORS

Spectra are not mysterious things known only to scientists. Rainbows are spectra of the Sun—we can see the visible-light part only, of course. When all the different frequencies or colors of light are mixed together, as in sunlight, our eyes are fooled into seeing only white light. We cannot see the different colors even though they are there. Raindrops have the ability to take apart white light into its component frequencies so we can then see the colors separately. The result is a display of colors that is the visible-light part of the spectrum of the Sun.

When sunlight strikes grass, the photons of most frequencies are strongly absorbed by the grass. Green light, however, is strongly reflected off into other directions. The spectrum of sunlight reflecting off of grass, therefore, is very strong in green light, and grass looks green. Any object tends to leave its fingerprints on the radiation that it emits or reflects, and so its spectrum is a very strong clue to what it is. Of course there are many green objects that have about the same color as grass, but if we can measure their spectra in great detail, we can easily tell whether they are grass or not. This is the principle astronomers use to try to interpret spectra in astronomy.

CONTINUOUS SPECTRA

Now let's be more specific. A solid or liquid object emits some radiation at essentially all frequencies; the higher its temperature, the more radiation it gives off, and the more the energy is concentrated at the higher frequencies. This case is called a "continuous spectrum" because the energy emitted changes in a smooth or continuous way with the frequency. Figure 3.6 shows two examples. What is interesting is that the continuous spectra of solids and liquids do not depend on the kind of object, only on its temperature. At room temperature, most of the radiation given off by solid or liquid objects has a frequency in the neighborhood of 10^{13} cycles per second, in the infrared. These rays are constantly being emitted and absorbed by walls, trees, the ground, and us. If we could see infrared rays, it would never seem to get dark, even at night or in a closed room. The visible light given off at room temperature is far too small for us to be able to see it. An object must be rather hot, close to 1000°F, before it emits enough light to be easily

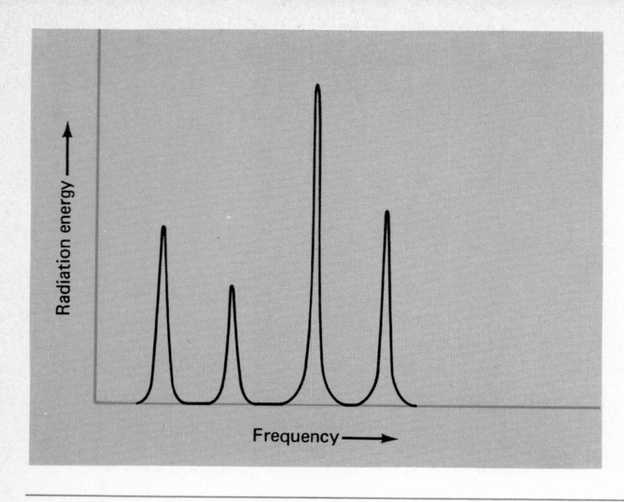

FIGURE 3.7
An emission spectrum.

seen. This again refers to radiation that is created in the object itself. We see most things around us only because they reflect sunlight or light from a lamp.

EMISSION LINES

The spectrum of a gas is usually more complicated than the spectra we just discussed, but the spectra of gases are also more important because most astronomical objects are gaseous. If a gas is thick enough or under high enough pressure, it will tend to have a continuous spectrum the way that a solid body does. When this is not the case, a radiating gas will have a spectrum that has a lot of energy at a number of special frequencies and almost nothing in between.

The narrow regions where most of the energy is emitted are known as *emission lines* (see Figure 3.7). The frequencies at which these emission lines occur depend on what the gas is made of. Each separate atom and molecule has its own special set of frequencies at which it likes to emit photons; when these have been found for a number of substances by experiments in the laboratory or by calculations, these substances can be readily identified if they occur in any emitting gas. For example, we know that hydrogen gas emits a certain pattern of emission lines, so when we see that pattern reproduced in an emission spectrum, we can be sure that hydrogen gas is responsible. The detailed appearance of the emission lines also depends on the temperature and the pressure of the gas. This makes things more complicated, but it also makes them more interesting, for it means that the spectrum can tell us that much more about the gas that is emitting it. It should be emphasized that, depending on the type of material and its temperature, the emission lines can occur in any part of the spectrum, from the gamma-ray region to radio frequencies.

ABSORPTION LINES

If a gas is too cool, it will not give off emission lines; however, this does not mean that a cool gas is of no importance in studies of spectra. All materials absorb radiation very strongly at the same frequencies that they emit strongly. If we send radiation through a cool gas, the gas will absorb some of the radiation at certain special frequencies; these special frequencies are exactly the same ones at which the gas has emission lines (when it is hot enough to have them). These narrow regions of frequency in which radiation is taken out by absorption are called *absorption lines*. Just as in the case of emission lines, the detailed appearances of the absorption lines depend on the type of absorbing gas as well as its temperature and pressure. It also depends, of course, on the presence of the background radiation that is being absorbed; if no radiation is passed through the

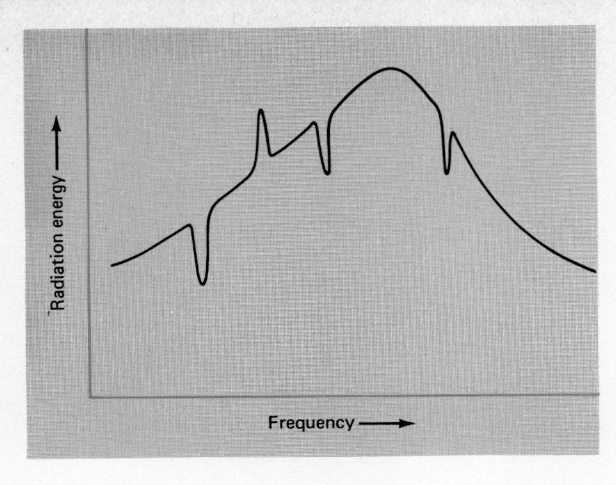

FIGURE 3.8
Absorption and emission lines on top of a continuous spectrum.

cool gas, there will be nothing to be absorbed and hence no absorption lines. Spectral lines, both in absorption and emission, are easily the most important sources of information that the astronomer has (see Figure 3.8). We will be referring to them constantly throughout the rest of the book.

REFLECTION

The reflection of sunlight from grass was mentioned earlier. This is an example of yet another type of spectrum, called a "reflection spectrum." Here the object does not produce its own radiation but simply reflects some of the radiation that is incident upon it. Such a spectrum depends both on the nature of the original source of radiation and on the reflecting properties of whatever is reflecting the radiation. This spectrum can be more difficult to understand than the other types of spectra, but it has important applications in astronomy. Examples are the reflection of sunlight by planets and satellites and the scattering of sunlight and starlight by very tiny particles which exist in space. We use reflection spectra all the time. We can usually tell a block of wood from a block of metal just by looking at them. These two objects do not look the same, and the reason is that each has its own way of reflecting light. All objects have their own reflection spectra, and we learn to recognize objects by these spectra.

SYNCHROTRON RADIATION

We have described how objects change part of their heat energy into radiation. Radiation produced in this way is called "thermal radiation." Not all radiation, however, comes from the heat energy of some object. Perhaps the most important discovery in astronomy in recent years is that there are tremendous amounts of nonthermal radiation being emitted into space by various kinds of objects. If electrically charged particles, such as electrons, are given very high speeds and move in a magnetic field, they emit a type of nonthermal radiation called "synchrotron radiation." As we noted in Chapter 1, many objects have spectra that indicate that the radiation they emit comes from the synchrotron process rather than from the thermal processes previously discussed.

An object that emits synchrotron radiation has a continuous spectrum, but it differs in a number of ways from the spectra shown in Figure 3.6. In particular synchrotron spectra have nothing to do with the temperatures of the emitting objects since this type of radiation does not come directly from the heat energy of the objects. Also, the details of how the photons are distributed over the different frequencies are quite distinct in the synchrotron case. An expert usually has no difficulty in distinguishing between a synchrotron and a thermal spectrum.

It is important to understand that a photon of a certain frequency is exactly like any other photon of the same frequency. There is no way of knowing from an examination of a single photon whether it was emitted by a synchrotron process, by a thermal process, or by a vibrating electron. Only by examining very large numbers of photons and seeing how they are distributed over the frequencies, that is, by studying spectra, can we hope to understand how these photons originated.

IMPORTANT WORDS

Electric field
Magnetic field
Electromagnetic wave
Wavelength
Frequency
Radio wave
Infrared ray
Visible light
Ultraviolet ray
X ray
Gamma ray
Photon
Emission
Absorption
Spectrum
Emission line
Absorption line

REVIEW QUESTIONS

1 How can an electron or a proton be made to emit radiation?

2 Why can't a neutron emit radiation in the same way?

3 What is an electric field?

4 What is the wavelength of a wave?

5 What is the frequency of a wave?

6 What are the six types of radiation?

7 How do our eyes distinguish different frequencies of light?

8 About how many times each second does red light vibrate?

9 How are wavelength and frequency related to the speed at which a wave is traveling?

10 How fast is the speed of light?

11 What is a photon?

12 What kinds of photons have the most energy? The least energy?

13 What are emission and absorption?

14 Why are ultraviolet rays, x rays, and gamma rays potentially dangerous?

15 Why don't the x rays and gamma rays emitted by the Sun cause us any harm?

16 What rays give us the warm feeling we have when we stand near an open fire?

17 Can we hear radio waves? Can we see TV rays?

18 How can collisions between atoms or molecules cause them to emit photons?

19 Why will an object cool off if left completely by itself?

20 How does the temperature of an object affect the kinds of photons the object radiates?

21 What is a spectrum?

22 What is a rainbow?

23 What makes grass look green?

24 How does an emission-line spectrum differ from a continuous spectrum?

25 How can the emission and absorption lines in the spectra of objects help us to know what the objects are made out of?

26 What does the term "thermal radiation" mean?

27 What is an example of nonthermal radiation?

28 Does a "synchrotron photon" look or act any differently from a "thermal photon" of the same frequency?

QUESTIONS FOR DISCUSSION

1 A freight train consisting of a long string of box cars is somewhat like a wave. The length of each car is the wavelength, and the number of cars that go by each unit of time (second or minute) is the frequency of the wave. Convince yourself that the speed of the train is equal to its wavelength multiplied by its frequency just as it is with a wave.

2 Why can a magnifying glass be used with sunlight to ignite paper?

3 If a house were coated with a strongly reflecting material in the summer and a strongly absorbing material in the winter, fuel would be saved. How?

4 A red star is usually cooler than a blue star, but a red book is not cooler than a blue book. What is the difference?

5 A piece of metal will become quite hot when placed in the sunlight, but a piece of glass won't. Why?

6 A gaseous substance is known to have a number of spectral lines at certain frequencies. Suppose the spectrum of a star shows, among other things, the same pattern of absorption lines as that substance except that the frequencies of each of the lines in the star are exactly one-half the frequencies of the corresponding lines in that substance. Do you suppose that substance exists in that star? Why or why not?

REFERENCES

Any elementary book on physics or on physical science will contain a discussion of radiation. A good introductory book is *Seeing Beyond The Visible* (Elsevier, New York, 1970), edited by A. Hewish.

Infrared rays are discussed by D. A. Allen in *Infrared: The New Astronomy* (Wiley, New York, 1975). An interesting story of the radiation that bombards the Earth from all directions is contained in *The Light of the Night Sky* by F. E. Roach and J. L. Gordon (Reidel, Dordrecht, Holland, 1976).

FOUR
INSTRUMENTS OF ASTRONOMY

Astronomers have developed some rather specialized instruments to help them gather observational data about the Universe we live in.

4.1 Observing

APPEARANCES

We can readily imagine the size, shape, and general make-up of familiar objects, such as a golf ball, a house, a tree, or a mountain. We have seen them many times from close up and from far away, and we have pretty good ideas of what they can or cannot do. But it is doubtful that anyone can honestly say that the Sun looks like a huge ball nearly a million miles across and about 100 million miles away. Sizes, distances, temperatures, masses, and so on, of objects in astronomy are so different from those that we are used to that our experience doesn't help us very much in understanding them. What is a million-mile-size ball supposed to look like anyway?

The appearances of things in astronomy are so misleading that common sense and everyday experiences have actually interfered with advances in understanding. For example, it just didn't seem reasonable that anything as heavy as the Earth could possibly be made to move, so throughout most of human history it was believed that the sky turned around the Earth once each day. Only by gathering large numbers of observed facts and by studying them over long periods of time have scientists gained much insight into the behavior of nature. An important part of this advance was made possible by the instruments that scientists have developed in order to improve the accuracy and efficiency of collecting facts and data.

EXPERIMENT AND OBSERVATION

Most scientific work is either *experimental* or *theoretical*. Theoretical work involves scratching the head and trying to understand, interpret, and predict the facts and data that have been or will be collected. Experimental work involves setting things up to see what will happen in a given set of circumstances. There is only a very limited amount of experimentation in astronomy, however, because the astronomer usually doesn't have any control over the object he is studying. The experimenter does have this control and actively sets up things so he can measure what he is interested in. The astronomer usually is forced to passively sit back and measure whatever the Universe chooses to reveal of itself. Thus in astronomy one speaks of theory and observation rather than theory and experimentation. Of course, there is some experimentation that can be carried out in astronomy, especially now that we can go into space, but passive observation is now and always will be the main source of astronomical data.

Some people like to accept only direct information. They cannot understand how we can know anything about the Sun, planets, or stars if we have never been there. (Mark Twain claimed to be surprised that astronomers could even find out the names of the stars, let alone their other properties.) Indeed our only source of information is the feeble radiation that manages to survive the long journey through space to reach Earth. Astronomers claim to be able to read that radiation and tell all or very much about the objects that emitted it. Magicians and frauds can make the same claim, but I hope to be able to convince you that astronomers can really do this and are not necessarily either tricky or fraudulent.

4.2 The telescope

The telescope is easily the most important instrument that the astronomer has. The main thing a telescope does is collect as much radiation as possible and bring it to a *focus*. Other instruments are placed near this focus to record or measure the radiation collected by the telescope.

MIRRORS AND LENSES

Either a *mirror* or a *lens* can be used to bring radiation to a focus. A mirror, of course, reflects the radiation back from its surface, while a lens bends the radiation that passes through it. When radiation passes from one substance to another, such as from air to glass, it changes its direction; this is known as *refraction*. By shaping a piece of glass in exactly the right way, we can make a lens that refracts radiation in a way that will bring it to a focus. Similarly it is possible to make a mirror into the proper shape so that it will reflect radiation to a focus. To do this a mirror must be curved and not flat like the ones in our homes.

Figure 4.1 shows how a lens and a mirror bring radiation to a focus and form an *image*. Consider first the lens. Radiation from point A (the point of the arrow) spreads out in all directions. Part of this will pass through the lens. If the lens is properly made, the radiation coming from point A that passes through the lens will be bent or refracted in such a way that all of it will pass through a second point, A', on the other side of the lens. Radiation from A is focused at A'; A' is called the "image" of A. Similarly radiation from point B is focused at B', and so for all points on the arrow between A and B. The line $A'B'$ is the image of the arrow AB.

The curved mirror at the bottom of the figure also forms an image $A'B'$ of the arrow AB. All the radiation from point A that strikes the surface of the mirror is reflected back through the point A'; radiation from B is focused at B' in the same way. The images formed by both the lens and the mirror are upside-down.

These images are quite real in the sense that placing a flat object at the position of $A'B'$ will cause a picture of the arrow to appear on the surface of that object. If the flat object is moved in or out, toward or away from the lens or mirror, the picture will go out of focus and become blurred. If we place photographic film at $A'B'$, we can take a photograph of the arrow. We can also put other lenses or mirrors near position $A'B'$ in order to magnify or do other things to the image. In the case of the mirror, any instruments placed at $A'B'$ to measure or record the image will block out some of the radiation coming from the arrow AB. If too much radiation is blocked out, very little will be able to get through to the mirror, and the image will be very faint. Often other mirrors

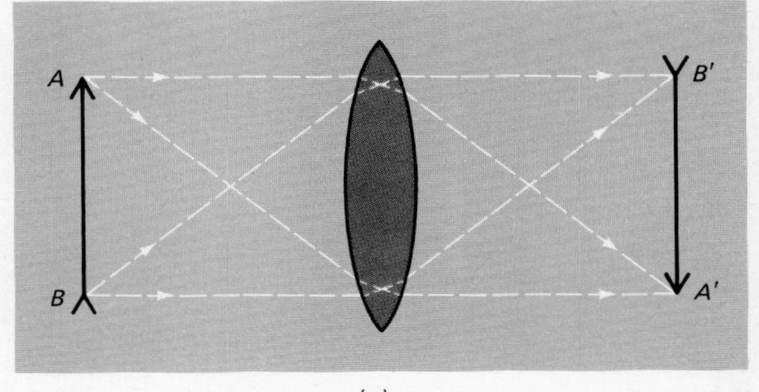

(a)

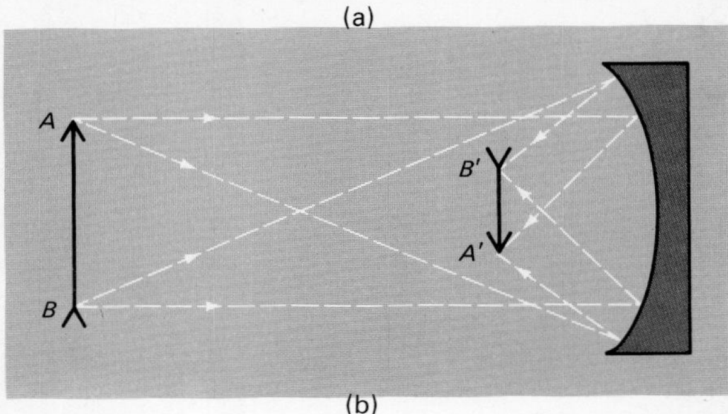

(b)

FIGURE 4.1

The lens and the curved mirror bringing light to a focus. In both cases $A'B'$ is the image of the arrow AB. (a) The lens. (b) The curved glass mirror.

FIGURE 4.2
The Lick 36-inch refracting telescope. The instrument is so well balanced that it can be easily moved by hand. (Lick Observatory photograph)

FIGURE 4.3
The Mount Palomar 200-inch reflecting telescope. The 17-foot mirror is at the bottom of the telescope. (The Hale Observatories)

are used to move the image out of the way of the incoming radiation.

REFLECTORS AND REFRACTORS

The most important part of a telescope is a large mirror or lens that collects radiation and brings it to a focus. Most astronomical objects are very faint, so astronomers want their telescopes as large as possible in order to collect more radiation. A large telescope is not necessarily long, but it has a large mirror or lens. A telescope is usually identified by a number giving its size and a word telling whether it uses a lens or a mirror. A 16-inch refractor is a telescope that has a lens that is 16 inches in diameter, while a 36-inch reflector has a mirror that is 36 inches across. The largest lens telescope in the world is the 40-inch refractor at Yerkes Observatory in Williams Bay, Wisconsin. Figure 4.2 shows the 36-inch refractor of the Lick Observatory in California. The largest mirror telescopes used for visible light are the 236-inch reflector in Russia and the 200-inch reflector on Mt. Palomar in California (see Figure 4.3). The model of an unusual telescope with six mirrors is shown in Figure 4.4. Mirror telescopes also can be specially made to collect radio waves from space. The largest of these is the 1000-foot *radio telescope* at Arecibo, Puerto Rico (see Figure 4.5).

The most common telescopes are those

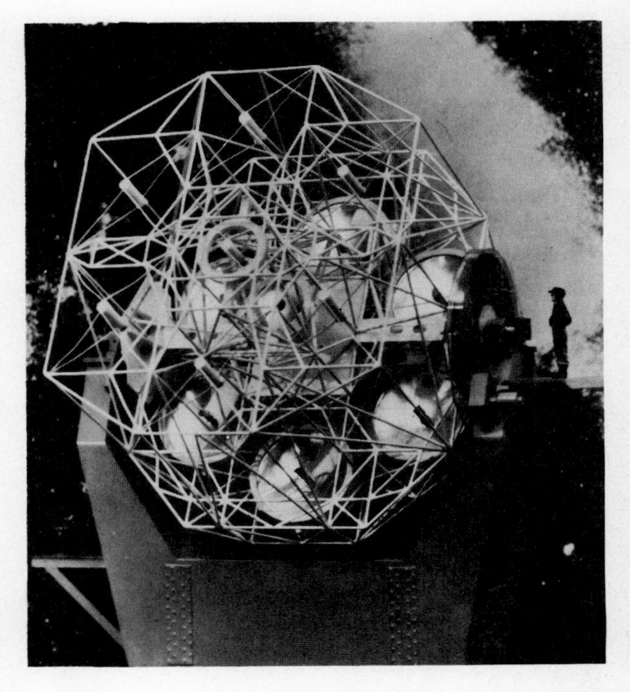

FIGURE 4.4
A model of the multiple-mirror telescope being constructed as a joint project by the Smithsonian Astrophysical Observatory and the University of Arizona. This unique telescope will have six 72-inch mirrors and is equivalent in light-gathering power to a 176-inch telescope. (Steward Observatory, University of Arizona)

FIGURE 4.5
An aerial view of NAIC's Arecibo Observatory upgraded 1000-foot radio/radar telescope. The National Astronomy and Ionosphere Center is operated by Cornell University under contract with the National Science Foundation.

designed to collect and focus visible light. The smaller ones of these, less than 12 to 15 inches in diameter, can be either lens- or mirror-type; however, it is much easier to make a large mirror accurately than it is to make a large lens since only the reflecting surface of a mirror, as opposed to both surfaces plus the interior of a lens, needs to be of high optical quality. For this and other reasons, all of the largest telescopes are of the mirror variety.

The mirror or lens of a telescope should not have any bumps or rough spots that interfere with its focusing. Radiation is not much affected by rough spots that are much smaller than the wavelength of the radiation. For an *optical telescope*, that is, one used for visible light, the mirror or lens must be smooth to within a few millionths of an inch. On the other hand, a radio telescope designed to collect radiation with a wavelength of one foot can have bumps as big as an inch without having its efficiency affected very much. Thus an optical telescope must be made much more carefully, and it is much easier to make very large radio telescopes (see Figures 4.5 and 4.6). An optical-telescope mirror is usually glass or quartz and is covered with a very thin aluminum surface to reflect light better. A radio-telescope mirror is made of thin metal foil or wire mesh since these materials reflect radio waves quite well.

FIGURE 4.6
The 300-foot radio telescope of the NRAO. (National Radio Astronomy Observatory)

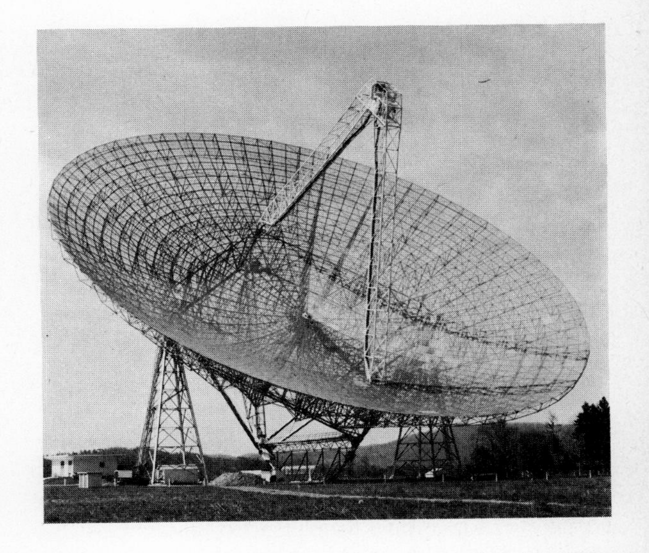

Radio telescopes can be made almost completely weatherproof so that they do not need to be protected inside observatory buildings. Figure 4.7 shows several radio telescopes that are used together in order to improve observational accuracy.

Many so-called radio telescopes do not really reflect or focus the radio waves; they just absorb the energy of the waves directly and change the energy into an electric signal in the way that the antennas in radio sets at home do. Because they don't focus the radiation,

FIGURE 4.7

Three radio telescopes that can be operated together to improve the efficiency of the measurements. (National Radio Astronomy Observatory)

these instruments are not really telescopes at all but simply radio receivers or antennas.

Astronomers naturally want to be able to point the telescopes to any part of the sky, so telescopes should be made movable. The Earth's rotation causes the stars to move across the sky from east to west, and the larger optical telescopes have a motor that slowly turns the telescope westward, following the stars. In this way the telescope can continue pointing at the same star for a long time. Radio telescopes are often too large to be easily moved, and many can be pointed only in a limited way. The mirror of the 1000-foot radio telescope at Arecibo is carved out of a natural depression in the ground and does not move at all. Only part of the sky can be measured with that instrument.

No telescope is any good if radiation cannot get to it. As we saw in Chapter 3, radiation having frequencies greater than about 10^{15} cycles per second cannot pass through the Earth's atmosphere. Such radiation cannot reach the ground from outer space, so it wouldn't do any good to set up telescopes on the ground for measuring ultraviolet rays, x rays, or gamma rays from space. We must get above the atmosphere, in rockets or artificial satellites, to be able to measure this very high frequency radiation from space.

4.3 Auxiliary equipment

DIRECT OBSERVATION

The main job of an observing astronomer is to take the radiation that has been collected by a telescope and convert it into useful information. There are three things that an astronomer can do with the collected radiation: (1) look at it, if it happens to be visible light or if it can be converted into visible light; (2) photograph it; or (3) send it through an instrument that will measure its energy.

To look through a telescope, a small lens called an *eyepiece* is placed near the image. The eyepiece acts like a magnifying glass, making the image larger for viewing. Astronomers do not often look through telescopes because they can usually get more accurate and useful information in other ways, but sometimes direct viewing is scientifically useful. Of course it can be interesting even when it is not scientifically useful.

PHOTOGRAPHY

Photography is a well-known part of astronomy. Certain chemical substances are sensitive to radiation, and they are put on the surface of a thin film or glass plate. If the film or plate is placed where radiation is being brought to a focus, such as at the position of $A'B'$ in Figure 4.1, the image will form on the chemical substance. Further chemical processes can then develop the image and make it permanent, and we then have a picture of whatever gave off or reflected the radiation. This is how a telescope is used as a camera. The permanent records available in the form of photographs of the sky are an extremely important source of vital information in astronomy.

One nice feature of photographs is that the response depends on how much radiation the film or plate is exposed to. If a galaxy is too faint to record on a short exposure, one makes the exposure longer, piling up more and more radiation until the image finally does come through. There is a limit on how far one can go in this way, but exposures of several hours with large telescopes will photograph stars and galaxies that are too faint to be seen by just looking through the same telescope.

It should be noted that all stars except the Sun are much too far away for us to be able to see or photograph their actual disks. The true sizes of the stars are much smaller than their images on a photograph because the light is diffused over a larger area. When one star has a larger image than another on a photograph, it doesn't mean that the first star is larger than the second but that it appears brighter.

Photography can also be used to measure the energy of radiation because a bright star will leave a stronger image on the plate or film than a faint star. There are different types of photographic plates sensitive to different kinds of radiation. A blue-sensitive plate, for example, will readily make a picture from the blue light to which it is exposed, but it will ignore the red light falling on it. Astronomers can photograph the same star twice, once with a blue-sensitive plate and again with a red-sensitive plate. By comparing the two pictures, they can determine whether the star radiates more red light or more blue light. This is an important clue in finding the temperature of the star. Astronomers can do the same thing by using *filters* with the telescope. A red filter allows

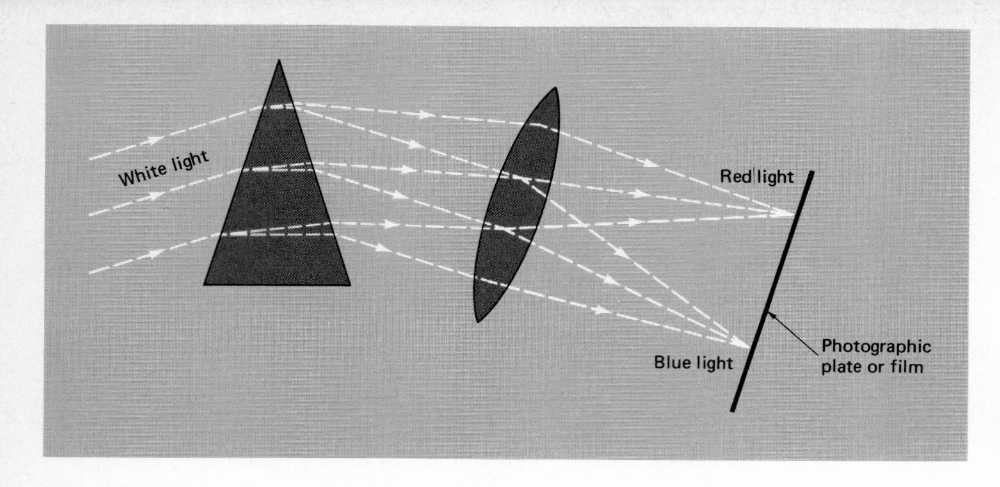

FIGURE 4.8
The spectrograph, used for measuring or recording spectra.

only red light to pass through it, and all other radiation is absorbed by the filter. Filters can be made for any frequency of radiation desired. Photography is by no means limited to visible light; photographic chemicals are available for radiation of all frequencies between infrared and gamma rays.

THE PHOTOELECTRIC CELL

There are a number of instruments designed to measure the energy of the radiation collected by a telescope. One such instrument is the *photoelectric cell,* sometimes called an "electric eye." The photoelectric cell has a special surface that gives off an electric signal when radiation strikes it. By amplifying and recording this electric signal, astronomers can measure the energy in the radiation very accurately. Photoelectric measurements are usually made with different kinds of filters so that energies at different frequencies of the radiation can be found. This is one way that the spectra discussed in Section 3.3 are obtained.

THE SPECTROGRAPH

The most important astronomical instrument next to the telescope itself is the *spectrograph.* As the name suggests, this instrument is used to determine the spectra of objects.

Figure 4.8 shows the main parts of a spectrograph. Radiation comes in from the left and passes through a glass prism. This example is designed for visible light only and is the most common kind, although spectrographs can also be made for use with other frequency regions. The prism separates the different frequencies of light and causes them to travel in slightly different directions. A lens then brings the light into focus, and each frequency has its own focus position. In Figure 4.8 a photographic plate is put at the focus points, and a picture is taken of the spectrum. One could also put a photoelectric cell at the focus instead of the photographic plate. By moving the cell back and forth (or by rotating the prism), the energy of the radiation could be measured one frequency at a time. The spectrograph takes apart the radiation so that some other instrument can measure the frequencies separately to obtain the spectrum.

The top part of Figure 4.9 represents a photograph of the spectrum of a star. The photograph is a negative, which means that intense radiation shows up dark and weak radiation is light. The spectrum is continuous, with absorption lines superposed. The bottom of the figure is a graph showing how much energy is in the spectrum at each frequency. There is an energy deficit at each absorption line.

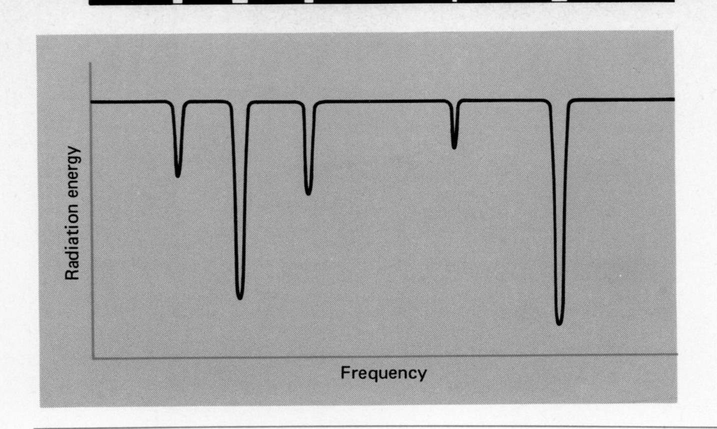

FIGURE 4.9
A stellar spectrum. At top is a simulated direct photograph of the spectrum, with the frequency of radiation changing from left to right. This photograph is a negative. The graph shows the same spectrum converted to radiation at the different frequencies. Note that the absorption lines show up light on the photograph and correspond to dips in the energy curve.

No single piece of equipment will record accurately radiation of all frequencies. Various kinds of instruments have been designed to measure special frequency intervals, and many kinds are needed if one wants the spectrum of an object over all frequencies.

In trying to find an instrument that will give accurate measurements in a given frequency of radiation, the scientist must find a substance that is sensitive to radiation of that frequency. The substance must give a reaction or a response when the given type of radiation strikes it, and that response must be measurable. The antenna of a radio telescope responds to radio waves by producing an electric signal, similar to a photoelectric cell, and the radio astronomer amplifies and accurately measures this electric signal. X rays and gamma rays can be made to cause electric pulses when they pass through certain pieces of equipment, so their detection also depends on the measurement of electric signals.

By far the major part of the energy reaching the Earth from outer space is in the form of visible light. There are many objects, however, that have a significant amount of energy in other regions of the spectrum. It is important to the astronomer to be able to measure all kinds of radiation, even if this requires carrying the instruments up and out of the Earth's atmosphere. Advances in astronomy depend strongly on those scientists who are constantly looking for improved ways of making these measures.

IMPORTANT WORDS

Experimental	Image
Theoretical	Eyepiece
Focus	Filter
Mirror	Photoelectric cell
Lens	Spectrograph
Refraction	

REVIEW QUESTIONS

1 Why don't our everyday experiences help us very much in understanding the sizes, distances, and other properties of astronomical objects?

2 How do instruments help in the advancement of science?

3 What is the difference between experimental and observational work?

4 Can any experimental work be carried out in astronomy?

5 What is the main purpose of an astronomical telescope?

6 What are the two types of objects that can bring radiation to a focus?

7 What does refraction mean?

8 Why do astronomers like their telescopes to be as large as possible?

9 How big are the largest lens and mirror telescopes in the world?

10 How big is the largest radio telescope?

11 Why does the mirror of an optical telescope need to be much more carefully made than that of a radio telescope?

12 Why are the larger optical telescopes provided with motors?

13 Why would it do no good to build an x-ray or a gamma-ray telescope on the surface of the Earth?

14 What are the three things an astronomer can do with the radiation collected by a telescope?

15 Can a telescope be used as a camera to take pictures?

16 How is a photoelectric cell used to measure radiation energy?

17 What is a spectrograph?

18 Why are electric and electronic equipment quite important to astronomers?

QUESTIONS FOR DISCUSSION

1 Name several ways in which astronomers can do experiments as opposed to just making observations.

2 What would happen to the image *A′B′* if the top half of the lens in Figure 4.1 were covered so light couldn't get through it?

3 Suppose a red filter and a blue filter are used together in a telescope for taking a photograph of some stars. What radiation would be recorded?

4 One problem with trying to measure infrared rays coming from space is that at room temperatures, the measuring equipment itself tends to emit in the infrared, interfering with the signal to be measured. How might this problem be alleviated?

5 Some x-ray and gamma-ray detectors are like little black boxes that go "ding" whenever an x-ray or gamma-ray photon passes through them. How could you set things up so that you could tell what directions the photons were coming from?

REFERENCES

The principles of image formation and the details of how telescopes work can be found in any elementary textbook on physics or optics. An excellent introductory book is *Tools of the Astronomer* by G. R. Miczaika and W. M. Sinton (Harvard University Press, Cambridge, 1961). An older book of the same type is *Telescopes and Accessories* by G. Z. Dimitroff and J. G. Baker (Blakiston, Philadelphia, 1945). A book that is more advanced but mostly understandable to the interested layman is *Telescopes*, edited by G. P. Kuiper and B. M. Middlehurst (University of Chicago Press, Chicago, 1960).

The interesting story of the development of the 200-inch telescope is given by D. O. Woodbury in

The Glass Giant of Palomar (Dodd, Mead, New York, 1939).

The story of radio astronomy is told by J. S. Hey in *The Evolution of Radio Astronomy* (Neale Watson, New York, 1973). A good article on radio telescopes and some of the things they are used for is "Giant Radio Galaxies" by R. G. Strom, G. K. Miley, and J. Oort in *Scientific American* [233 (Aug. 1975), 26].

Two good books for those who wish to build or buy their own telescopes and use them are *All About Telescopes* by Sam Brown (Edmund Scientific Company, Barrington, New Jersey, 1967), and *Telescopes for Star-Gazing* 2nd ed., by H. E. Paul (American Photographic Book, Garden City, New York, 1966). An unusual book for those who want to build their own radio telescope is *Radio Astronomy for the Amateur* by Dave Heiserman (Tab Books, Blue Ridge Summit, Pennsylvania, 1975).

PART TWO
THE SOLAR SYSTEM

We begin the study of the Universe with the solar system, which contains the astronomical objects nearest to us. Chapter 5 introduces the solar system and presents a brief historical account of our progress in understanding it. Chapter 6 is devoted to the planet on which we live, and Chapter 7 describes the Moon, our nearest neighbor in space. The other planets are the subject of Chapter 8, while we cover the minor members of the solar system in Chapter 9. Part Two closes with Chapter 10, in which we look at an outline of the origin of the solar system and examine the scientific evidence bearing on the existence of life elsewhere in the Universe.

FIVE
BACKGROUND AND HISTORY

Although the solar system is only a minute part of the known Universe, it does contain the most conspicuous objects: the Sun, the Moon, the brighter planets, and, of course, the Earth itself. In this chapter we will acquaint ourselves with the general properties of the system and see how our knowledge of it has developed over the centuries.

5.1 General information

THE CENTRAL SUN

The solar system is the system of the Sun. It includes the Sun itself plus all of the objects held captive by the gravity of the Sun: planets, satellites, asteroids, comets, meteors, and the very tenuous matter that exists in the space between the planets. The Earth we live on is one of the nine planets that move in nearly circular paths around the Sun.

As Figure 5.1 indicates, the planets are all quite small when compared to the Sun. In fact, the Sun contains nearly 1000 times as much matter as all of the planets put together, while all of the other objects in the solar system contain a negligible amount of mass by comparison. The Sun is very large, very massive, and very hot, and it emits very large amounts of visible light and other kinds of radiation. It is easy to see why the Sun dominates the whole solar system.

The Sun is a star. The stars we see at night appear faint because they are at tremendously large distances from us. The Sun would also seem very faint if it were as far away from us as even the nearest of the other stars. The Sun and other stars are hot enough to be self-luminous; we can see them because they emit their own light, like a lamp or a candle.

PLANETS AND SATELLITES

Since the planets are not hot enough to emit their own visible light, how do we see them? We see them the same way we can see a house or a book or any other nonluminous thing: They reflect other light back to us. The Sun is the lamp that lights up the solar system and allows us to see the planets, comets, etc. Five of the planets, Mercury, Venus, Mars, Jupiter, and Saturn, are all close enough to the Sun and us that they appear quite bright in the night sky. They seem like bright stars and are easily seen without a telescope. As a result these planets have been well known throughout history. Uranus is so far away that it is just barely visible to the naked eye. It was undoubtedly seen many times throughout history and mistaken for a faint star, but it was not

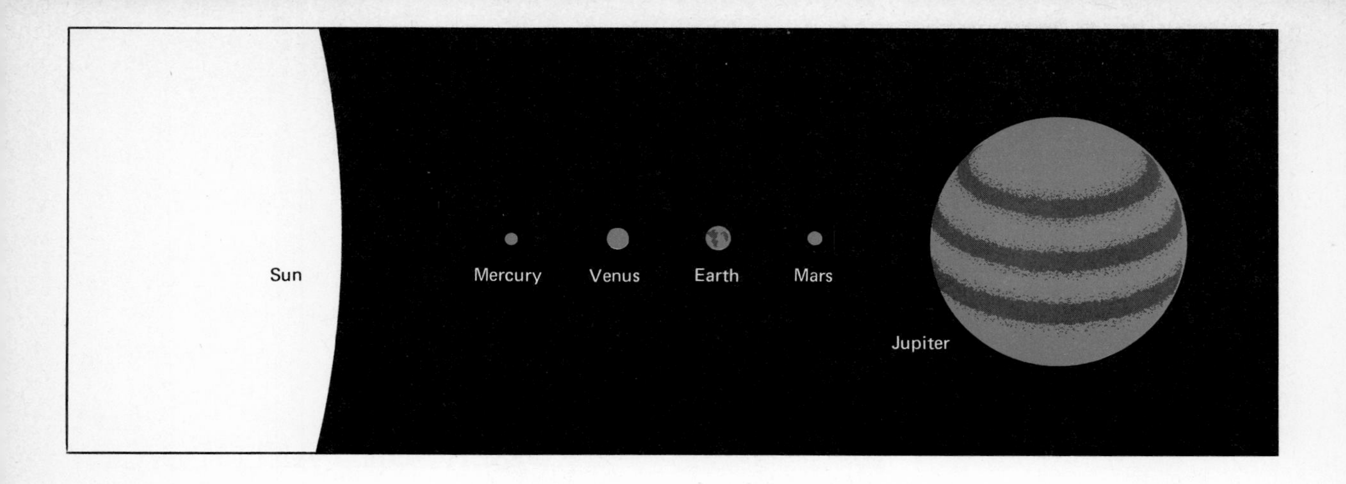

discovered to be a planet until 1781. Neptune and Pluto, the two outer planets, are much too faint to be seen without a telescope. Neptune was discovered in 1846, Pluto in 1930.

There are two kinds of planets: the small, dense, rocky ones and the giant, low-density, gaseous ones. The small planets include the four inner ones—Mercury, Venus, Earth, and Mars—plus the outer planet Pluto. The four giant planets are Jupiter, Saturn, Uranus, and Neptune. Some of the properties of the planets are listed in Table 5.1. (More detailed information about them is given in Appendix 4.) The Moon and Ceres, the largest asteroid, are included in the table for comparison.

Table 5.1 lists both the actual average densities and, for the inner planets, the densities the materials would have if they were relieved of the high pressures in the interiors. It is the zero-pressure densities that we compare when trying to identify the substances the planets are made of. The giant planets are

TABLE 5.1

Planetary data

Planet	*Distance from sun (millions of miles)*	*Diameter (miles)*	*Mass (Earth = 1.0)*	*Density (water = 1.0)*	*Density at zero pressure*	*Period around Sun*	*Number of satellites*
Mercury	36	3030	0.055	5.4	5.3	88 days	0
Venus	67	7520	0.82	5.2	3.9	255 days	0
Earth	93	7930	1	5.5	4.0	1.0 yr	1
(Moon)	93	2160	0.012	3.3	3.3	—	0
Mars	142	4220	0.11	3.9	3.8	1.88 yr	2
(Ceres)	257	590	0.00017	2.2	2.2	4.60 yr	0
Jupiter	484	89,000	320	1.3	—	11.9 yr	13
Saturn	887	75,000	95	0.7	—	29.5 yr	10
Uranus	1780	32,000	15	1.2	—	84 yr	5
Neptune	2790	31,000	17	1.7	—	165 yr	2
Pluto	3670	6000(?)	0.1(?)	?	?	248 yr	0

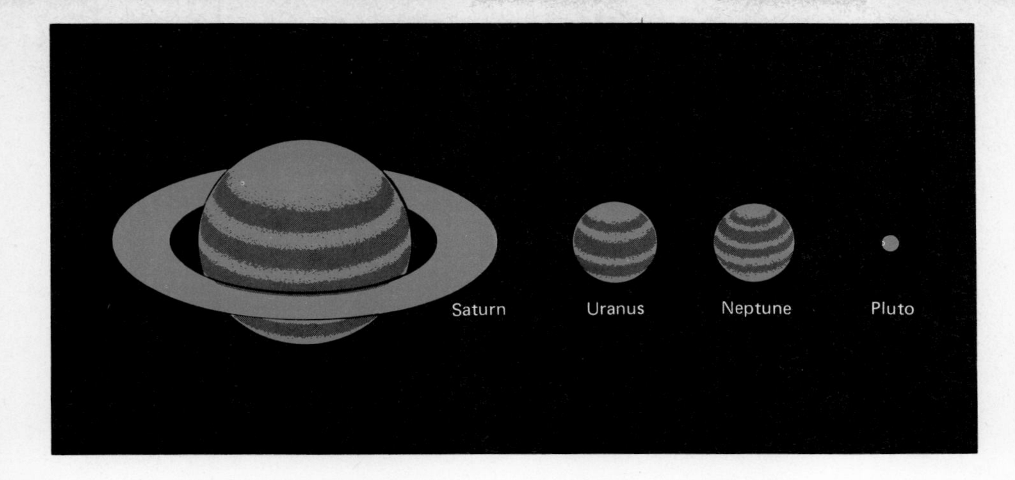

FIGURE 5.1
The relative sizes of the Sun and the planets.

largely gaseous, so densities at zero pressure have no meaning. (Figures 5.2 and 5.3 show the top and side views of the planetary orbits.)

Just as the planets move in paths around the Sun, so there are *satellites* or moons that move in paths around the planets. The Earth has only one satellite, the Moon (not counting the many artificial satellites that have been put into orbit in recent years), while Jupiter has 13 satellites. Mercury, Venus, and Pluto have no known satellites.

It is interesting to consider a scale model of the solar system. As suggested by Figures 5.2 and 5.3, the orbits of the planets around the Sun are

FIGURE 5.2
Top view of the orbits of the planets about the Sun. (a) Inner planets. (b) Outer planets.

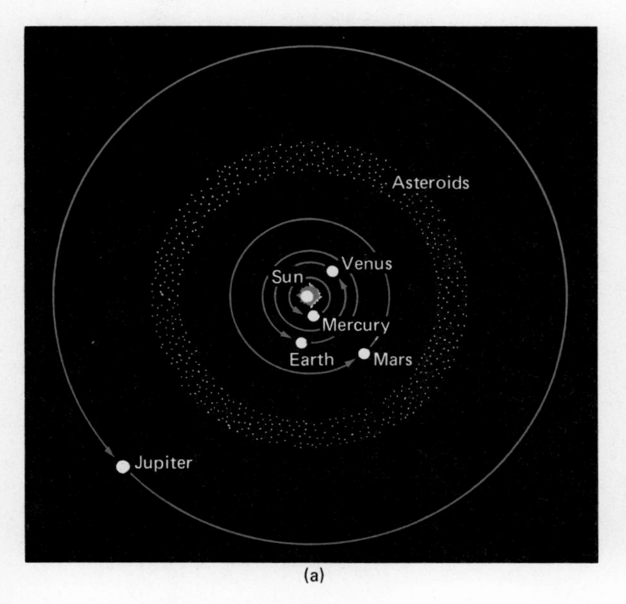

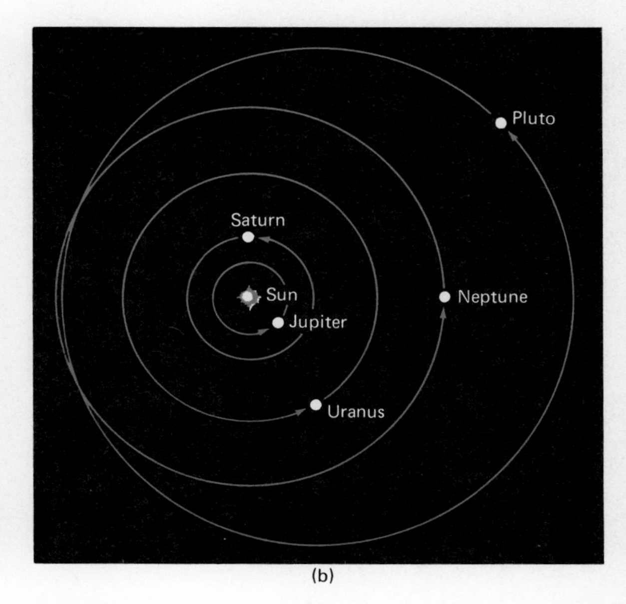

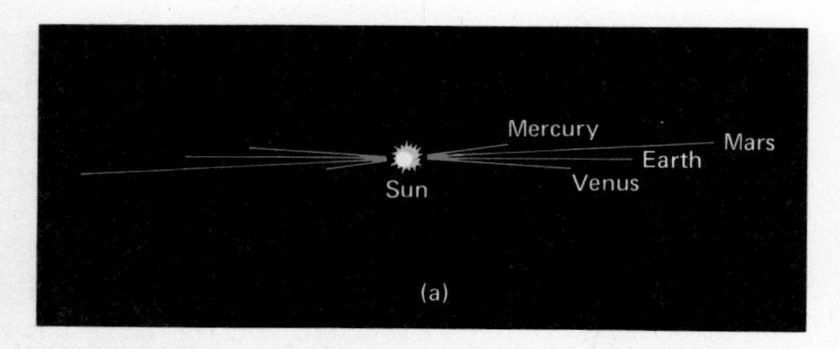

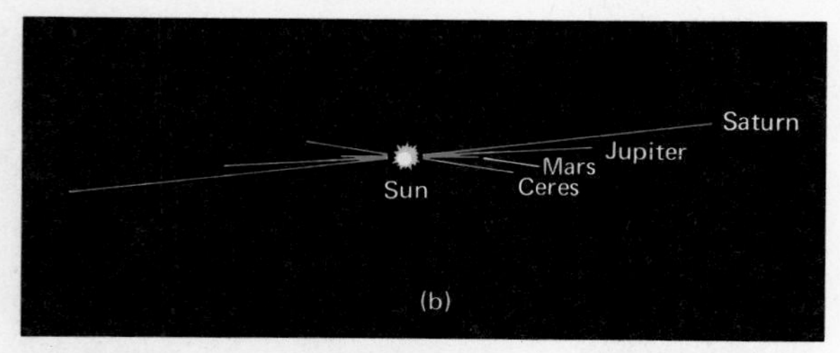

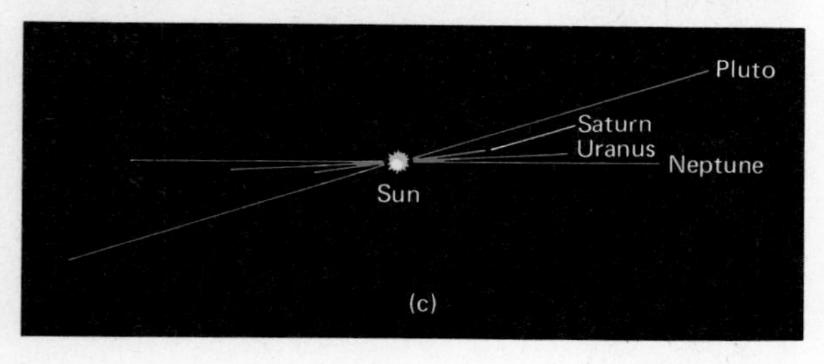

FIGURE 5.3
Side view of the orbits of the planets. (a) Inner planets. (b) Intermediate planets. (c) Outer planets.

all very nearly circles and all lie very nearly in the same plane. This means that the solar system is very flat. If a scale model of it were placed in a box 2 feet wide and 2 feet long, the box would need to be only 7 inches deep in order to contain all of the orbits. Practically all of this depth would be needed only for the planet Pluto because the other planets would have room to spare if the box were just ½ inch in depth.

Suppose now that we want a model of the solar system in which the sizes of the Sun and planets are made to the same scale as their distances apart. We choose a ball 1 foot in diameter to represent the Sun, and we place this ball in the center of a large field. Then the innermost planet, Mercury, would be represented by a grain of sand 14 yards away from the Sun. Venus and Earth would each be tiny balls about 1/10 inch in diameter at distances of 26 yards and 36 yards, respectively, from the center. The Moon would be a grain of sand about 3 inches from the Earth. Mars would be a slightly larger grain of sand 55 yards from the central Sun. The giant Jupiter would be 1¼ inches in diameter and 187 yards from the center, while Saturn would be 1 inch in size and nearly 1/5 mile from the Sun. The planets would not be strung out in a straight line but would be in random directions from the Sun. Uranus and Neptune would be small marbles 4/10 and 6/10

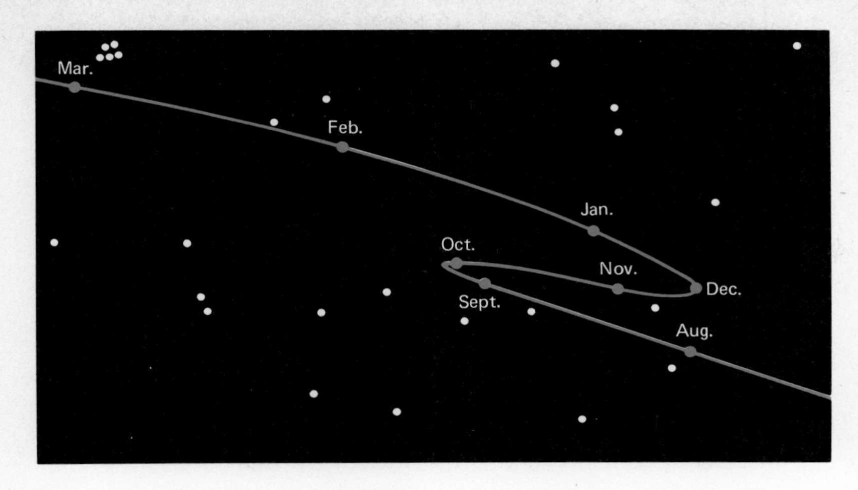

FIGURE 5.4
The motion of Mars in the sky during 1973 and 1974. Some background stars are shown. The cluster of stars in the upper left corner is the Pleiades (PLE-a-deez) in the constellation of Taurus.

mile, respectively, from the Sun, while Pluto would be a grain of sand 8/10 mile away from the ball representing the Sun. The field would have to be over 1½ miles across to hold the orbits of the planets on this scale.

Our large field is filled with the solar system, yet it seems nearly empty. This is generally the way of things astronomical: There is a lot of empty or nearly empty space between the objects of interest. A few grains of sand and marbles are spread over a field nearly 2 miles across. The planets are in forced isolation from each other. But if the solar system seems empty, consider the space between the Sun and the next nearest star. On the same scale as our field, the nearest star to the solar system would be another 1-foot-size sphere more than 5000 miles away—not even any grains of sand in between. It is possible to get a lonely feeling when studying astronomy.

5.2 The solar system over recorded history

MOTIONS OF THE PLANETS ON THE SKY

The stars appear as though they are stuck on a huge dome that rotates around us once each day. The stars keep their same positions on the dome day after day, year after year. There are seven bright objects, however, that do not remain fixed on the dome of the sky. As they rise and set each day, they are seen to roam slowly from one part of the dome to another, that is, they move in the sky among the background stars. These seven objects were called planets and included the Sun, the Moon, Mercury, Venus, Mars, Jupiter, and Saturn. Today the term "planet" has a somewhat different meaning, but it is related in an obvious way to the original definition.

How could the wanderings of the planets be accounted for in ancient times? The observed motions of the Sun and Moon on the sky are relatively simple: The Sun travels slowly and rather smoothly among the stars and goes around the dome of the sky once a year (that is, it takes a year for the Sun to return to the same group of stars again), and the Moon does the same in about a month. The other planets, in contrast, have much more complicated motions across the sky. They usually travel eastward among the stars, but occasionally they back up and move westward for a short time (see Figure 5.4). Today we know that the planets move around the Sun and that we observe them from the moving Earth; thus rather complicated wanderings across the sky are to be expected.

It was easy in olden times to assign the irregular motions of the planets to the whim of

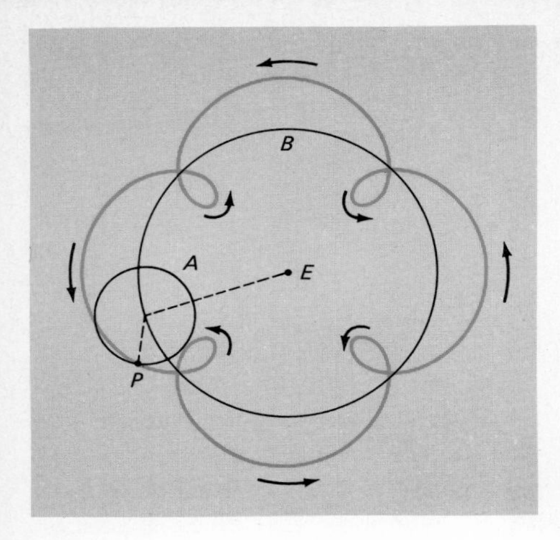

FIGURE 5.5
Epicycle motion.

the gods. Eventually natural causes were sought by a few Greek philosophers and scholars, but the truth did not come easily. Around 250 B.C. Aristarchus (air-is-TAR-kus) suggested that the main part of the observed planetary motions could be easily understood if one assumed that the Earth and the other planets moved in circles around the Sun. While this idea is basically correct, almost everyone rejected it for the simple reason that the Earth did not seem to be in motion. Why didn't they feel the strong wind from this motion? Why didn't the stars appear to change positions as the Earth carried them around the Sun? (Refer to Section 12.1.) Perhaps the strongest objection to Aristarchus' point of view was the belief that the heavy, "earthy" bodies always fell down due to gravity, so only the light, ethereal objects could belong in the sky. It was easy for people to give motion to the heavenly bodies lacking substance, but they thought that the Earth itself was much too ponderous to move. They considered the correct picture, but they rejected it as contrary to common sense.

PTOLEMY

Since the Greeks generally did not believe the Earth could move, they would require the planets and Sun to move around the Earth. A number of elaborate models were brought forth in an attempt to explain the observed planetary motions around an Earth fixed at the center. The most accurate of these models was a system of *epicycles* perfected by Ptolemy (TOL-e-mee) around A.D. 150. Ptolemy's model for planetary motions is of great historical interest because people seriously considered it through late medieval times.

An epicycle is a circle moving on another circle. In Figure 5.5 the small circle or epicycle *A* moves around the larger circle *B* at the same time that it is turning around its own center. This combination of two motions causes the point *P* on the epicycle to trace out the rather complicated dotted curve shown here. The precise nature of this curve can be changed by varying the size of the epicycle and the rate of its rotation. The interesting thing is that a curve like the one shown in the figure is rather similar to the motions of the planets as observed on the dome of the sky. If the Earth is placed at the center of the large circle, *E* in the figure, we can find a size for the epicycle and a rotation rate that will cause the point *P* to follow fairly well the observed motions of any planet.

Ptolemy used accurate observations of the planets, including those of the famous astronomer Hipparchus who lived three centuries earlier, to check out his model of planetary motions. He found that the observed motions

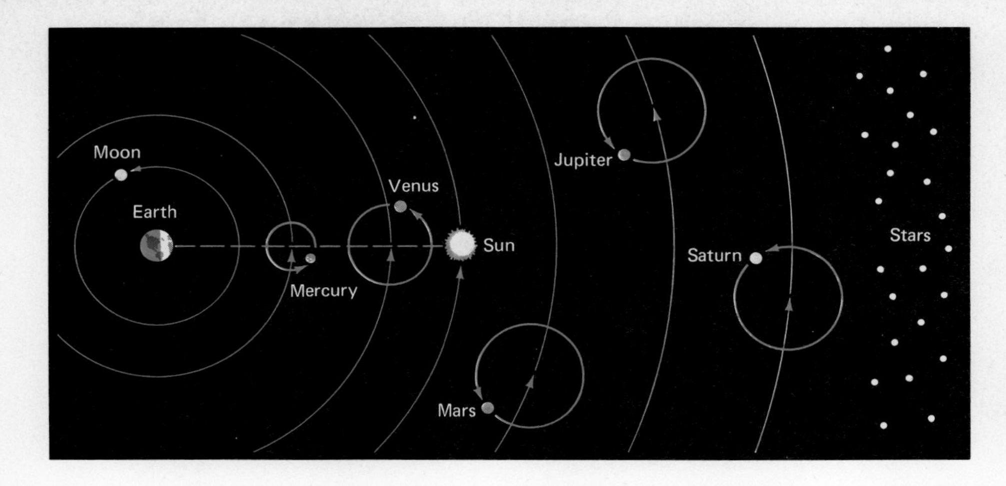

FIGURE 5.6
The planetary system of Ptolemy.

are much too complicated to be accurately represented by a single epicycle for each planet. In fact many epicycles, small circles with smaller circles on them, each having its own motion, were needed for each planet before a totally satisfactory fit with the observations could be attained. But a good fit was indeed found, and this was a major triumph of the patience and ingenuity of the Greek scientist.

Ptolemy's system of epicycles was very complicated, and it took a major effort to use it to predict new positions of the planets. In the thirteenth century, Alfonso X, King of Castile (Spain) and patron of astronomy, was shown the intricacies of Ptolemy's system. The king commented that if he had been present at the creation, he would have recommended a much simpler system. And the correct system is indeed so much simpler: Put the Sun near the middle instead of the Earth, make the paths slightly oval or *elliptical* instead of circular, and presto! There is a perfect fit without any epicycles. But to the ancients it was "obvious" that the Earth could not move, so it had to be at the center. Furthermore, there was a strong philosophical argument against any curve except the circle. It was self-evident to the Greeks that the circle is the perfect figure; would celestial objects be burdened with imperfection, such as an egg-shaped orbit? The Ptolemaic system, complicated though it was, satisfied both the practical and the philosophical needs of the ancients.

It is not my intention here or elsewhere to belittle the efforts of people of the past. We are all subject in some degree to blindness and irrationality in thought and action, although there is a popular misconception that scientists are immune to this. Whether the ancients "should have" pursued a Sun-centered system more thoroughly and introduced elliptical orbits is not the question. The facts are that they were very close to finding the correct model of the solar system, and they had the necessary data and mathematical tools[1] with which to find it, but they didn't. Ironically enough when Kepler introduced elliptical orbits in the seventeenth century, he was a little hesitant because he felt that a system that simple would surely have been discovered by the ancient Greeks.

THE MIDDLE AGES

For the next 1000 years after Ptolemy, scientific thought was in a sort of hibernation. No

[1]It is very curious that the Greek Apollonius (about 200 B.C.) made one of the earliest studies of the properties of the ellipse: He is the same person who introduced the epicycle for explaining the observed motions of the planets!

FIGURE 5.7
Copernicus. (From *A Short History of Astronomy*, by Arthur Berry. Dover Publications, Inc., New York, 1961)

advances in astronomy occurred, although there were occasional observations of significance. In fact the Western World even lost what knowledge the ancient Greeks had, but fortunately the Arabs preserved a number of the ancient records. In the twelfth century Western scholars rediscovered these records, and a new era of learning began.

Twelfth- and thirteenth-century European scholars were delighted by the newly translated works of Aristotle (AIR-is-tot-l), Ptolemy, and others. It is natural that they would see in these ancient writers a confirmation of the current Christian beliefs and an extension of these beliefs into science. Soon ancient Greek science and Christian theology were united into a single structure that was a marvel of reason and self-consistency. This intellectual structure, due largely to Thomas Aquinas, eventually became so universally and enthusiastically received that it was converted into dogma—statements of fact not to be questioned. The new age of learning became stifled and did not become a new age of enlightenment.

Aristotle lived about 350 B.C., and he taught and wrote on practically all areas of knowledge. It certainly was not his fault that some 1500 years later he would be held up as the last word in scientific matters, not to be questioned even by experimental evidence.

Aristotle thought that the Earth is motionless at the center of the Universe, and this was accepted in medieval times along with his other beliefs. Not many scholars gave a detailed study to Ptolemy's complicated system of epicycles, but they granted that it was an interesting possibility. After all, it did satisfy the major requirement that the system be Earth-centered.

COPERNICUS

In 1543, the year of his death, Nicolaus Copernicus (ko-PER-ni-kus) of Poland published a work in which he revived Aristarchus' old view that the Sun, rather than the Earth, is at the center of the planetary motions. Like other scholars of his day, Copernicus argued primarily from philosophical principles rather than from scientific data. There were certain details of Ptolemy's system that he did not like, not because they were contrary to observations, but because they were not sufficiently pleasing to his mind. It is interesting to note that Copernicus was quite willing to accept the notion that the Earth moves around the Sun, but he also claimed that "the intellect recoils with horror" from the suggestion that the motions might be in any form except perfect circles.[2] Since the planets do not in fact move

[2]G. Holton, *Introduction to Concepts and Theories in Physical Science*, Addison-Wesley, Cambridge, 1952, p. 129.

FIGURE 5.8
Kepler. (From *A Short History of Astronomy*, by Arthur Berry. Dover Publications, Inc., New York, 1961)

in perfect circles, Copernicus was forced to introduce epicycles of his own in the planetary orbits in order to get good agreement with the observed planetary motions. Thus the system of Copernicus was not really much if any simpler than the system of Ptolemy. It did represent a great change in thinking, though, and that is what is important.

Copernicus was a devout Catholic who believed that his work in no way ran contrary to the teachings of the Church. During his lifetime he discussed his ideas with many officials of the Church, and he often received favorable comments. Well after his death, however, most Church officials became convinced that the suggestion that the Earth moves was an attack on their own authority. Copernicus was posthumously labelled a heretic, and his book was made forbidden reading (which it remained until the mid-nineteenth century). Copernicus unwittingly became a symbol for the question of whether scientific truths could be dictated by the authority of the Church. In this way he had a very strong influence on the growing scientific renaissance.

TYCHO AND KEPLER

While many persons argued about the religious merits of Copernicus, Tycho Brahe (TIE-ko BRA-he) of Denmark made the observations that would eventually provide the correct answers. Tycho lived before the invention of the telescope, yet he was able to devise instruments and measure the positions of stars and planets with an accuracy far greater than any of his predecessors.

In 1600 the German Johannes Kepler (see Figure 5.8) came to work with Tycho. After Tycho's death in the following year, Kepler inherited Tycho's observations. Kepler knew that these observations were far more accurate than any previous ones and therefore would provide a good test to find the correct model for the solar system.

The Copernican system made a good but not perfect fit with Tycho's observations of the planet Mars. This probably surprised and disappointed Kepler, who had long been a dedicated supporter of Copernicus. Kepler then rejected the system of Copernicus and continued looking for the correct model, or at least one that would give a satisfactory fit to the data. In doing this, Kepler demonstrated one of the earliest examples of belief in the mechanical and mathematical nature of the Universe.

Many times in the past the model of Ptolemy had been found to be in error in predicting the positions of the planets. The model required periodic adjustments in order to keep its errors from becoming very large as time went on. These errors and required adjustments were not taken as evidence of the model

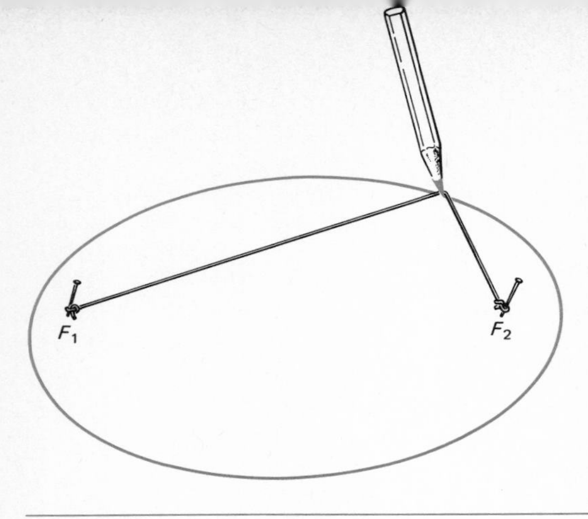

FIGURE 5.9

The ellipse. You can draw an ellipse by tying the ends of a string to two pins. Hold the string tight with a pencil and trace out the ellipse. The two points where the pins are stuck are the two focus points (F_1 and F_2) of the ellipse. The closer together the pins are, the more nearly like a circle is the ellipse.

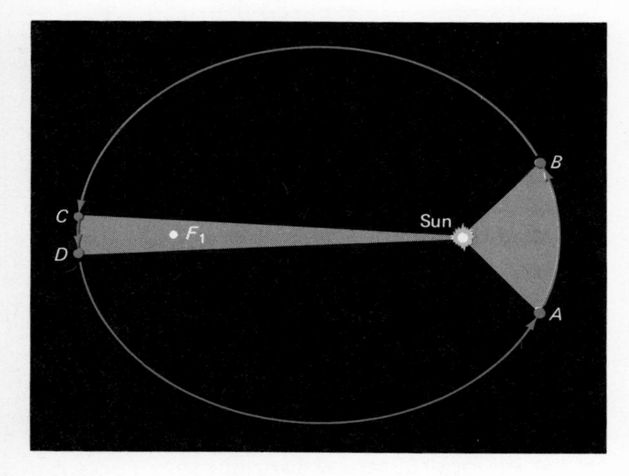

FIGURE 5.10

Kepler's second law. The ellipse is the orbit of a planet with the Sun at one focus. The other focus (F_1) is empty. If the planet moves from point A to point B in one day, the area covered by the line from the Sun in that time is shown shaded. Later the same planet moves from points C to D in one day, and the area covered in this interval is also shaded. According to Kepler's second law, the two shaded areas must be equal. It is apparent that this requires that the planet move more slowly in its orbit when it is farther from the Sun. Actual orbits of the planets are more nearly circular than this figure indicates.

being wrong, for who could expect mathematical exactness? Close was good enough, for did not angels push around the planets at the whim of God? It is curious that Kepler, who was an astrologer and a mystic, would be among the first to give experimental data complete authority over philosophical prejudice in deciding the validity of a scientific theory.

Kepler abandoned circular motion and, after years of trial and error, found that elliptical orbits are needed to fit the data. The Sun is not at the center of the ellipse but at a point called the "focus" of the ellipse. An ellipse is an oval; it can be perfectly round (a circle is a special ellipse), it can be very long and narrow (a straight line is also a special kind of ellipse), or it can be between these extremes. The planets all have orbits that are very close to being circles, while many comets have elliptical orbits that are very long and narrow.

Kepler studied the orbits of the planets for many years, and what he found is summarized in the three famous laws of planetary motion:

1 Planets move in elliptical orbits with the Sun at one focus, a point offset a certain amount from the center. (See Figure 5.9.)

2 A planet moves in its orbit in such a way that a line drawn between it and the Sun sweeps out equal areas in equal lengths of time.

3 In comparing two or more different planets, the square of the length of time taken

FIGURE 5.11
Galileo. (From *A Short History of Astronomy*, by Arthur Berry. Dover Publications, Inc., New York, 1961)

to go once around the Sun is proportional to the cube of the average distance to the Sun.

The second law states that a planet moves faster when it is at a point in its orbit nearer the Sun than it does when it is farther away from the Sun (see Figure 5.10).

The third law is a mathematical relation that we can easily check with the planetary data given in Table 5.1. (Or see Appendix 4.) We expect planets in large orbits to take longer to go around the Sun than those in smaller orbits; the third law says how much longer it should take.

Let us compare the planets Neptune and Mars. According to the table, Neptune is about 2790/142 or 19.6 times as far from the Sun, on the average, as is Mars. The third law says if we should cube this number, getting 7530, and take the square root of it, 87, this will tell us how much longer Neptune takes to go around the Sun than Mars. The table indicates that this number is 165/1.88 = 88, showing that the law does work.[3] Pluto and the Earth are another example. Pluto takes 248 times as long as the Earth does to travel around the Sun. The third law states that if we square this number, getting 61,500, and then take the cube root, 39.5, this will be how much farther Pluto is from the Sun than the Earth is. The table indicates that this is 3670/93 = 39.5, again confirming the third law.

Each of Kepler's three laws can be expressed as a mathematical equation. No longer could the planets be considered to be whimsically wandering across the dome of the sky. Science had become quantitative. This does not mean that Kepler's laws are exact; they are not, although they do come much closer to exactness than any previous theory had.

GALILEO

While Kepler was studying planetary orbits in Prague, another man was stirring up trouble for the Church and for himself in Italy. This man was Galileo (gal-i-LAY-o). In 1609 Galileo heard of a Dutchman named Hans Lippersheim who combined two lenses in such a way as to make a useful telescope. Galileo himself experimented with many different lenses and made many telescopes. He was among the first to make astronomical use of the telescope, and in 1610 he published the results of some of his observations. Among the things he discovered were four of the satellites of Jupiter, the phases of Venus, mountains and craters on the Moon, spots on the Sun,[4] and a peculiar shape

[3]Eighty-eight is not exactly equal to 87, but the numbers in the tables are not exact, and neither are Kepler's laws.

[4]Sunspots had been reported many times before Galileo reported them, but the naked-eye observations were infrequent and not convincing.

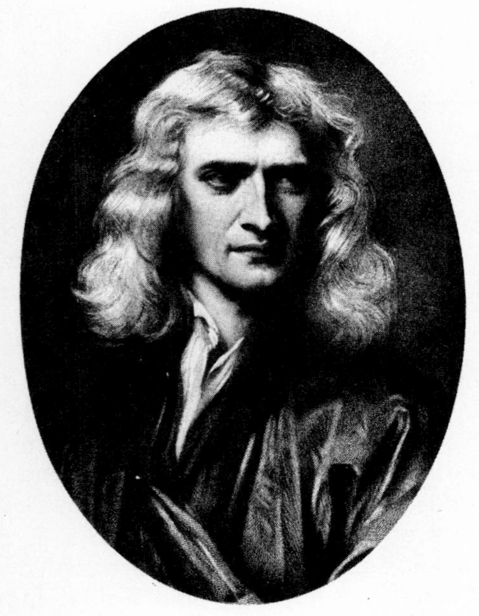

FIGURE 5.12

Newton. (From *A Short History of Astronomy*, by Arthur Berry. Dover Publications, Inc., New York, 1961)

to Saturn that was later found to be a ring around the planet. Most of these discoveries were embarrassing to the Church (see Question 7 at the end of Chapter 1), and Galileo's telescope got the reputation of being tainted by the devil. This attitude did not stop the use of the telescope, and at long last astronomy had received what even today is its most useful instrument.

In addition to his work with the telescope, Galileo also carried out many mechanical experiments with objects and found errors in a number of theories that were then currently held. Since these theories had the sanction of the Church, these activities also made for difficulties between Galileo and the authorities. Finally Galileo strongly and openly supported the Copernican system of planetary motions, and this eventually proved to be too much. For a number of years the Church made threats and Galileo promised to mend his ways, promises he was not able to keep. In 1633, at the age of 69, he was called to Rome to stand trial for his heresies. He was forced to make a public renunciation of the system of Copernicus, and for the remaining years of his life the Church restricted his activities and kept close watch to see that he didn't continue in his old ways. Galileo became the most famous martyr to the cause of scientific freedom.

NEWTON

Kepler's laws tell us quite a bit about the orbits of the planets, but they are too specific to give us much understanding of how or why they seem to be valid. Are they simply the result of a whim of God, or can they be understood in terms of more basic properties of matter? Do the planets have one set of regulations governing their motions while the matter with which we are familiar has a different set? Or is there a universal set of rules that governs the behavior of matter everywhere? If there is such a universal set of rules, can we hope to discover these rules and understand them? Up through the mid-seventeenth century, these were very puzzling questions; yet by the end of the 1600s they all had been answered, largely by one man—Isaac Newton (see Figure 5.12).

Newton was born in England in 1642, the year after the death of Galileo. He found in England an atmosphere that was nearly free of the religious oppression to which Galileo had been subjected. He also had many friends and acquaintances educated in science with whom he could discuss and debate ideas, a situation strongly helping the development of science. It was around this time that a number of scientific societies were established in Europe.

Newton's contributions to the understanding of planetary motions consist of two parts: his laws of motion and his law of gravity. These are universal laws that apply to all matter, regardless of where it is. The laws of motion are precise statements about how forces affect the motions of objects; the law of gravity is a precise statement about one of the basic forces between masses.

The laws of motion state, in essence, that an object will move in a straight line with constant speed if left to itself. When a body behaves differently from this, it is because it has come under some outside influence: A force has been exerted on it. Any change in motion, such as speeding up or slowing down or swerving to one side or another, depends directly on the strength of the force that is applied. For a given strength of force, the changes in motion are greater for an object of small mass, less for an object of large mass. Lastly, forces always occur in equal and opposite pairs: If you push on a building, the building pushes back equally on you. Some of these ideas had been developed earlier by Galileo, but Newton was the first to formulate them in a precise way.

Bodies around us tend to come to rest, so a speed of zero would seem to be the natural state of things. This is what Aristotle thought, but Newton said that it was not true; objects stop because some force is acting on them. In most cases it is the force of friction caused by

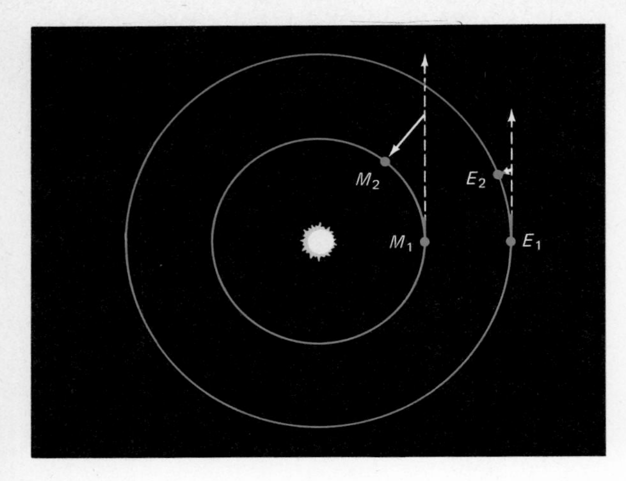

FIGURE 5.13
The gravitational force of the Sun. In the time that the Earth moves from E_1 to E_2, Mercury moves from M_1 to M_2. If it were not for the gravitational pull of the Sun, the planets would move along the straight lines shown by the arrows. The pull of the Sun keeps the planets on their appoximately circular paths. Note that the deviation of Mercury from straight-line motion is greater than that of Earth, indicating that the pull on Mercury is stronger than the pull on the Earth.

different objects rubbing on each other. If friction could be completely removed, a body pushed horizontally on a flat surface would keep sliding on the surface indefinitely.

When we apply the laws of motion to the planets, we see immediately that some force must be exerted on them since they do not move in straight lines. Where could this force come from? The force is directed toward the inside of the orbits since that is the direction in which the planets deviate from straight lines. There is nothing outside to push the planets inward; is there anything inside that could be pulling on them? Yes, there is the Sun. Thus we suspect the Sun of pulling on the planets and holding them in their elliptical orbits instead of letting them go off in straight lines as the laws of motion say they "want to do."

What is this magical force with which the Sun holds onto the planets? If forces occur in pairs and if the Sun pulls on the Earth, then the Earth must pull back on the Sun with an equal force. How can the Earth exert such a force without our knowing about it? Yet we do know about it, for the force of gravity appears to be exerted on all things. The same force that holds us onto the surface of the Earth and gives us weight also holds the Moon in orbit around the Earth.

If the Earth pulls on the Sun with the same force as the Sun pulls on the Earth, why is it the Earth that does all of the moving? It isn't; they both move. Remember, however, that the reaction of a body to a force is larger if the body has a small mass. The Sun has far more mass than the Earth; therefore the same force that brings the Earth around in its 93-million-mile orbit causes only a slight wobble in the Sun. When an ocean liner collides with a row-boat, they both feel the same force; it is only that the force does not have the same effect on both objects.

How strong is the force of gravity of the Sun? It must be strong enough to hold the planets in their orbits. We can actually measure the orbits and see how much they deviate from straight lines, and this indicates the strength of the gravity of the Sun. The innermost planet, Mercury, has the greatest deviation from straight-line motion, and so the force on it must be the greatest. In this way we can measure directly how the gravity of the Sun weakens with greater distance from it (see Figure 5.13).

Newton studied the motion of the Moon and of objects falling on the surface of the Earth, but his procedure was essentially the same as the one just outlined. He found the force needed to produce the observed motions, and he stated his results in the famous law of gravity: Any two bodies exert an attractive force on each other that becomes stronger with the product of their masses and weaker as the square of the distance between them.

The laws of motion and the law of gravity together completely determine the possible orbits that planets can have. These laws gave Kepler's laws of planetary motion a firm physical and mathematical foundation. Newton actually made several modifications to Kepler's laws that make them more accurate and more general. The planets behave the way they do because of the type of force exerted on them and because of the basic properties of all matter, not just the planets. When different forces are acting, as in a bouncing ball or a stone falling through water, the motions are different; however, the motions can still be determined from Newton's laws of motion if we know what forces are being exerted.

Newton's laws of motion and law of gravitation can, of course, be expressed as mathematical equations. In order to obtain a maximum amount of information from these equations, we must be able to solve them; that is, we must know something about mathematics. Sometimes solving the equations is easy; more often it is very difficult and requires much mathematical knowledge. When forces other than gravity are important, the equations may become even more difficult. It is essential to be skilled in mathematics in order to determine the behavior of matter. Newton helped to develop some of the mathematics needed to solve his own equations. It is no coincidence that from Newton's time on, the development of science and of mathematics went hand-in-hand.

Planetary motions as we understand them today are not very much different from the way Newton understood them three centuries ago. A number of details have been filled in, and the laws of motion have been twice modified in very significant ways by relativity theory and by quantum theory; however, relativity has only very small effects on the motions of the planets, and quantum theory is here totally unimportant.

Newton has probably received more praise than any other scientist. Alexander Pope wrote:

> *Nature and nature's law lay hid in night*
> *God said, Let Newton be, and all was light.*

But we should not take ourselves or our heroes too seriously. Referring to the major revolution in physics caused by A. Einstein's theory of relativity, J. C. Squire added:[5]

> *It did not last. The devil howling: Ho!*
> *Let Einstein be! restored the status quo.*

[5]Quoted in C. Payne-Gaposchkin and K. Haramundanis, *Introduction to Astronomy*, 2nd ed., Prentice-Hall, Englewood Cliffs, New Jersey, 1970, p. 204. Used by permission.

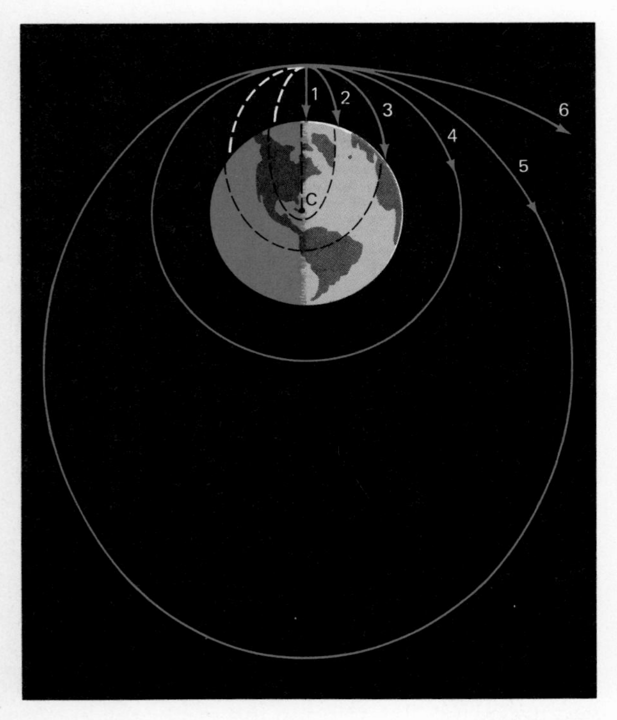

FIGURE 5.14
Orbits around the Earth. The center of the Earth (C) is a focus for all of the orbits.

5.3 Ellipses and hyperbolas

THROWING A STONE

Let us examine the path taken by a stone or other object that we can throw from near the surface of the Earth. We must pretend that the air has no resistance, for that would add a frictional force that changes the problem. When we throw the stone, we put it into orbit around the Earth, and we wish to see what kinds of orbits are produced.

If we simply drop the stone, it falls straight down to the surface as in line 1 in Figure 5.14. If the Earth did not get in the way, the stone would fall to the center (C), as indicated by the dotted continuation of the line. As mentioned before, this straight line is a special kind of ellipse.

Suppose we throw the ball horizontally instead of just letting it drop. Experience shows that it travels a certain distance and then hits the ground, as in curve 2. Curve 2 is actually part of an ellipse, shown by the dotted line in the figure. If the Earth did not get in the way, the stone would follow this ellipse all the way around the Earth and come back again to the point from which it was thrown.[6] Throwing the

[6]We are often told that a falling body has a type of path called a parabola; this is approximately true, but it is more accurately an ellipse.

stone harder causes it to move in a wider ellipse, curve 3, and it travels farther before hitting the surface of the Earth. Again it would continue completely around the Earth in this elliptical path if the ground did not stop it. By throwing the stone hard enough, we will cause it to travel in the circle indicated by curve 4 in the figure. In this case the ground no longer gets in the way, so the stone does complete its orbit around the Earth and come back to the starting point. Unless something stops it, the stone will keep moving in this circle indefinitely. In practice, of course, the resistance of the air will stop it.

By throwing the stone even harder, we will cause the stone to follow the larger ellipse, number 5 in the figure. It moves a good distance away from the Earth, but then it is brought back again to the starting point by gravity. Note that all ellipses are closed paths, coming back to the point of origin.

If we keep throwing the stone harder and harder, its elliptical path will become larger and larger, carrying it farther and farther from the Earth before it is brought back again by gravity. Eventually a velocity will be reached in which the stone will go off and not come back at all. When the stone is thrown faster than a certain critical velocity, called the *velocity of escape*, the gravity of the Earth is not strong enough to pull it back again; the stone escapes completely from the Earth. Curve 6 represents this case. The escape velocity from the Earth is about 7 miles per second or 25,000 miles per hour.

When the stone travels faster than the escape velocity, its path is no longer an ellipse; it is a type of curve called a *hyperbola* (hy-PER-bo-la). The hyperbola is an open curve that will go off indefinitely without coming back. Figure 5.15 illustrates some different kinds of ellipses and hyperbolas that a stone or a planet can have for an orbit. The path for an object with exactly the velocity of escape is a special type of hyperbola called the *parabola* (pa-RAB-o-la).

ORBITS OF PLANETS

We see that when two bodies move under the influence of their gravities, there are two types of curves they can follow, ellipses and hyperbolas. The difference between them is that elliptical motion is *periodic*. That is, it keeps repeating itself indefinitely. Hyperbolic motion is a one-time affair: The bodies meet once and then move off, never to meet again. It is understandable that the planets all have elliptical orbits: If a planet ever had a hyperbolic orbit, it would have left the solar system long ago.

What is it that decides whether an orbit

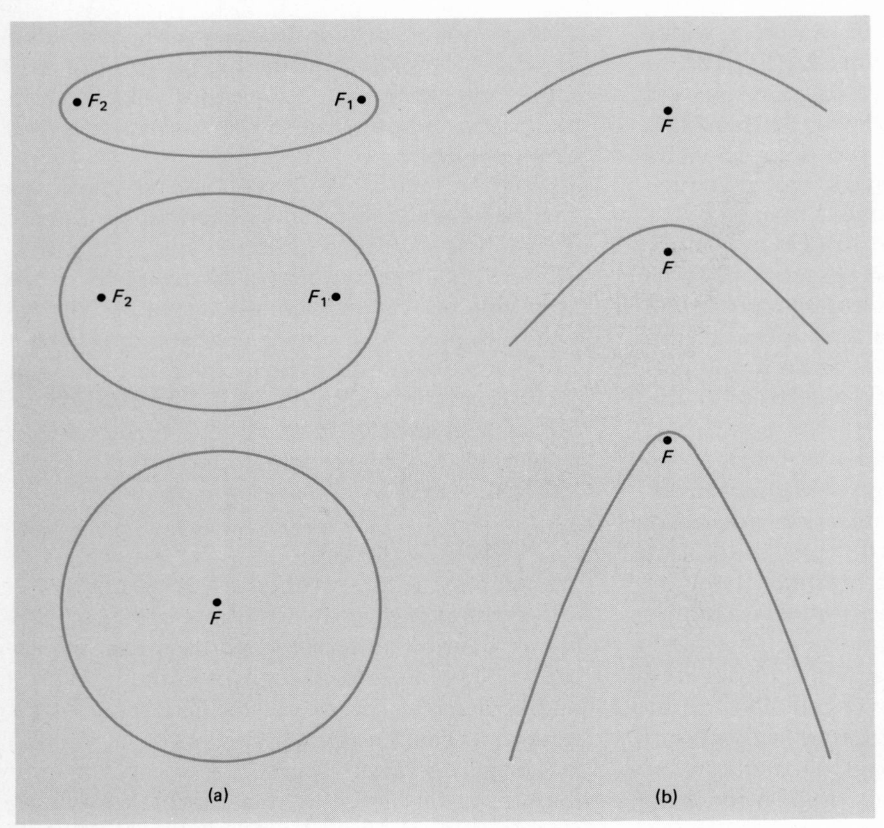

FIGURE 5.15
Different shapes of ellipses and of hyperbolas. In each case the focal point or points are indicated by F. (a) Ellipses. (b) Hyperbolas.

will be an ellipse or a hyperbola? Energy. A planet moving in an elliptical orbit around the Sun does not have enough energy to escape from the gravity of the Sun. An object following a hyperbolic orbit around the Sun has too much kinetic energy for the Sun's attraction to hold it back. If we want a rocket ship to go to the Moon, we must require the engines to build up enough kinetic energy to put the rocket in a hyperbolic (or near hyperbolic) orbit around the Earth.

If more than two bodies exert gravitational forces on each other, the paths become much more complicated than simple ellipses and hyperbolas. Since the Sun has far more mass than any of the other bodies in the solar system and since the planets are quite far apart, the main force each planet feels is from the Sun. To a very good approximation, each planet may be considered to be alone with the Sun, and this approximation leads to elliptical orbits.

For very high accuracy, however, the forces between the planets must be taken into account. This leads to complicated paths that are nearly though not exactly elliptical. The same is true of artificial satellites traveling around the Earth. They move primarily under the gravity of the Earth, but they are also disturbed by the Sun, the Moon, and the resistance of the tenuous upper atmosphere through which they must move. Another complication affecting their paths is caused by the Earth's not being exactly spherical. All of these factors can produce significant deviations from elliptical motion.

IMPORTANT WORDS

Satellite	**Tycho**
Aristarchus	**Kepler**
Ptolemy	**Galileo**
Epicycle	**Newton**
Ellipse	**Velocity of escape**
Aristotle	**Hyperbola**
Copernicus	

REVIEW QUESTIONS

1 What does the solar system contain?

2 What object dominates the solar system?

3 What are the two types of planets?

4 Which planets were not known in ancient times?

5 How flat is the solar system?

6 Is the solar system crowded with bodies?

7 What were the seven bodies called "planets" by the ancients?

8 Why did ancient people not like the idea of the Earth moving?

9 How did Ptolemy explain planetary motions?

10 How did Copernicus modify Ptolemy's ideas?

11 What did Tycho Brahe do that eventually helped to show that Ptolemy was wrong?

12 How did Kepler arrive at his notions about planetary orbits?

13 Who first used the telescope in astronomy?

14 What is the law of gravitation?

15 What are Newton's laws of motion concerned with?

16 Are the laws of motion exact, or have they been modified by more recent theories?

17 When you throw a baseball, is it really in orbit around the Earth?

18 What is meant by the velocity of escape?

19 What are the two different types of orbits?

20 What can you say about the energy of an object that is in a hyperbolic orbit around the Sun?

21 Are the orbits of the planets exactly elliptical?

QUESTIONS FOR DISCUSSION

1 Referring to Figure 5.2, explain why we cannot see Mercury or Venus at midnight.

2 Suppose that the Sun were to be turned off for one second and then turned back on again. Would we notice anything by looking at the planets and stars?

3 Discuss the relative merits of the following three statements:

Ptolemy's system of epicycles is obviously artificial, very cumbersome, and based on strong prejudice. It is ridiculous to believe it represents the actual solar system.

Ptolemy's system is a marvel of mathematical and engineering skill that predicts the planetary motions quite well. It served a useful function.

Ptolemy's system is aesthetically pleasing and adequately represents planetary motions in the sky; therefore, it must be true.

4 One could argue that Copernicus deserves no more credit than others of his day since he also was motivated primarily by philosophical bias. He was just lucky that his biases turned out to be closer to reality than those of other people. Do you agree?

5 Suppose that Kepler somehow had been able to prove that mathematical equations or models can never give excellent agreement with the observed positions of the planets so people would have to be satisfied with just being rather close. What difference would this make in science?

6 If we shot a projectile off the Earth with a speed greater than the escape velocity, is it possible that it would ever come back to the Earth?

REFERENCES

A good introduction to the solar system is contained in the first two chapters of F. L. Whipple's *Earth, Moon, and Planets*, 3rd ed. (Harvard University Press, Cambridge, 1971).

The historical part of the chapter is covered in

detail by J. L. E. Dreyer in *A History of Astronomy from Thales to Kepler* (Dover, New York, 1953), by A. Berry in *A Short History of Astronomy* (Dover, New York, 1961), and by A. Pannekoek in *A History of Astronomy* (Allen & Unwin, London, 1961).

Interesting personal accounts of the lives of the chief actors in our historical drama are given by A. Koestler in *The Sleepwalkers* (Grosset & Dunlap, New York, 1963), by P. Moore in *Watchers of the Stars* (Putnam, New York, 1974), and by R. Thiel in *And There Was Light* (Knopf, New York, 1957).

How modern space ships use the law of gravity is described in "Navigation Between the Planets" by W. G. Melbourne in *Scientific American* [234 (June, 1976), 58].

SIX
THE EARTH

After the general background of the last chapter, we begin our study of the individual members of the solar system with the Earth. This is appropriate because, small though the Earth may be in the objective view, it is our home and we know it best.

6.1 The structure of the Earth

AROUND THE EARTH

The world we live on is a large ball about 8000 miles in diameter. We would like to understand, at least in broad outline, how the Earth is put together and how it probably got that way. This information is of interest in itself, and it also can help us understand the other planets for which we have much less accurate data.

As seen from space (see Figure 6.1), the clouds are the most obvious part of the Earth. The clouds float in the thin atmosphere that separates the surface of the Earth from outer space. This atmosphere is extremely important to us and all living creatures. We need its air to

FIGURE 6.1

A photograph of the Earth in the crescent phase taken in 1967 from a position near the Moon by the spacecraft Lunar Orbiter I. (NASA)

FIGURE 6.2
A mosaic of photographs of the United States at night taken from an Air Force satellite at an altitude of about 500 miles. The fitting is not perfect: Tucson, Arizona, shows up twice. (NASA)

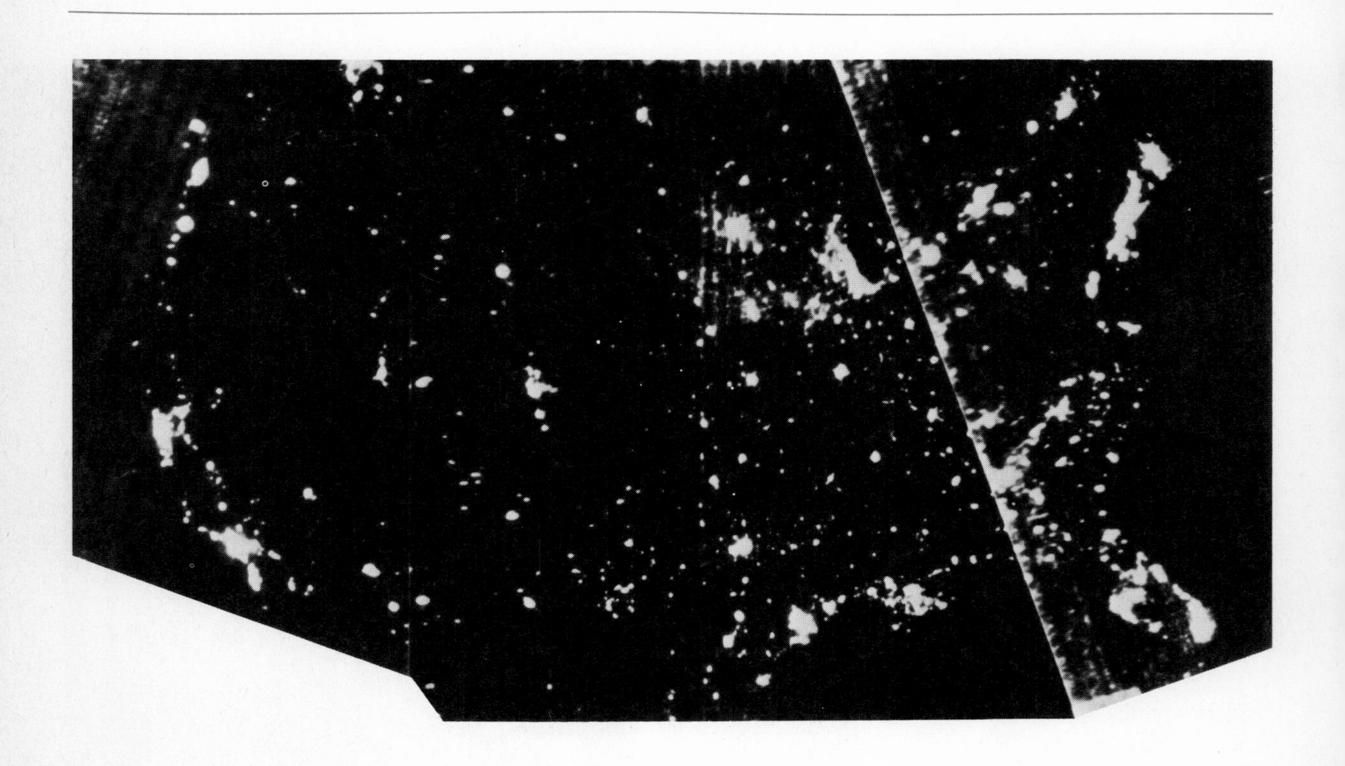

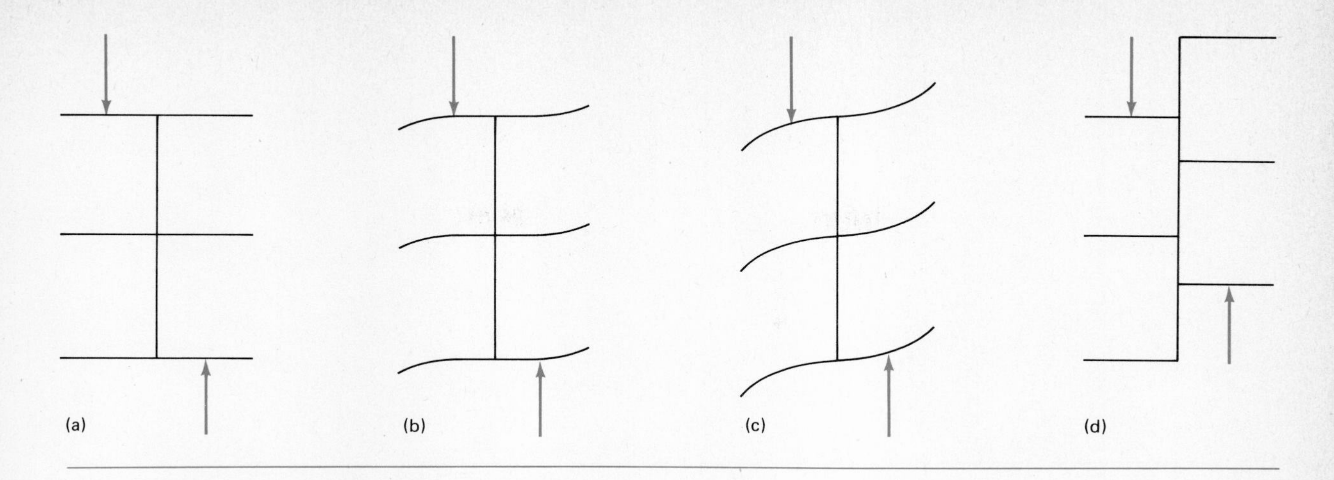

FIGURE 6.3
A schematic drawing of an earthquake.

breathe, and we need it to protect us from the dangerous high-energy radiation from the Sun. The air is only a very minor part of the Earth as a whole, however, since it makes up only one part in a million of the Earth's total mass.

The Earth is the only planet known to have life. The question sometimes comes up about how obvious this life is as viewed from outer space. Figure 6.2 is a composite photograph of the United States taken at night from a satellite. The coastlines and the lights of the large cities are clearly visible.

INSIDE THE EARTH

We know the surface of our planet rather well, but the interior is not accessible to us directly; we must obtain information about it indirectly. One such important piece of information is the average density. The mass of the Earth is known from the strength of its gravity. Its size and therefore its volume can be accurately measured, and we find the density of the Earth, its mass divided by its volume, to be 5.5 times the density of water. Now, this is a very large density. Most of the rocks on or near the surface of the Earth have a much lower density, usually between two and three times water. This means that the deep interior of the Earth must be made out of material that is quite dense in order to bring the average for the Earth as a whole up to 5.5. Part of the difference between surface and interior material is due to compression: The deep interior is naturally under a very high pressure, and this helps to increase the density. Another part of the difference is that the interior of the Earth is made of different types of materials than the surface layers. The surface rocks are not typical of the matter that exists deep within the Earth.

Our main source of information about the interior of the Earth is *earthquake waves*. An earthquake is a sudden shift in the rocks within the Earth. It is caused by pressures that build up until the rock is at the breaking or shifting point (see Figure 6.3). An earthquake releases a large amount of energy, sending earthquake waves in all directions through the rocks. These waves are similar to sound waves which pass through the air and allow us to hear noises. Geologists have placed a large number of stations on the surface of the Earth to measure the strength and times of arrival of earthquake

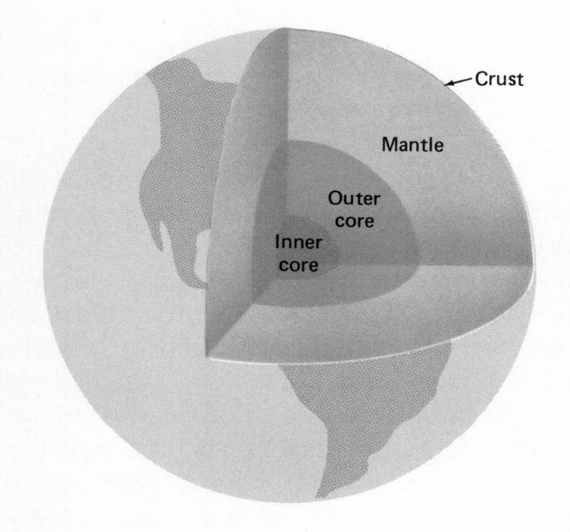

FIGURE 6.4
The interior of the Earth.

waves. By collecting data from many stations for many earthquakes, geologists can tell much about the material properties of the interior of the Earth. To study the interior of the Earth, scientists can also create artificial earthquakes through underground explosions.

Studies of earthquake waves show that the Earth has an internal structure as shown in Figure 6.4 and in Table 6.1. The *crust* is the thin outer layer, the only region we can explore directly. Its average thickness is only 10 miles, although it is much thicker under the continents than under the oceans. The *mantle* is the thickest and most massive part of the Earth. The *cores* are small but of very high density. The outer core is a liquid. This can be determined because there are two different kinds of earthquake waves, one of which cannot pass through a liquid; scientists have observed that this kind of wave cannot pass through the outer core. The crust and mantle are composed of various kinds of rocks, while the cores are thought to be mostly metal iron. The molten outer core of iron is probably the source of the magnetic field that the Earth possesses. As already mentioned, the density of each layer is increased by the high internal pressures that compact the material. Table 6.1 lists what the estimated densities of the materials would be if they were brought to the surface and relieved of these high pressures.

EARLY MELTING

How did the Earth get separated into the various zones containing different types of materials? Why isn't everything mixed together? As long as the material is a solid, it would be difficult for the different parts to separate; however, if the material is at least partly in the liquid state, then it would be natural for the heavier parts to sink to the bottom while the lighter materials rise or float to the top. This is

TABLE 6.1

Interior of the Earth

Part	*Thickness (miles)*	*Average density (water = 1.0)*	*Mass (tons)*	*Density for zero pressure (water = 1.0)*
Crust	10	2.8	2.6×10^{19}	2.8
Mantle	1790	4.5	4.5×10^{21}	3.3
Outer core	1380	10.9	2.0×10^{21}	7
Inner core	780	12.3	1.2×10^{20}	7

[Data from W. M. Kaula, *An Introduction to Planetary Physics,* Wiley, New York, 1968.]

apparently what has happened: As we go inward from the crust to the inner core, we pass regions of successively higher density. We here have direct evidence that much, if not all, of the interior was once molten, although today it is all solid except for the outer core. As we shall see, part of the mantle is actually in a sort of plastic state that is between a rigid solid and a freely flowing liquid.

Where did the heat that melted the interior come from? One source of heat is the process of formation itself. As the material that formed the Earth came together, it released a large amount of gravitational potential energy. The particles falling in under their own gravity collided with each other, heating themselves up. This is the same process that makes the stars hot. (Refer to Section 15.3.) There is far more than enough potential energy released in this way to melt the entire Earth, but another important question is whether this process took place rapidly or slowly. If the Earth formed rather slowly, the heat would have had a chance to be lost by radiation as fast as it was produced, and the result would have been a cold, solid Earth. If the formation was rapid, however, most of the heat would have been trapped inside before it had a chance to escape by radiation from the surface. In this case a hot, molten Earth would have been the result after all of the "dust had settled." It is not known at present whether a hot or a cold Earth was produced at its formation.

Even if the Earth formed cold and solid, there is another process that could make it hot: radioactivity. It is mentioned in Section 2.1 that the nuclei of certain types of atoms are unstable and tend to break apart spontaneously with the release of large amounts of energy. If radioactive substances are sufficiently abundant, the energy they release over a period of time could heat the surrounding materials and cause them to melt. Uranium, thorium, and a certain type or isotope of potassium are the main radioactive substances that occur in rocks. Although these elements are quite rare, they are still abundant enough to produce the required heating of the interior of the Earth. Through the process of the formation of the Earth or by means of radioactivity or both, the interior was heated to melting. This allowed the heavier and the lighter materials to separate and form the different zones indicated in Figure 6.4. Since that time, the mantle and the crust have lost enough heat to become solid again. The inner core is still very hot, possibly as hot as the surface of the Sun (10,000°F), but the very large pressures force it into the solid state.

We might wonder how the mantle and crust could have lost enough heat to resolidify, since radioactivity continues to be present in

TABLE 6.2
Radioactive nuclei

Parent	Daughter	Half-life (billions of years)
^{235}U	^{207}Pb	0.71
^{238}U	^{206}Pb	4.5
^{232}Th	^{208}Pb	14
^{40}K	^{40}Ar	12

the rocks. There are two reasons for this. First, although uranium and thorium are very heavy atoms, the molecules they and potassium form with other atoms tend to gather in the lighter rocks. Most of the Earth's radioactivity has thus risen to or near the surface, allowing much of the heat that it generates to escape into space and halting further heating of the deep interior. The second point is that radioactivity destroys the uranium, thorium, and potassium atoms, and there is no place on Earth where these atoms are being replaced. The amount of radioactivity decreases with time, so there is much less heat produced today than when the Earth was young.

RADIOACTIVE DATING

Radioactivity appears to have had an important influence on the history of the Earth, and it also provides us with an important tool for understanding this history. A radioactive nucleus breaks apart into smaller nuclei after a certain length of time. If the smaller nuclei are also unstable, they will eventually break into still smaller ones. The stable nuclei produced in this way will, of course, remain as they are. Each atom that undergoes radioactive decay will eventually be changed into stable nuclei of various kinds. As time goes on, the abundance of the original radioactive atoms (the parent) goes down, while the stable end products (the daughters) are becoming more plentiful. By measuring the amounts of parent and daughter elements and by knowing how fast the parent decays into the daughters, we can determine how long the radioactivity has been taking place. Actually this only tells us how long the rock has been solid, because in the liquid state the parent and daughters can become separated.

The main kinds of radioactivity used to date rocks are listed in Table 6.2. Here U is uranium, Th is thorium, K is potassium, Pb is lead, and Ar is argon. The number at the upper left of the element symbol is the atomic weight of the nucleus. Thus ^{235}U is the uranium isotope with a total of 235 protons and neutrons in the nucleus. The *half-life* tells about how long it takes the parent to decay into the daughter nucleus. For example, if we have a group of ^{232}Th nuclei, about half of them will have changed into ^{208}Pb after 14 billion years. There are other daughter nuclei produced by the radioactive substances listed in the table, but most of them are not easily used in radioactive dating studies.

When rocks from the crust are examined for their radioactivity, it is found that they have ages ranging up to a maximum of about 3.7 billion years. This means that the oldest rocks on the surface of the Earth became solid about 3.7 billion years ago. Most rocks are much younger than this, and ages under 1 billion years are common. What does this mean regarding the age of the Earth? Can we assume that the Earth is 3.7 billion years old?

The Earth must be at least 3.7 billion years old, but it could be much older. If radioactivity and the formation process caused much of the rock material to melt and to remain melted for very long, the ages we find for the rocks will be less than the time since the Earth formed. The early stages of the Earth have erased the records of what those stages were like, much as all traces of a snowman are lost when the snow melts.

There is no known way to determine the age of the Earth from the Earth materials alone. Meteorites (these are discussed in Section 9.3) can help because they are pieces of rock and metal that arrive on the Earth from outer space. The meteorites show radioactive ages that range up to a maximum of 4.6 billion years. If the meteorites and the Earth were formed at the same time, this would also be the age of the Earth.

The element lead has a number of stable isotopes, some of which are produced by the radioactivity of other elements and some of which are not. By studying the abundances of the different isotopes of lead and of uranium, we find that Earth material and meteorite material are different from each other today, but they were the same about 4.6 billion years ago. This suggests that the material that eventually formed the Earth and the material that made the meteorites were once the same, but they separated 4.6 billion years ago and have had different histories since. This is the basis for the belief that the Earth has an age of 4.6 billion years.

Over four and a half billion years ago the Earth material separated from the rest of the solar system, or at least from the material that formed the meteorites, and took on the approximate size and shape that our planet has today. The surface must have been molten for at least part of the time during the next billion years, since we can find no rocks older than about 3.7 billion years. But why is it that most rocks are much younger than this? If the surface solidified 3.7 billion years ago, why aren't all surface rocks 3.7 billion years old? There are yet other processes that strongly affect the surface of the Earth. To understand them, we need to examine the Earth's surface and near-surface layers more closely.

CONTINENTAL DRIFT

Geologists have discovered that the surface of the Earth is composed of a number of huge blocks or *plates* (see Figure 6.5). They are some 60-miles thick and may extend thousands of miles horizontally. The plates consist of the

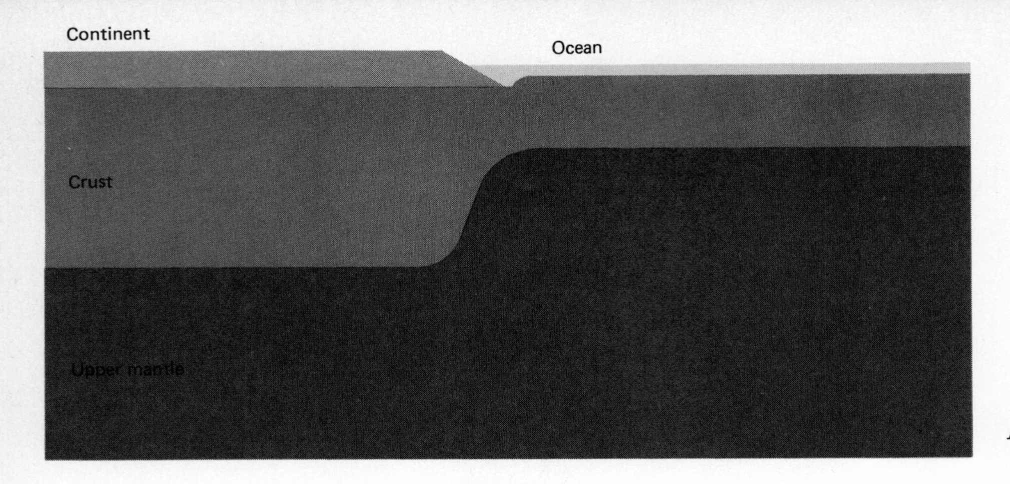

FIGURE 6.5
A surface plate on the Earth.

crust, including continents and oceans, plus a layer of hard rocks in the upper part of the mantle. The plates are floating on a denser but softer region of the mantle, a region that is somewhere between a rigid solid and a freely flowing liquid in consistency.

The most interesting thing about the plates is that they do not hold still but move around on the Earth. They act like rafts, carrying the continents around with them, reshaping the oceans, and giving rise to the term *continental drift*. The motions of the plates on the Earth are quite appreciable—up to several inches per year, an amount easily measured. There are many other lines of evidence that confirm that continental drift has been taking place for at least several hundred million years. This evidence includes studies of magnetism, radioactivity, and fossil records in rocks. It seems incredible that such huge, massive objects as the continents can actually be carried from one part of the Earth to another as a river carries a raft, and prior to the 1960s geologists themselves considered this impossible. Today, however, it is an accepted fact (see Figure 6.6).

It is not known for certain why the plates are moving, but a strong possibility is that slow currents in the soft mantle rocks carry them along, the way that the current of a river carries driftwood. Any solid will flow like a viscous fluid if it is subjected to steady forces for a long enough period of time. It is possible that excess heat deep within the Earth causes this flow in the mantle and the consequent drift of the continents (see Figure 6.7).

Earthquakes may be caused by two plates sliding past each other. When two plates collide head on, they could crumple each other

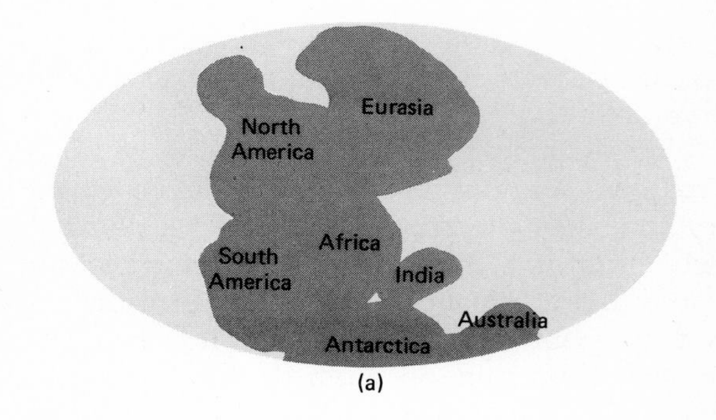

FIGURE 6.6
Changes in the positions of the continents over the past 200 million years. (a) Two hundred million years ago. (b) One hundred million years ago. (c) Today. [Based on "The Breakup of Pangaea" by R. S. Dietz and J. C. Holden, *Scientific American* 223 (Oct. 1970), 30.]

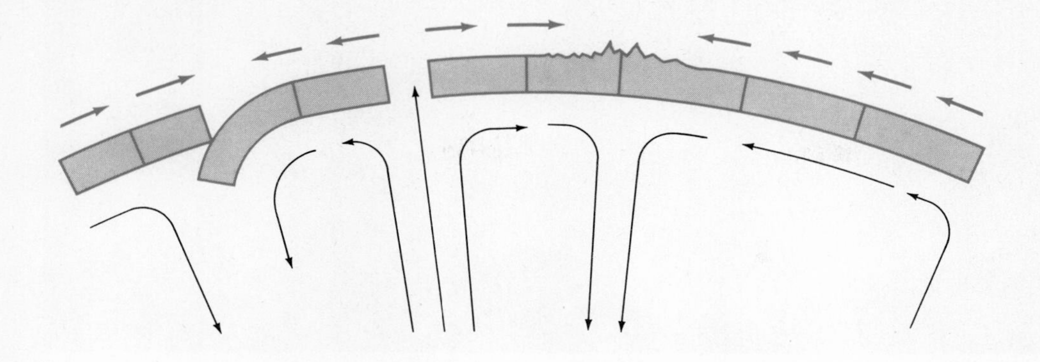

FIGURE 6.7
Mantle currents and plate motion.

and form mountain ranges, or one might be forced under the other, deeper into the mantle. In the latter case, the submerged plate is heated and mixed in with the softer mantle rocks. At places between separating plates, the soft rocks from the interior are being pushed upward, forming new surface material. In this way the upper parts of the Earth, including the crust and the upper part of the mantle, are slowly being mixed and recycled. Now we can understand why few surface rocks are as old as 3.7 billion years: Most surface material gets recycled into the hot, semimolten mantle in a shorter time than 3.7 billion years.

It has been known for some time that the air around us can not be the original atmosphere of the Earth. The reason is that there are a number of gases, such as neon, that are quite common in the Sun and in the rest of the Universe and that one would expect to be common in the original atmosphere of the Earth. Actually, such gases are extremely rare on the Earth. We conclude that the Earth lost its first atmosphere and that it was probably blown away by the very strong energy emission of the Sun early in its life.

Where did the Earth get its present atmosphere? The hot, molten material (magma), which comes up to the surface from the mantle in volcanoes (see Figure 6.8) and between receding plates, releases large amounts of various gases, including water vapor. These gases undergo many complicated chemical reactions with the surface materials, resulting in the nitrogen and oxygen atmosphere plus the surface water that make the Earth so attractive to life. The large amount of free oxygen in the atmosphere is probably due to the action of

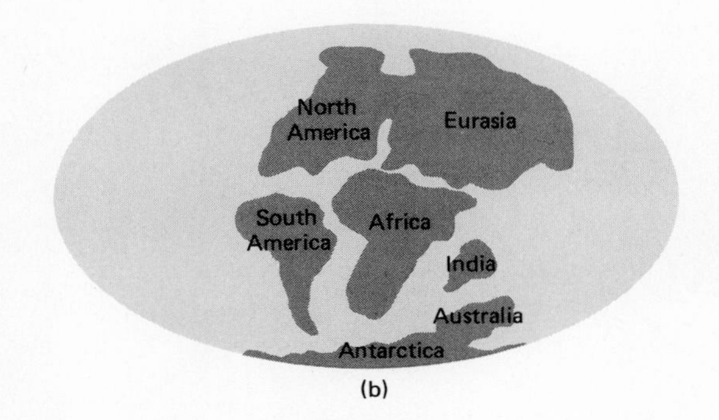

(b)

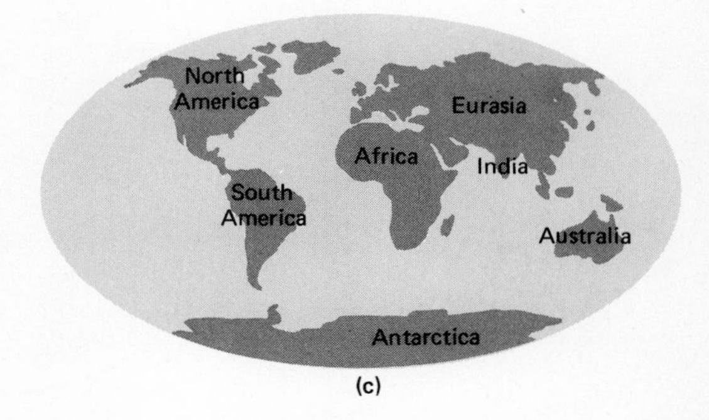

(c)

plant life over several billion years. Actually, the atmosphere, the oceans, and the surface rocks are probably all continually recycled by chemical reactions and plate movements that mix the surface materials and the upper mantle rocks over long periods of time.

6.2 The Earth in motion

ROTATION OF THE EARTH

Scholars since the time of Aristotle have known that the Earth is spherical, although this knowledge did not always drift down to the rest of the people. We also know that the Earth turns around once a day on an axis that passes through the Earth at two points; we call those points the *poles* (see Figure 6.9). The rotation carries all objects on the surface of the Earth in a direction we call *east;* the opposite direction we call *west*.

It is apparent that as long as we remain on the surface of the Earth, we can see only half of the sky at any instant; the Earth itself blocks off the view of the other half since that half is below the horizon. Figure 6.9 shows that as rotation carries us around, we have a view of

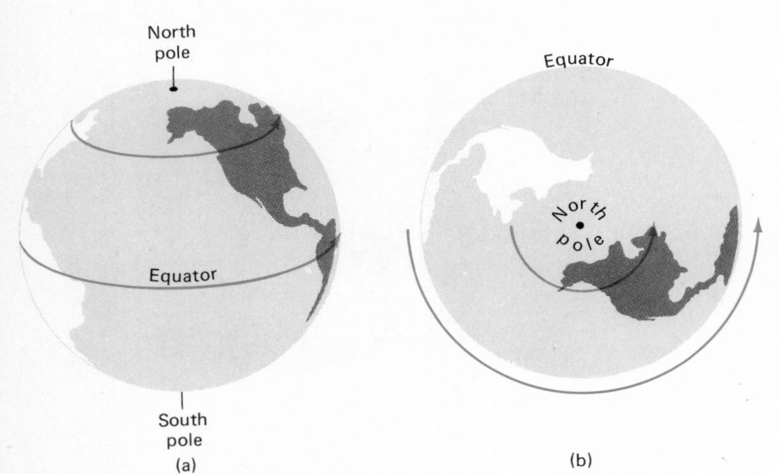

FIGURE 6.9

The eastward rotation of the Earth. (a) View from above the equator. (b) View from above the North Pole.

FIGURE 6.8

The eruption of the volcano Halemaumau in Hawaii in 1968. (Robert G. Strom)

different parts of the sky at different times of the day. The person on the equator is exposed to the entire sky throughout one day, but people at far northern or southern latitudes will have a section of the sky which they can never see. The entire southern half of the sky is always below the horizon for someone standing on the North Pole.

The rotation of the Earth produces the same effect as if we held still and the sky rotated around us in the opposite direction, that is, westward. This is, of course, what the ancients thought was happening. Figure 6.10 shows how rotation changes the view of the sky for someone at middle latitudes in the Northern Hemisphere.

Of course, it makes a big difference in the appearance of the sky whether or not the Sun is up, in other words, whether it is daytime or

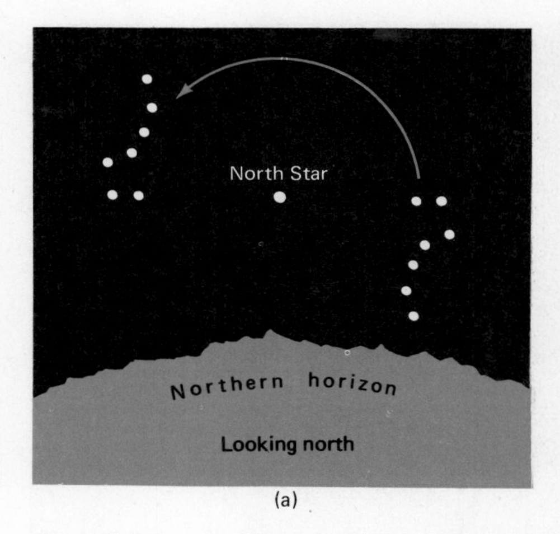

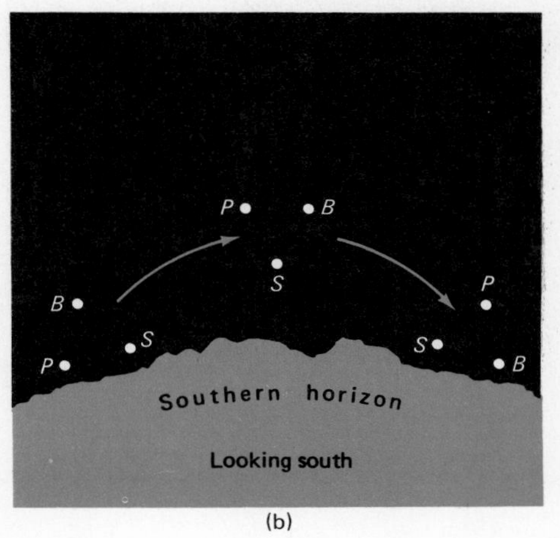

FIGURE 6.10

The apparent westward rotation of the sky. Looking north, we see two views of the seven stars in the Big Dipper as they appear to swing around the North Star. Looking south, the letters *B*, *P*, and *S* represent the three bright stars Betelgeuse, Procyon, and Sirius, which form a conspicuous triangle in the winter skies.

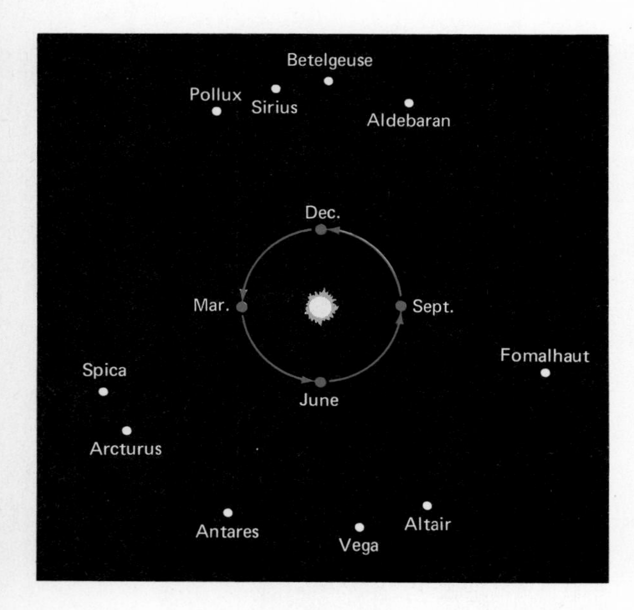

FIGURE 6.11
The motion of the Earth around the Sun. The stars are very much farther away than the proportions in the figure would indicate.

nighttime. It is too bad, but the Sun brightens up the sky so much that all stars are lost in its daytime glare. This is due to the scattering of sunlight by the atmosphere of the Earth. The only stars that we can see are those above the horizon when the Sun is down. Fortunately the motion of the Earth in its elliptical orbit around the Sun causes the Sun to appear to move among the stars; if a star is too close to the Sun to be visible now, wait a few weeks, the Sun will appear to move out of the way, and the star will become visible.

ORBITAL MOTION

Figure 6.11 shows how the orbital motion of the Earth makes it appear as though the Sun moves around the dome of the sky once each year. In early July, the Earth is in a position where the Sun is seen nearly in line with the bright star Sirius, so that star is not visible to us. Antares, Vega, and Altair are in the opposite part of the sky and are easily seen on a clear night. In September, the Earth has moved to the point in its orbit at which Spica and Arcturus become lost in the glare of the Sun, Fomalhaut is easily seen, and Sirius (along with Pollux, Betelgeuse, and Aldebaran as shown) is coming out from behind the blinding Sun. During the months of December and January, Pollux, Sirius, Betelgeuse, and Aldebaran help to brighten the winter sky at night while Altair is nearly hidden from view. In spring, Fomalhaut is gone while Spica and Arcturus command the night skies. As summer again approaches, the Sun appears to creep up on Aldebaran, Betelgeuse, Sirius, and Pollux as it did the year before, and as it will next year and in future years.

Note that the effects described here are worldwide. Around December 1, the Sun is nearly lined up with the star Antares, and this holds true for anyone on the Earth. When you look at the Sun on December 1, you will know that Antares is near it in the sky (although of course you cannot see it), and it makes no difference whether you are in the United States, China, or Australia. Opposite sides of the Earth see opposite parts of the sky at any given time. The United States sees now what China saw 12 hours ago, and vice versa.

Figure 6.11 does not show the complications caused by different north-south positions on the Earth or in the sky. It appears from the figure that in early January the Sun is in line with Vega; however, Vega is actually far north of the Sun in the sky on that date. To show this, a three-dimensional view is needed (see Figure 6.12).

In Figure 6.12, the Earth is the point at the center, and the large sphere is the imaginary dome of the sky. Some of the bright stars are shown on this dome. The large circle around the sky is the path that the Sun appears to take as seen from the Earth. In other words, it is the extension of the plane of the Earth's orbit around the Sun. This circle is called the *ecliptic,* and the times of the year for which the Sun is at certain positions are indicated. Here we

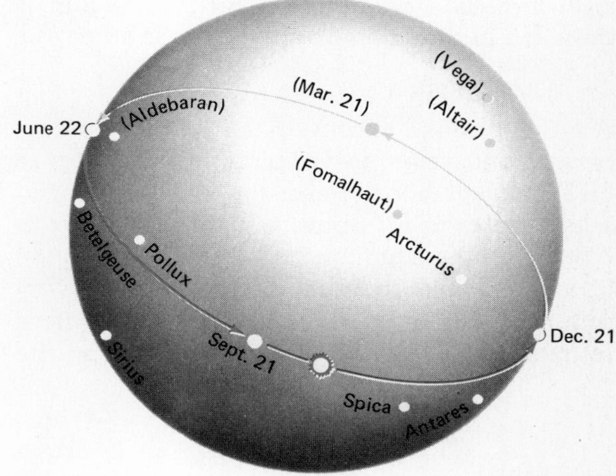

FIGURE 6.12
The Sun's apparent motion along the dome of the sky. The Earth is at the center of this imaginary representation.

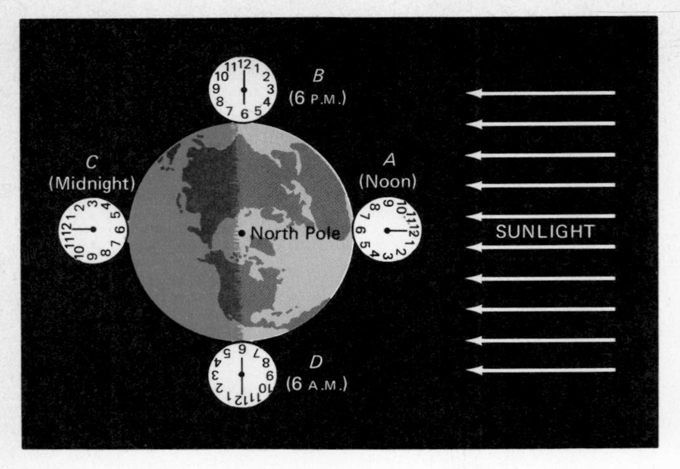

FIGURE 6.13
Timekeeping on different parts of the Earth.

note that as seen from the Earth, the Sun is quite far in direction from Vega in January.

TIMEKEEPING

The Sun and the rotation of the Earth provide a convenient system for keeping time (see Figure 6.13). Call it noon when the Sun is directly overhead or as far above the horizon as possible. This is the situation for a person at *A* in the figure. A person at *B* is one-fourth of the way around the world to the east of *A*. This is where rotation will take *A* in another 6 hours so the time for *B* should be 6 hours later than noon or 6 P.M. (one-fourth of a day ahead). A person at *B* is seeing the sunset. Halfway around the world from *A* is *C*, whose time should be one-half day, or 12 hours, ahead of (or behind!) *A*. The time for someone at *C* is midnight. Finally, someone at *D* is watching the Sun rise at 6 A.M., the position where *A* was 6 hours earlier.

FIGURE 6.14
Time zones on the Earth.

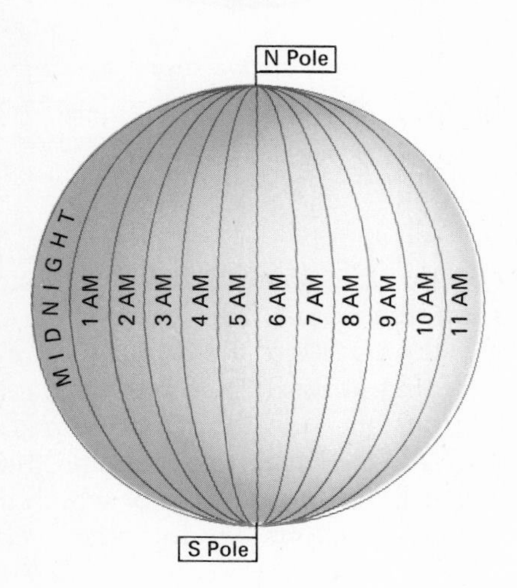

We can see from this illustration that we do not change our time by moving north or south but by moving in an eastward or a westward direction. For convenience the Earth has been divided into 24 long, narrow strips or zones that extend from the North Pole to the South Pole. These are called *standard time zones*, and by definition everyone in a given zone has the same time (see Figure 6.14). Each zone is 1 hour later than the zone west of it and, naturally, 1 hour earlier than the zone to its east. The boundaries of the time zones are bent in many places to fit political boundaries. The use of *daylight saving time*, setting the clock 1 hour ahead of standard time, is a common seasonal practice.

THE SEASONS

The rotation axis of the Earth is not perpendicular to the orbit around the Sun but is tipped by an angle of about 23½ degrees. During June, the northern half of the Earth is leaning toward

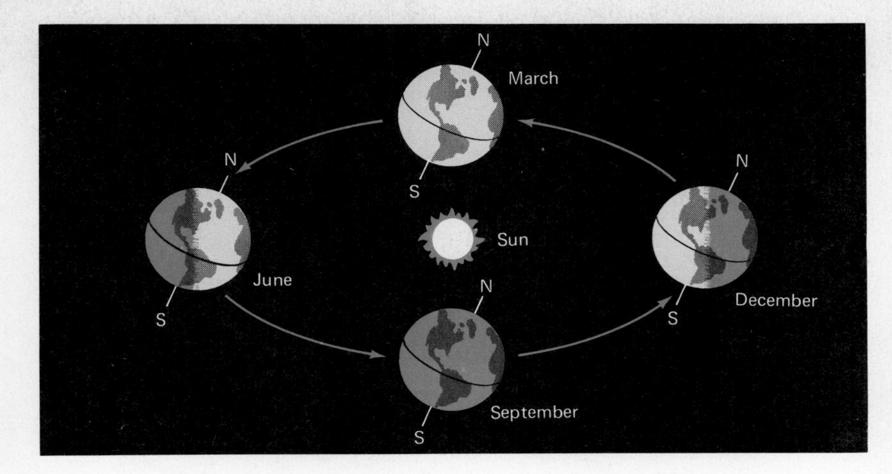

FIGURE 6.15
The seasons. Sizes and distances in this figure are, of course, not to scale.

the Sun and the southern half is leaning away; in December, the two are reversed (see Figure 6.15). It is apparent that the hemisphere tipped toward the Sun receives more than its share of light and heat from the Sun, and it does so at the expense of the hemisphere tipped away. It is this increased exposure to the sunlight that brings on summer, while winter occurs in the half of the Earth turned away from the Sun. Northern latitudes thus have summer starting in June and winter starting in December, while these seasons are reversed for regions south of the equator. The Earth is actually a little closer to the Sun in January than at other times of the year, but the amount is not enough to have much influence on the seasons.

IMPORTANT WORDS

Atmosphere	Standard time zone
Earthquake waves	Daylight saving time
Crust	Radioactive dating
Mantle	Half-life
Outer core	Plates
Inner core	Continental drift
West	Pole
Ecliptic	East

REVIEW QUESTIONS

1 What fraction of the mass of the Earth is contained in the atmosphere?

2 How do we know that the deep interior of the Earth is more dense than the rocks on the surface?

3 What is our main source of information about the interior of the Earth?

4 What are the four main layers of the solid part of the Earth?

5 How do we know that the outer core is liquid?

6 How do we know that much of the Earth must have been in the liquid state in the past?

7 What could have made the interior of the Earth liquid?

8 Why is there less radioactive heating today than in the distant past?

9 How does radioactivity help us find the ages of rocks?

10 How old are the oldest rocks found on the Earth?

11 What is continental drift?

12 How do surface rocks get recycled into the semi-molten mantle?

13 Where did the Earth's atmosphere come from?

14 How are the directions east and west defined?

15 Why can't we see all of the sky at one time?

16 Will a person on the equator or a person on the North Pole see more of the sky over one day?

17 Why is the star Sirius easily seen during the

winter but not generally visible during much of the summer?

18 Do we change time by moving north or south? East or west?

19 Why does the Northern Hemisphere receive more light and heat from the Sun during June than during December?

QUESTIONS FOR DISCUSSION

1 When I was in the seventh grade, my science textbook said that the Earth could not have a liquid core. As proof it suggested that we try to spin a raw egg on its end on a table. We did, and it wouldn't spin. When the egg was hard-boiled, however, so it had a solid core, it spun easily. Since the Earth is known to spin around once a day, this was considered proof that the core of the Earth must be solid.

What flaws can you find in this "proof"?

2 There are many radioactive substances known besides those listed in Table 6.2. For example, ^{14}C is an isotope of carbon that is radioactive and has a half-life of about 6000 years. Why can't ^{14}C be used in determining the ages of rocks on Earth?

3 Suppose that North America was attached to Europe about 200 million years ago. Calculate roughly how fast these two continents would have to be separating from each other in order to reach their present positions. Does this agree at least approximately with the rate of continental drift given in this chapter?

4 Does the recycling of the air and water through mixture with the rocks in the mantle offer us any help for our pollution problems?

5 The United States and China are both in the Northern Hemisphere, and stars that are visible in the United States are also visible in China. Australia is in the Southern Hemisphere, and Australians see a different set of stars, although there is much overlap. Explain.

6 Look at Figure 6.9 and decide whether stars would rise and set for a person standing on the North Pole or the South Pole.

7 Which of the stars in Figures 6.11 or 6.12 is most nearly opposite the Sun in May? What is most nearly in line with the Sun in May?

8 In Figure 6.13, is person *C* 12 hours ahead of *A* or 12 hours behind? Does it make any difference? If it is November 28 for *A*, should *C* be just starting or just ending November 28?

The international date line (IDL) is a line running north–south between the poles such that, by definition, a point slightly west of it is 24 hours later than a point just to the east of it. This line is placed in the Pacific Ocean. Do you see the need for the IDL in answering these questions? (See question 9.)

9 Referring to the question 8, suppose the IDL runs through person *B* in Figure 6.13. Would person *C* be 12 hours ahead of or behind person *A*? How about if the IDL goes through person *D*?

10 Suppose the rotation axis of the Earth were in the plane of its orbit around the Sun. Draw a figure similar to Figure 6.15 to represent this. How would summer and winter be affected?

REFERENCES

The material in this chapter can be found in greater detail in most elementary textbooks on geology and astronomy. A good source is Chapters 5 and 6 of *Earth, Moon, and Planets*, 3rd ed., by F. L. Whipple (Harvard University Press, Cambridge, 1971). Beautiful pictures of the Earth from space are contained in "Skylab Looks at the Earth" in the *National Geographic* 146 (Oct. 1974), 470.

"Continents Adrift" is a collection of articles on the structure of the Earth from *Scientific American* (Freeman, San Francisco, 1972). Other articles from *Scientific American* include "The Earth's Mantle" by P. J. Wyllie [232 (March 1975), 50]; "The Earth" by R. Siever [233 (Sept. 1975), 82]; "The Subduction of the Lithosphere" by M. N. Toksöz [233 (Nov. 1975), 88]; and "Convection Currents in the Earth's Mantle" by D. P. McKenzie and F. Richter [235 (Nov. 1976), 72]. A well-illustrated article is "This Changing Earth" by S. W. Mathews in *National Geographic* [143 (Jan. 1973), 1].

SEVEN
THE MOON

Next we examine the Moon, our nearest neighbor in space. Its motions and aspects as seen from the Earth are the topics of Section 7.1, and its properties as an astronomical body are covered in Section 7.2.

7.1 Motions of the Moon

ORBIT

The Earth was once thought to be the center of the Universe. After the dust settled from centuries of argument, we can still agree with the ancients that the Moon goes around the Earth—it is all that the Earth has left. Figure 7.1 is a diagram of the Earth and Moon, with the sizes and distance drawn to scale.

The Moon travels in a nearly circular orbit around the Earth as the Earth goes around the Sun. The result as viewed from the Sun would be something like the epicyclic motion that we saw in Figure 5.5. Figure 7.2 shows the orbits of the Moon and the Earth. Neither the distances nor the sizes are drawn to scale. You can easily see the winding motion of the Moon as it follows the Earth around the Sun.

PHASES OF THE MOON

Like the planets, the Moon shines only by reflecting the light of the Sun. The half of the Moon facing the Sun is lit up as is the daylight

FIGURE 7.1
The Earth and the Moon with their sizes and distance between them drawn to scale.

FIGURE 7.2
The orbits of the Moon and the Earth. The sizes and distances are not drawn to the correct scale.

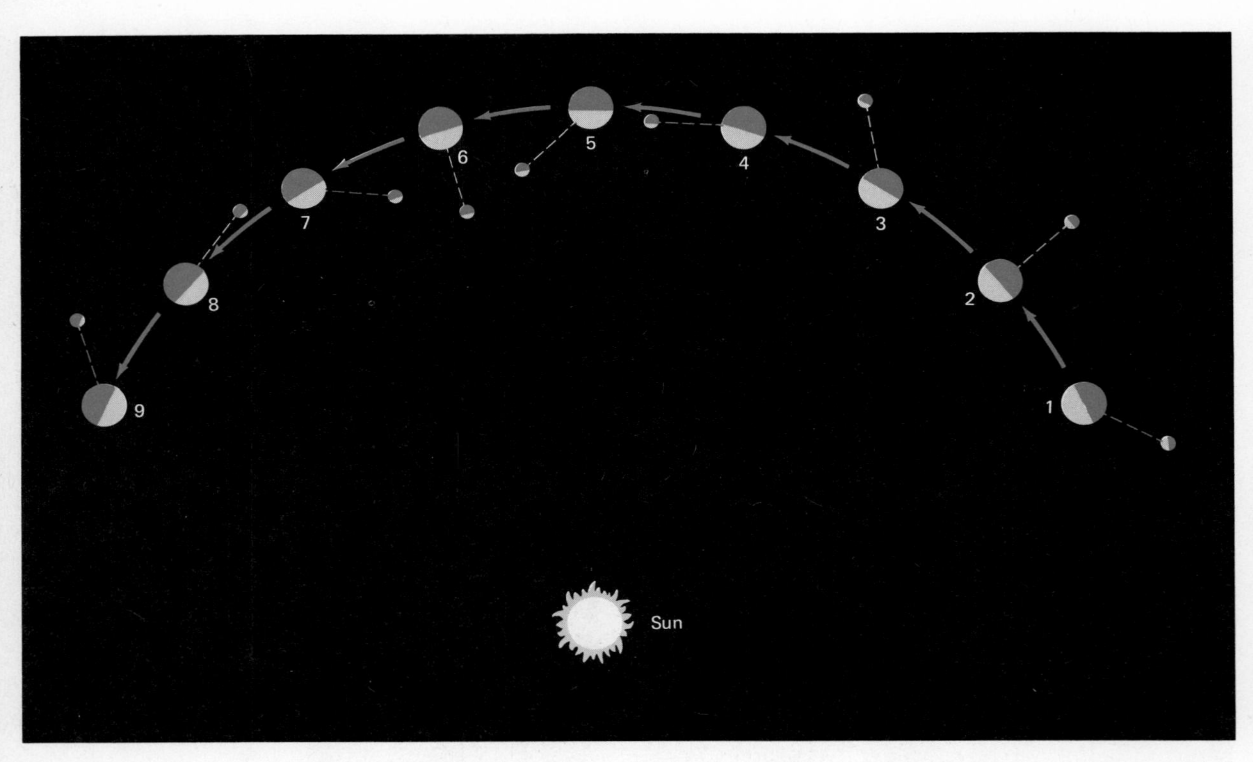

side of the Earth; the other half is dark and invisible, or nearly so.

When we see the Moon almost opposite the Sun in the sky, as in position 2 of Figure 7.2, the daylight half of the Moon is facing us. The whole disk is visible, and the Moon is in the *full phase*. The opposite case occurs when the Moon as seen from the Earth is in nearly the same direction as the Sun. This is illustrated by position 6 of Figure 7.2. Here the dark half of the Moon is toward us, and we cannot see it. This phase is called the *new phase*. When the Moon is not exactly new but nearly new, as in positions 5 and 7 in the figure, only a thin *crescent* Moon is visible (see Figure 7.3). Half-way between the full and new Moon are what are called the *quarter phases* (see Figure 7.4). Note that the phases are worldwide: If the Moon appears full to one person, it appears full to anyone on Earth who can see it. Of course, an astronaut who goes beyond the Moon will see the phases differently.

The phases of the Moon depend on how close in the sky the Moon appears to the Sun. In the new phase, the Moon must be very close to the Sun as seen from the Earth, while in the full phase, it is opposite the Sun. The full Moon rises about sunset and sets at around sunrise, although there can be exceptions at extreme northern and southern latitudes. Figure 7.5 shows the positions of the Moon in the sky at sunset for different phases.

FIGURE 7.3
The crescent Moon. (Lick Observatory photograph)

FIGURE 7.4
The first quarter Moon. (Lunar and Planetary Laboratory, University of Arizona, copyright pending)

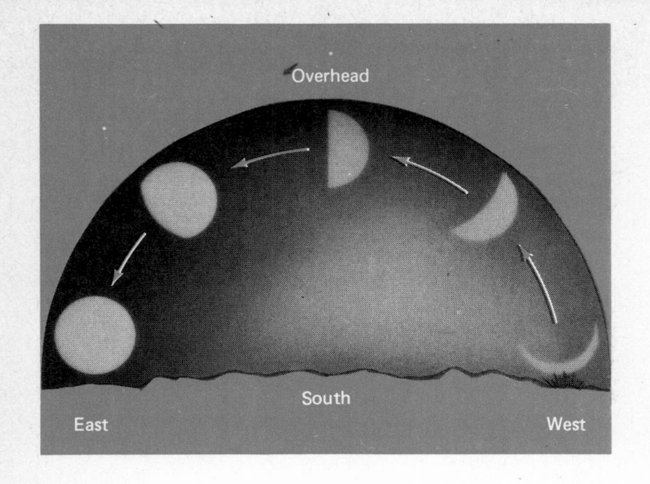

FIGURE 7.5

(Left) The Moon at sunset. The different views represent changes in the Moon over a period of about two weeks.

FIGURE 7.6

(Below) Spring and neap tides. The tides are greatly exaggerated. (a) New moon, spring tide. (b) First quarter phase, neap tide. (c) Full moon, spring tide. (d) Third quarter phase, neap tide.

TIDES

The Moon, along with the Sun, is responsible for tides. The Moon's gravity pulls hardest on the near side of the Earth and least hard on the far side. This causes a tidal force that tries to stretch out the Earth in the direction of the Moon. The solid part of the Earth is so rigid that it does not stretch very much, but the fluid parts—the atmosphere and the oceans—are drawn out noticeably in the directions toward and away from the Moon. We don't notice the atmosphere tides, but the ocean tides are obvious to anyone who spends time on the seacoast.

The Sun has a weaker tidal pull on the Earth than the Moon does, but it is also important. When the Sun and Moon are in a straight line with the Earth, that is, either at full or new Moon, the two are trying to stretch out the Earth in the same direction. By acting together, they raise extra large tidal bulges called *spring tides*. When the Moon is in either quarter phase, it tries to elongate the Earth in one direction while the Sun tries to do the same thing in a perpendicular direction. The two efforts partly cancel, and the tides are weaker than usual. These weaker tides are known as *neap tides* (see Figure 7.6).

As you would expect because of its larger mass, the Earth exerts a stronger tidal force than the Moon. The Earth's tidal force acting on

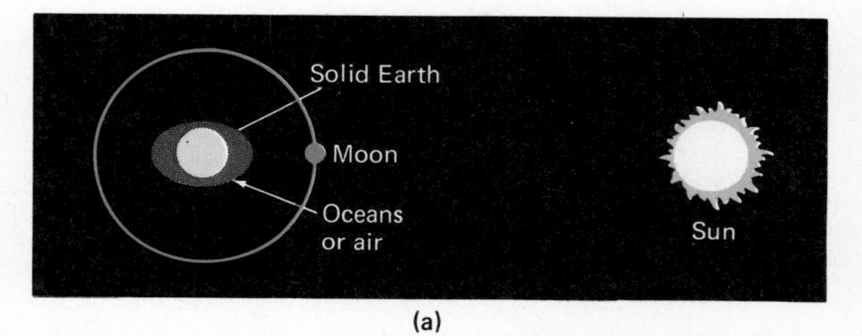

(a)

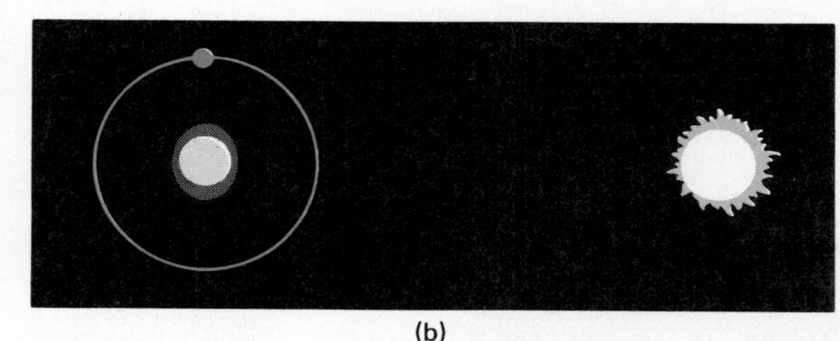

(b)

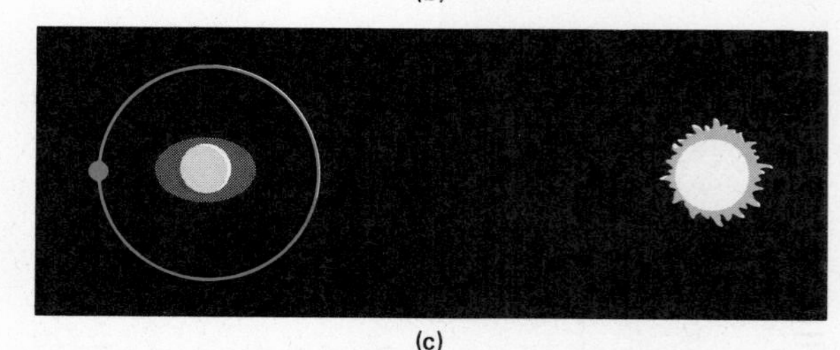

(c)

(d)

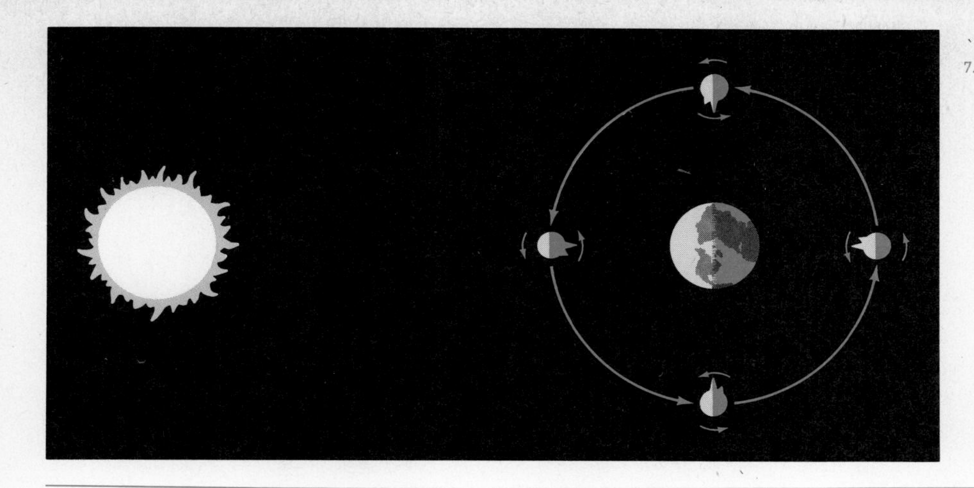

FIGURE 7.7
The rotation and revolution of the Moon.

the Moon has succeeded in slowing down the rotation of the Moon. This force causes the Moon to turn once on its axis in exactly the same time that it takes to orbit the Earth, or about 27 days. This causes the Moon to always keep the same side turned toward the Earth (Figure 7.7), and before the Space Age no one had seen the back side of the Moon.

Tidal forces tend to slow down the Moon's rotation until it has the same period as the orbital motion. The reason is that then the tidal bulge is fixed in one place on the Moon. In the same way, the weaker tidal force of the Moon is slowing down the rotation of the Earth, causing the day to gradually get longer. The lengthening of the day amounts to only a tiny fraction of a second per century, but in the distant future the Moon will finally succeed in forcing the Earth to always keep the same side facing our satellite.

ECLIPSES

When the Moon is full, it is nearly opposite the Sun. If it were exactly opposite the Sun, it would be in the Earth's shadow. This phenomenon does occur sometimes, and is known as an eclipse of the Moon or *lunar eclipse* (see Figure 7.8). The Moon is eclipsed on the average of once or twice a year. The Moon actually never completely disappears from view during a lunar eclipse because a little sunlight manages to get through the Earth's atmosphere

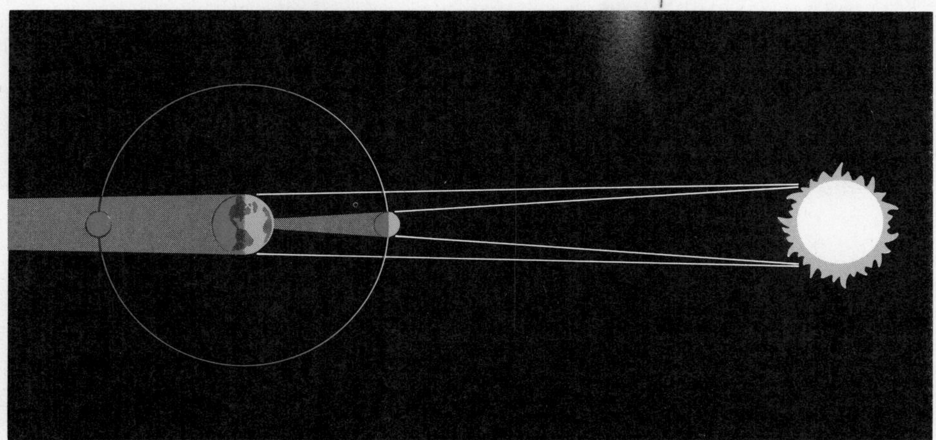

FIGURE 7.8
Solar and lunar eclipses.

FIGURE 7.9
A photograph of the full Moon with some of the features identified. (Lunar and Planetary Laboratory, University of Arizona, copyright pending)

and onto the Moon. The eclipse is said to be total if the entire Moon gets into the Earth's shadow and partial if only part of the Moon is shaded.

A *solar eclipse* occurs when the Moon comes exactly between the Earth and Sun, blocking off part of the sunlight to the Earth (also see Figure 7.8). This, of course, can take place only at new Moon. Most of the time the Moon is not exactly lined up with the Sun at new phase, and no eclipse takes place. There are at least two solar eclipses every year, usually more.

If the Moon's orbit were in exactly the same plane as the Earth's orbit around the Sun, then there would be two eclipses every month: a solar eclipse every new Moon and a lunar eclipse every full Moon. Actually, the Moon's orbit is inclined to the Earth's orbit by a small angle of about 5 degrees, and it is much more rare that the Sun, Moon, and Earth are lined up closely enough for an eclipse to occur. To look at it differently, as the Moon moves across the sky, it will cause a solar eclipse whenever it covers the disk of the Sun, either partially or completely. Most of the time, the Moon passes the Sun in the sky a little bit above or below the disk of the Sun, and no eclipse occurs. Only rarely will it come close enough to a "direct hit" to cause an eclipse.

The Moon is much smaller than the Earth, and its shadow can cover only a small part of the Earth's surface. Thus a solar eclipse is not visible to a very large part of the Earth's population. It is particularly rare to witness a total solar eclipse, when the disk of the Sun is completely covered by the Moon. (This phenomenon is shown in Figure 11.2.) Total solar eclipses occur every year or two, but they are visible only from such a limited area that few people have seen one. The outer solar atmosphere then becomes visible, and the view is quite spectacular.

7.2 The structure of the Moon

SURFACE FEATURES

We see the face of the Moon so often that it ought to look familiar. Figure 7.9 shows this face with some of the features identified. The dark areas are called *maria* (singular *mare*), which is Latin for "seas," because that was what Galileo thought they were when he saw them through his small telescope. Today they are known to be large lava flows that have long since hardened.

On a closer look, the craters, mountains, and other features show up well (see Figures 7.10 through 7.14). Most of the craters were formed by meteoric impact throughout the long

Plato
Mare
Imbrium
Apennine
Mts
Mare
Serenitatis
Aristarchus
Mare
Crisium
Oceanus
Procellarum
Copernicus
Mare
Tranquillitatis
Kepler
Mare
Fecunditatis
Mare
Nubium
Mare Nectaris
Mare
Humorum
Tycho

FIGURE 7.10
The Apennine Mountains with Mare Imbrium to the upper left. (Lick Observatory photograph)

FIGURE 7.11
Central region of the Moon showing parts of Mare Serenitatis (upper right), Mare Tranquillitatis (lower right), and the Hyginus Rille as the line near the center. (Lunar and Planetary Laboratory, University of Arizona, copyright pending)

FIGURE 7.12
Part of Mare Nubium, with the straight wall running some 60 miles long and about 5000 feet high. (Lunar and Planetary Laboratory, University of Arizona, copyright pending)

FIGURE 7.13

Mare Tranquillitatis. The Apollo 11 landing site is near the lower left edge. (Lunar and Planetary Laboratory, University of Arizona, copyright pending)

FIGURE 7.14

The crater Tycho taken from Lunar Orbiter V in 1967 from an altitude of 135 miles above the Moon. With close inspection the major features in Figures 7.10 through 7.14 can be identified in the full Moon photograph in Figure 7.9. (NASA) history of the Moon, although some of these craters are of volcanic origin. Such craters also form on the Earth, but the action of the wind and water erases them in a few million years or so. The Moon has no wind or water, so its face keeps the scars for much longer periods of time.

The craters do not cover the face of the Moon uniformly. The maria, for example, have much fewer craters on them per unit area than the highlands, as the other areas are called. This is evidence that the maria are relatively young parts of the lunar surface, as only the craters formed since the lava hardened are visible. The maria probably were made when large meteoric bodies crashed into the Moon. These crashes released from the interior large amounts of molten lava that flooded the general area and buried the older craters. The craters formed after the maria solidified are seen today, while older craters partially filled by the lava are also visible.

PAST HISTORY

Moon rocks brought back by the astronauts can be dated through their radioactivity, just as it is possible to date the Earth and meteorite rocks discussed in Chapter 6. This radioactive dating tells us how long ago the rocks solidified from molten material. Ages of different parts of the Moon's surface have been found in this way. The maria range in age from a little over 3 billion years to about 4 billion years. The oldest parts of the lunar surface are slightly over 4 billion years old.

The numbers of craters observed on different parts of the surface tell us how many craters were formed at different ages. Suppose, for example, that region *A* is known to be 3.8 billion years old, while region *B* solidified 3.6 billion years ago. If region *A* has twice as many craters per unit area as *B*, then we know that just as many craters were formed between 3.6 and 3.8 billion years ago as have been formed during the past 3.6 billion years. These numbers would indicate that the meteoric bodies that cause craters were much more common in the distant past than they are today.

By studying the abundances of craters on parts of the surface of known ages, it has been found that few craters have been formed during the past 3 billion years. This means that most of the chunks of matter that cause craters by colliding with the Moon (and with the Earth and other inner planets) had been cleaned out of the inner solar system by 3 billion years ago. Going further back in time, we find that the Moon's surface shows a rapidly increasing rate of crater formation. About 4 billion years ago, the debris bombarded the Moon so strongly that the surface was almost completely broken up, and few Moon rocks have survived from earlier times. Records of the first half-billion years of the Moon's history have been destroyed. Was this an avalanche of material suddenly released from the outer parts of the solar system?

It seems more likely that this was the final phase of the origin of the Moon and the inner planets. The main parts of these bodies were probably formed about 4.6 billion years ago, but a lot of debris, in the form of solid pieces of all sizes, was left over and continued to rain down on them. As time went on, more and more of this material was swept up by the large bodies, and by 3 billion years ago, the interplanetary space was nearly cleaned up.

Early in its existence the Moon was so violently bombarded that the outer parts, at least, were melted. It is uncertain whether the deep interior was ever completely melted. As with the Earth, melting could have been caused by the process of formation, if it occurred quickly enough, or by radioactive heating, if there were enough of this. When the bombardment lessened enough, the surface of the Moon cooled and solidified. Further collisions with interplanetary debris caused craters in the solid surface. Occasionally a very large chunk of material would hit the Moon hard enough to fracture the surface and allow the still molten material underneath to flow out. In this way the maria could have been formed.

The Moon is made out of rather different kinds of materials than the Earth. Iron, in particular, is much less plentiful than on Earth, and the Moon possesses little or no dense metallic core. In certain respects, though, the materials making up the Moon resemble the kinds of material that make up the mantle of the Earth.

Due to its small gravity, the Moon has lost into space any atmosphere it might have had in the past. Note in Figure 7.15 that the sky on the Moon appears dark, even in broad daylight. This is because the Moon has no air to scatter the sunlight.

The Moon shows no evidence of mantle or crustal activity, such as continental drift on the Earth. There are occasional "moonquakes" detected by equipment that was set up by astronauts, but the Moon is much more nearly a dead world geologically than the Earth.

ORIGIN

Scientists had hoped that the rocks and other data brought back by the astronauts would tell us all about the origin of the Moon (see Figure 7.16). Much was learned from these data, but how the Moon was formed is still a mystery.

There are three main theories of the Moon's formation:

1 The Moon was once a part of the Earth and was thrown off by rapid rotation.

FIGURE 7.15

Apollo 17 on the Moon. Note the dark sky, even in broad daylight. This is due to the lack of air on the Moon. (NASA)

FIGURE 7.16
This large boulder looks like the many we see everyday on the Earth, but it happens to be a Moon rock. (NASA)

2 The Earth and Moon formed together as a double planet and have stayed that way.

3 The Moon was formed in some other part of the solar system and was later captured by the Earth.

The first theory accounts well for most of the differences in composition between the Moon and the Earth, but there is some question as to whether the Earth could possibly lose a large blob and have it turn into the Moon as it is today. A variation on this theory has the Earth throw off a rather large envelope of material, part of which completely leaves the Earth and part of which stays in orbit around the Earth, eventually forming the Moon. According to the second theory, the Earth and Moon should be made out of nearly identical material, yet we have seen that this is not the case. The capture of the Moon as required by the third idea is a difficult and unlikely event at best. A variation on this theory is that a little at a time, material was captured and brought into orbit around the Earth. This material gradually coalesced and formed the Moon.

If the Moon were expelled by the rotation of the Earth or if it were formed as part of a double planet with the Earth, then we might expect its orbit to be aligned with the Earth's equator, but this is not the case. Instead the Moon's orbit is nearly aligned with the Earth's orbit about the Sun, which makes an angle of about 23½ degrees with the equator. This lends credence to some form of the capture theory. It is interesting to note that most of the satellites in the solar system, the so-called "regular ones," are all closely aligned with the equators of their parent planets. The irregular satellites are not so aligned. It is generally believed that the regular satellites were formed with their parent planets, while the irregular ones are thought to have been captured long after they had been formed in their own orbits around the Sun.

We see that there is evidence against all of the main theories of how the Moon could have formed, and the experts in the field seem to agree less now than they did in earlier times when they had much less data. Still, we can't argue with the fact that the Moon does exist. One interesting point is that all other satellites are insignificant specks in comparison to the planets to which they are attached, but the Moon is by no means negligible compared to the Earth. The Moon is among the largest satellites in the solar system, and it moves around one of the very small planets. There are a number of other ways in which the Moon seems different from other satellites, and it is possible that the process by which it was formed is completely different from that of the other satellites of the solar system.

IMPORTANT WORDS

New phase	**Neap tides**
Crescent phase	**Lunar eclipse**
Quarter phase	**Solar eclipse**
Spring tides	**Maria**

REVIEW QUESTIONS

1 What kind of motion does the Moon have as it follows the Earth around the Sun?

2 Why does the Moon show phases?

3 How does the Moon produce tides?

4 What is the difference between spring and neap tides?

5 What are the two types of eclipses?

6 Why are the dark areas on the Moon called maria?

7 How were most of the lunar craters formed?

8 Were most of the craters on the Moon formed recently or long ago?

9 Why are no Moon rocks found that are much older than 4 billion years?

10 How did the outer layers of the Moon become melted?

11 Does the Moon have continental drift?

12 Does the Moon have a large metallic core?

13 What are the three main theories about the origin of the Moon?

14 Is the Moon typical of the other satellites in the solar system?

QUESTIONS FOR DISCUSSION

1 As the sun is setting, you look to the west and see the Moon. Could it possibly be the full Moon? Why? What phase must it be in?

2 If the Moon is full, about what time will it rise? Why?

3 A witness to a robbery that occurred during the night claims that he saw what happened by the light of the full Moon. If an eclipse of the Sun took place a week before the robbery, could the witness be telling the truth? What if the eclipse had been lunar?

4 Draw a diagram showing how, if the Moon did not rotate on its axis, all sides of the Moon would be exposed to the Earth at one time or another.

5 How often would a person living on the Moon see the Sun rise? How often would this person see the Earth rise?

6 Examine Figures 7.10 through 7.14 and find some areas that must be younger than others. Look at some examples of two craters overlapping each other. Can you tell which was formed first?

7 How does the theory that the Moon broke off of the Earth explain the main differences in composition between the Earth and the Moon?

REFERENCES

The Moon is an important subject in all elementary astronomy books. Special mention is made of Chapters 7, 8, and 9 of F. L. Whipple's *Earth, Moon, and Planets*, 3rd ed. (Harvard University Press, Cambridge, 1971), and Chapter 7 of Z. Kopal's *The Solar System* (Oxford, London, 1973).

There is a large literature on our knowledge of the history and structure of the Moon. This includes "The Moon" by J. A. Wood [*Scientific American* 233 (Sept. 1975), 92]; "The Interior of the Moon" by D. L. Anderson [*Physics Today* 27 (March 1974), 44]; "Whence the Moon?" by A. L. Hammond [*Science* 186 (Dec. 6, 1974), 911]; and "Surface Geology of the Moon" by F. El-Baz [*Annual Review of Astronomy and Astrophysics* 12 (1974), 135]. A good article on the ways in which craters give us information about the history of the Moon and planets is "Cratering in the Solar System" by W. K. Hartmann [*Scientific American* 236 (Jan. 1977), 84].

Well-illustrated works on the Apollo program include "To the Mountains of the Moon" by K. F. Weaver [*National Geographic* 141 (Feb. 1972), 233]; "Apollo 16 Brings Us Visions of Space" [*National Geographic* 142 (Dec. 1972), 856]; and "Apollo 17," a series of articles in *National Geographic* [144 (Sept. 1973), 289]. An outstanding book on the Apollo program is *Apollo Expeditions to the Moon*, edited by E. M. Cortright (NASA SP-350, U.S. Government Printing Office, Washington, D.C., 1975).

EIGHT
THE PLANETS

Here we examine the planets one-by-one, working our way out from the Sun. Each planet has its own pecularities, its own special points of interest, and its own puzzles.

8.1 The inner planets

Mercury and Venus go around the Sun inside the orbit of the Earth (see Figure 8.1). They are alternately on the near side and the far side of the Sun. When on the near side, they are close to us, but they have their night sides toward us and appear in crescent or new phases. When they are on the far side of the Sun, they are more distant, and they appear between quarter and full phase. Like the Moon, Mercury and Venus go through a complete set of phases. Because their orbits are interior to the Earth's, they are never seen very far from the Sun in the sky. If you see a bright planet overhead at midnight, it cannot be Mercury or Venus.

MERCURY

Mercury is not only the nearest planet to the Sun, but it is also the smallest. Because of its closeness to the Sun, it is very hard to see and study. The Mariner 10 spaceship showed that it is nearly airless and well covered with craters, looking much like the Moon (see Figures 8.2 through 8.4).

Mercury's density is about the same as the Earth's, nearly 5½ times the density of water. (Refer to Table 5.1 or Appendix 4.) The Earth and Mercury are the two bodies with the greatest densities in the solar system. Part of the Earth's large density is due to compression of the material under the large pressures in the deep interior. Mercury is a smaller and less massive planet, and the pressures in its interior are much less than those in the Earth. Thus, the density of Mercury is not significantly increased by compression, and we conclude

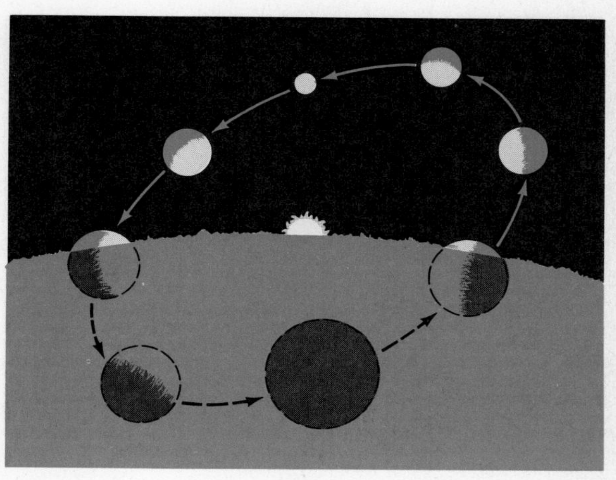

FIGURE 8.1
The orbit and phases of Mercury or Venus as seen from the Earth at sunrise.

FIGURE 8.2

Mercury in the crescent phase as photographed by Mariner 10 in 1974. The similarity to the Moon is quite striking. (NASA/JPL)

that Mercury has a different composition than the Earth. (Compare the actual densities with the zero-pressure densities in Table 5.1.) It is probable that Mercury has a very large, dense iron core and that its mantle is much less extensive than the Earth's.

The heavily cratered surface of Mercury reminds us of the Moon, and it has probably had a history of meteoric bombardment similar to the Moon's. This evidence shows that the meteoric particles were not confined to the vicinity of the Earth and Moon but existed over the entire inner solar system.

FIGURE 8.3

A mosaic of photos taken by Mariner 10 of the surface of Mercury. Note the large circular basin at the bottom center. (NASA/JPL)

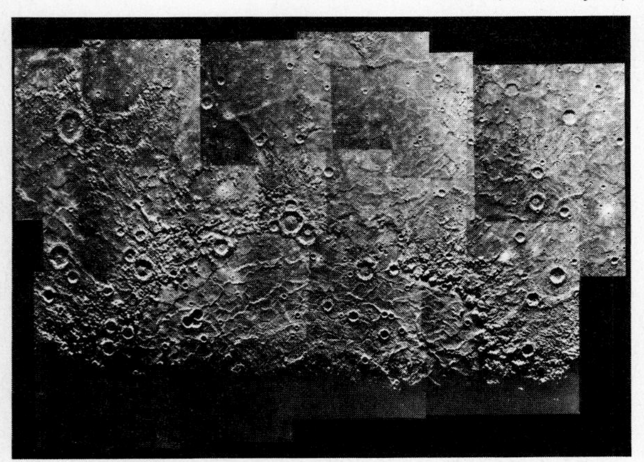

Mariner 10 detected a small magnetic field and a tiny trace of an atmosphere on its close pass by Mercury in 1974. Like the Moon, Mercury has too little gravity to hold onto its gases for very long, so they should rapidly escape into space and leave the planet without an atmosphere. The extremely tenuous atmosphere actually observed must consist either of particles ejected by the Sun or of gases that are being continuously released through the surface of the planet; these gases reside for a short time near the surface before breaking the gravity barrier and escaping into space. Helium is the only gas detected so far. It is possible that volcanism continues today on Mercury, or the helium could come from radioactivity of surface rocks.

Years ago Mercury was reported to rotate exactly once on its axis during the time it takes to go around the Sun, thus keeping the same face toward the Sun as the Moon does to the Earth. Surface markings on Mercury are very hard to make out from the Earth, and this report was not adequately checked. It did correspond to what astronomers expected, since the strong tidal force of the Sun could make Mercury rotate in this way. In the 1960s these old observations were shown to be wrong: Radar signals reflected off of Mercury indicated that the planet actually rotates on its axis exactly 1½ times as it goes once around the

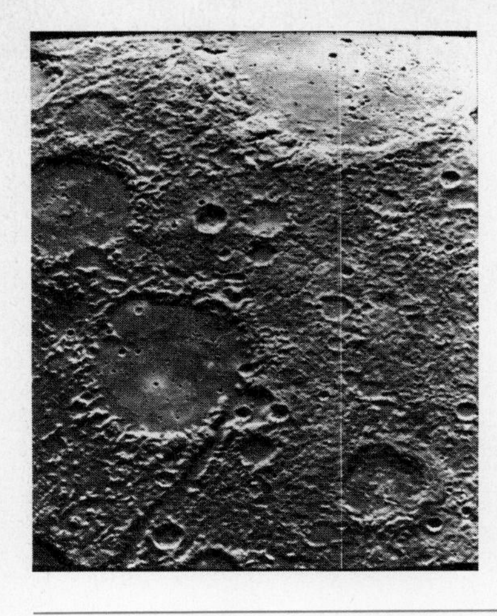

FIGURE 8.4

A close-up of the surface of Mercury taken by Mariner 10 and computer-enhanced at the Jet Propulsion Laboratory. The picture covers an area of 136 by 180 miles. (NASA/JPL)

Sun. This type of rotation can also be produced by the tidal force of the Sun. It is interesting to note that for many years, astronomers (including me) gave an erroneous value to the rotation of Mercury and argued convincingly that that value had to be correct.

VENUS

After the Sun and Moon, Venus is the brightest object in the sky. If you see an object in the west shortly after sunset or in the east just before sunrise, an object that looks much brighter than any star or planet ought to look, you are probably seeing Venus. It is often visible to the naked eye in broad daylight, although it is not easy to find. The phases of Venus, seen from Earth, are shown in Figure 8.5.

Venus is very much like the Earth in size and mass, and it has a very thick atmosphere surrounding it. The atmosphere is too thick to

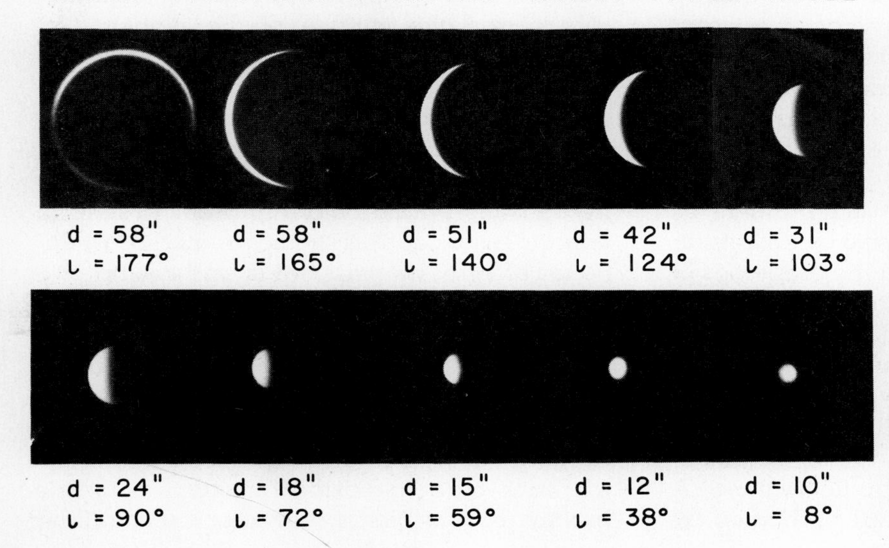

FIGURE 8.5

The phases and different sizes of Venus in different parts of its orbit, as viewed from the Earth. Compare this with Figure 8.1. (New Mexico State University Observatory)

see through, so we cannot observe the surface directly. Visible light only penetrates to the tops of high cloud layers before being scattered. For many years astronomers have wondered whether the surface of Venus is like the Earth in having large amounts of water and oxygen and generally being favorable to the support of life. The answer finally came through studies of radio waves from Venus, which can penetrate the atmosphere, and from a series of Venus space probes. Russian space shots actually parachuted instrument packages onto the planet and photographed parts of its surface. The atmosphere was found to be mainly the gas *carbon dioxide* with very little water and even less oxygen. The surface temperature is about 800°F. We cannot expect life to exist under these extreme conditions. Figure 8.6 is a close-up photograph of Venus taken by Mariner 10 in 1974.

Why did Venus, so like the Earth in so many ways, develop such hostile conditions on its surface? The answer lies in the properties of carbon dioxide gas (CO_2). This gas is nearly transparent to visible light, but it absorbs infrared radiation very strongly. As we learned in Section 3.3, a very hot object, such as the Sun, emits a lot of visible light, while a planet, being much cooler, emits radiation mainly in the infrared and radio regions. A planet with no atmosphere will be heated by the sunlight it absorbs, and it will be cooled by the infrared and radio waves it emits back into space. The planet will arrive at a balance between the amount of energy it receives from the Sun and the amount it radiates into space, and this balance will fix its temperature at a certain value: The more energy it receives from the Sun, the more energy it must radiate in order to be in balance, and so the hotter it will be.

Now let us put a thick CO_2 atmosphere around this planet. Some of the sunlight will be scattered by the atmosphere, but much of it will penetrate deeply enough to heat the atmosphere and the surface of the planet. The infrared radiation emitted by the surface now has trouble getting out through the CO_2 atmosphere and into space because the CO_2 absorbs it so strongly. The planet has lost part of its ability to cool itself, and its surface will heat up more than before in order to keep in energy balance. This is known as the *greenhouse effect:* The glass walls and ceiling of a greenhouse will allow sunlight in to heat the interior, but they won't let the infrared radiation emitted by that interior back outside; this is what keeps greenhouses quite warm.

Venus would be somewhat hotter than the Earth simply by being closer to the Sun, but it is the thick carbon dioxide atmosphere that is mainly responsible for the extreme heat of the

FIGURE 8.6
Venus, taken in ultraviolet radiation by Mariner 10 in 1974. (NASA/JPL)

surface. The Earth has a very small amount of CO_2 in its atmosphere, so why does Venus have so much?

It is believed that both Venus and the Earth have built up their atmospheres from gases released from the interior through volcanoes. Volcanoes pour CO_2, along with other gases, into the atmosphere, but the Earth, unlike Venus, is able to remove it as fast as it appears. Part of the gas combines chemically with surface rocks, part of it is dissolved in the oceans, and part of it is changed into oxygen and other things by the action of plant life. At the higher original temperature of Venus, the rocks were not able to remove CO_2 from the Venus atmosphere as fast as it was produced by volcanoes. Some liquid water might have formed on Venus early in its history, but the build-up of CO_2 eventually drove the surface temperature so high, through the greenhouse effect, that practically all of the water evaporated into steam. When steam rises high in the atmosphere, ultraviolet radiation from the Sun breaks it down into oxygen and hydrogen; the hydrogen then escapes into space (neither Venus nor the Earth are able to hold onto hydrogen gas, although they can keep heavier gases), and the oxygen combines chemically with the surface rocks. In this way Venus has lost most of its water and gaseous oxygen and has built up a thick atmosphere of nearly pure CO_2. Venus probably has modest amounts of nitrogen, the main ingredient in the Earth's atmosphere. The rather small difference in original temperature between Venus and Earth has produced, over billions of years, very large differences in surface and atmospheric conditions. If life ever did get started on Venus, it probably soon lost its struggle for survival.

Venus rotates very slowly on its axis, taking 243 days to turn around once. The slow rotation could perhaps be attributed to the tidal force of the Sun, as in the case of Mercury; however, this explanation will not account for the fact that Venus is actually rotating backward: While practically all objects in the solar system are rotating in an eastward direction, Venus (and Uranus) rotates in a westward direction. If Venus originally rotated in the normal direction, as seems likely, tidal action by the Sun could slow down the rotation, but it could not reverse the direction. The Earth may very well be involved in this. The different motions are such that the same side of Venus is always turned toward the Earth whenever the two planets are nearest each other. Either this is a great coincidence, or the feeble tidal force of the Earth on Venus is responsible for this phenomenon.

The rotation of the surface of Venus was measured by reflecting radar waves off the planet. These are a type of radio wave that can pass through the atmosphere of Venus. It is interesting that while the surface of Venus rotates very slowly, the upper atmosphere is observed to rotate around the planet in only four days. There must be very strong winds on Venus.

MARS

Mars is the planet that has captured the interest and imagination of the general public. Life on Mars has been the topic of conversation and study ever since the so-called "canals" were reported about a century ago. A comparison of Plate 16 with Figure 8.7 will illustrate how much the space age has increased our knowledge of Mars.

Mars has a very thin atmosphere that, like Venus's, is mainly carbon dioxide. The atmosphere is too thin to have a strong greenhouse effect, and the surface is quite cold on the average; however, it does get as warm as around 70°F at the point directly under the Sun. There are polar caps that change in size with the seasons and are probably a mixture of water ice and frozen carbon dioxide (dry ice). Most of the canals have proven to be nonexistent, although some of them could be the very

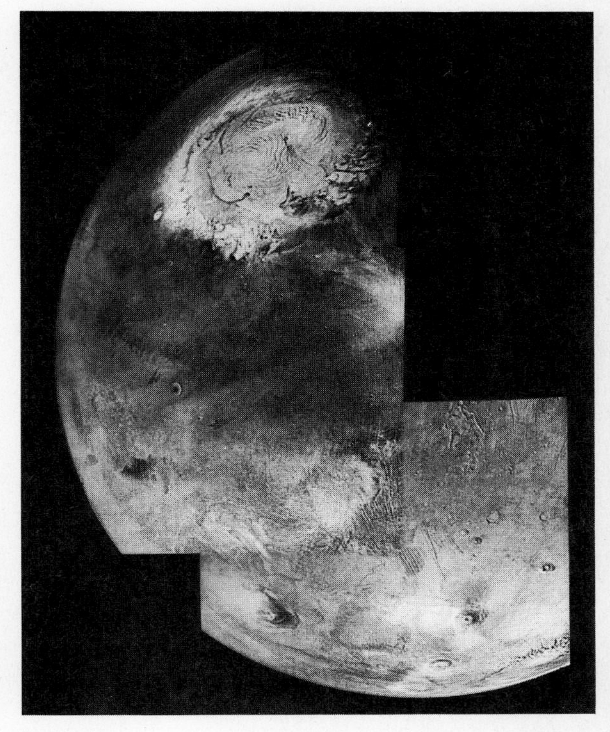

FIGURE 8.7
The northern hemisphere of Mars as photographed by Mariner 9 in 1972. The north polar ice cap shows considerable detail, and several huge volcanoes are visible toward the bottom. (NASA/JPL)

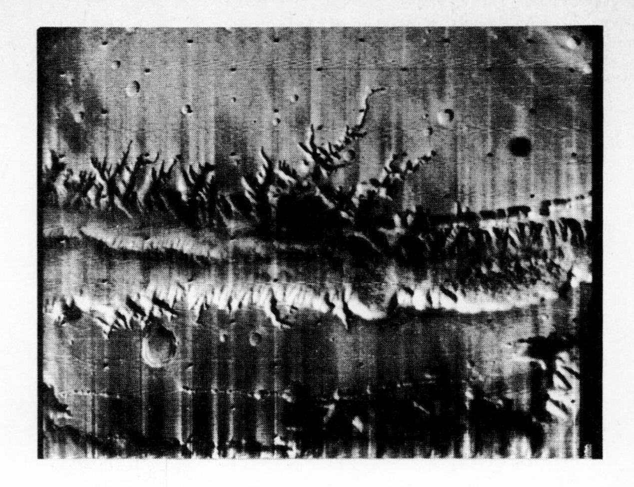

FIGURE 8.8
A tremendous canyon carved out of the south equatorial regions of Mars. The picture covers an area about 235 by 300 miles. (NASA/JPL)

large canyons that are now known to exist (see Figure 8.8).

Volcanoes (in Figure 8.7), canyons (in Figure 8.8), sand dunes (see Figure 8.9), craters (see Figure 8.10), and what appear to be dry stream beds (also Figure 8.10) have been found in close-up pictures taken from various Mariner and Viking spacecraft. Viking 1 and 2 actually landed on the surface of Mars in the summer of 1976 (see Figures 8.11 and 8.12).

It is obvious that a lot of surface activity has taken place on Mars. The volcanoes and canyons are far greater than the corresponding structures on the Earth. The huge volcano Olympus Mons, for example, is over 300 miles across at its base and reaches about 15 miles high. This and other volcanoes are probably the sources of the Martian atmosphere.

How are the dry stream beds to be interpreted? It is difficult to think of a plausible explanation other than running water for their origin. At the low temperatures and atmospheric pressures prevailing today, liquid water cannot exist on Mars: It would quickly freeze or evaporate. A much denser atmosphere is needed to hold liquid water on the surface. Perhaps Mars had a more extensive atmosphere and lots of surface water at some time in the past. Some scientists speculate that Mars might periodically undergo major changes of climate, alternating warm and wet

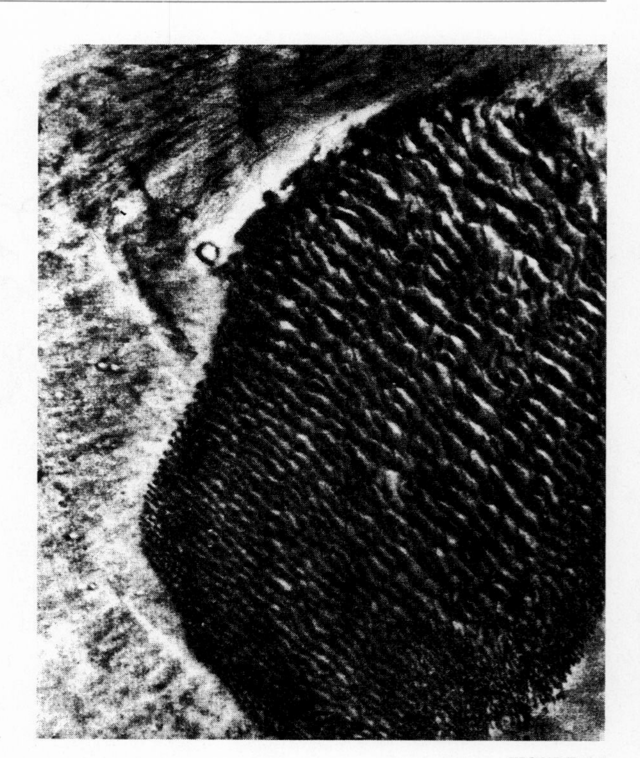

FIGURE 8.9
A large sand dune found by enlarging a photo from Mariner 9. (NASA/JPL)

FIGURE 8.10

A rather heavily cratered region of Mars with a channel of some sort. Photographed by Mariner 9. (NASA/JPL)

FIGURE 8.11

(Upper photograph below) Surface detail from Mars, taken by Viking 2 shortly after landing on September 3, 1976. The rocks range in size up to about 8 inches. (NASA)

FIGURE 8.12

(Lower photograph below) The rocky desert of Mars, taken by Viking 2. The tiny channel running from the upper left toward the lower right makes it appear that it rained the previous week. (NASA)

FIGURE 8.13
Phobos, the larger of the two satellites of Mars, as photographed by Viking Orbiter 1. The large crater at the top is about 3 miles across. (NASA)

intervals with cold and dry ones, like the present. During the dry spells the water could be frozen out underground or in the polar caps. The ground around the polar caps has a layered structure like a sandwich, suggesting that alternate layers of material were deposited during alternating climatic conditions. There are a number of possible causes of a variable climate, including variable amounts of volcanic activity and varying amounts of energy from the Sun. These points today are speculative.

It is possible that life developed on Mars in the past when conditions might have been more favorable. Certain life forms could have adapted themselves to the slowly changing conditions on the planet, and they might have survived until today. Unfortunately, the experiments conducted by the Viking landers on Martian soil did not resolve the question of life on Mars. Certainly, though, if Mars were teeming with life, the way the Earth is, these explorations would have found it.

In the early eighteenth century Jonathan Swift wrote his famous tale *Gulliver's Travels*, in which he told how astronomers in the fictitious land of Laputa discovered two satellites of Mars. In 1877 two satellites were discovered (see Figure 8.13), and they have properties that in some respects are remarkably close to those mentioned by Swift. This is sometimes used as evidence for the assertion that the satellites were actually discovered much earlier than believed, but the telescopes available in Swift's day were far too poor for the moons to have been seen. This is just a very interesting coincidence.

Mercury, Mars, and the moons of Mars all show the same types of craters so conspicuous on the Moon. It seems that the entire inner part of the solar system was subjected to a heavy bombardment of particles soon after the planets and satellites were formed about 4.6 billion years ago. This bombardment might be part of the process of formation of the planets. The crater formation slowed down considerably after the first half-billion years or so, and it has continued at a low pace ever since. Earth also shows some impact craters, although the older ones have been erased by surface erosion as well as interactions between the surface plates. Radar signals have even detected craters on Venus. The chunks of matter that are responsible, varying in size up to perhaps 100 miles across, are probably related to the asteroids. (Refer to Section 9.1.)

8.2 The outer planets

The outer planets consist of the four giants, Jupiter, Saturn, Uranus, and Neptune, plus Pluto. As Figure 5.1 showed, the giants are far larger than the other planets, and they are

much more massive. They are similar to the
Sun in that they are made primarily of the light
…es hydrogen and helium, as opposed to the
…, rocky material of the smaller planets.
…easons for these differences will be dis-
… Section 10.1.)

… giant of the giants. It has 2½
…s as all the other planets
…atellites, more than any
… fastest on its axis: The
… 10 hours long. The
… to be noticeably
…und, as can be

…red radia-
… planet
…unted
…her
…of

…er

…, and many
…iscovered by the
…e planet (Figures
…s are parallel to its
… due to rising and
… temperatures that
…way. The rapid rota-
…s excess internal heat
… current in the gas. The
…e shape and detailed
…hat we are not seeing a
…mous red spot is a long-

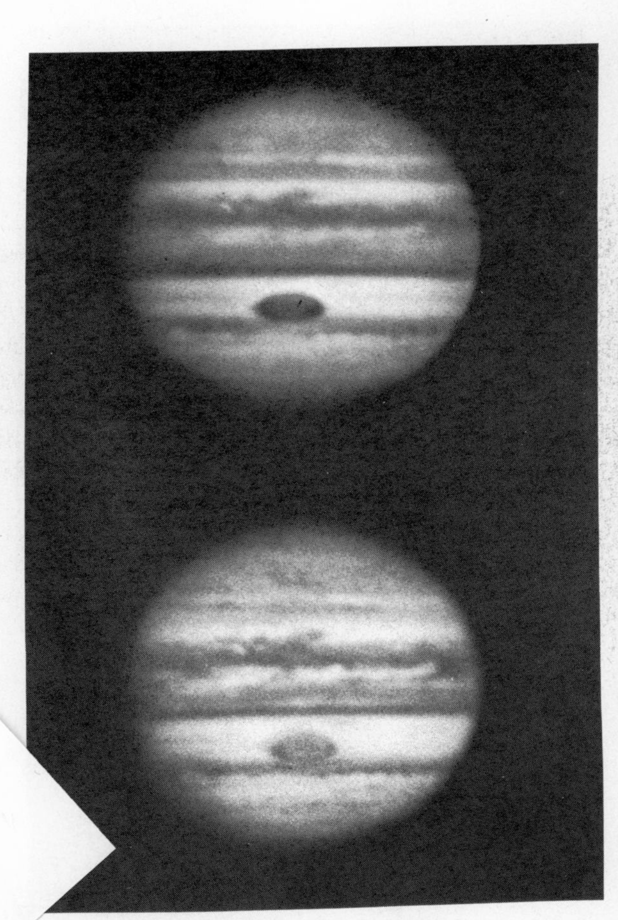

FIGURE 8.14
Jupiter, as photographed with the 61-inch reflecting telescope at Catalina Observatory. The top picture was taken with blue light, the bottom with red. (Lunar and Planetary Laboratory, University of Arizona, copyright pending)

FIGURE 8.15
Jupiter, as photographed by the Pioneer 10 spacecraft in 1973. Note the greater detail seen here as compared with the Earth-based photographs in Figure 8.14. (NASA/University of Arizona)

FIGURE 8.16
Jupiter, as seen from Pioneer 11 as it approached the planet from south of the equator at a distance of about one-half a million miles. (NASA/University of Arizona)

to Section 16.2.) In spite of Jupiter's internal heat, the outer layers which we can see are quite cold, some −200 to −250°F.

The inside of Jupiter probably is mainly in the liquid state. The extremely large pressures inside the planet squeeze hydrogen into being a liquid. At the much higher temperatures inside a star, the hydrogen is gaseous, even at far higher pressures. It is possible that Jupiter possesses a small, rocky core made of heavier materials.

As seen through a telescope, Jupiter reveals a banded structur
detailed features have been d
Pioneer 10 and 11 flybys of th
8.15 and 8.16). Jupiter's band
equator and are probably
sinking gases of differen
align themselves in this
tion of the planet and it
could induce this type o
bands slowly chang
structure, indicating
solid surface. The fo

FIGURE 8.14
Jupiter, as photographed with the 61-inch reflecting telescope at Catalina Observatory. The top picture was taken with blue light, the bottom with red. (Lunar and Planetary Laboratory, University of Arizona, copyright pending)

much more massive. They are similar to the Sun in that they are made primarily of the light gases hydrogen and helium, as opposed to the heavy, rocky material of the smaller planets. (Some reasons for these differences will be discussed in Section 10.1.)

JUPITER

Jupiter is truly the giant of the giants. It has 2½ times as much mass as all the other planets combined. It has 13 satellites, more than any other planet, and it spins fastest on its axis: The day on Jupiter is less than 10 hours long. The rapid rotation causes Jupiter to be noticeably flattened instead of perfectly round, as can be seen in Figure 8.14.

Measurements made of the infrared radiation emitted by Jupiter show that the planet gives off more energy than can be accounted for by the heating effects of the Sun. In other words, Jupiter must have its own supply of energy. When it was in the process of formation, it contracted under its own gravity from a larger size. This action produces heat, and Jupiter still has much of this internal heat left. Jupiter could be continuing the contraction and generation of heat even now. If the center of the planet could have gotten hot enough, nuclear reactions would have taken place, and Jupiter would be a star instead of a planet. Jupiter would need about 80 times as much mass as it actually has to become a star. (Refer

FIGURE 8.15

Jupiter, as photographed by the Pioneer 10 spacecraft in 1973. Note the greater detail seen here as compared with the Earth-based photographs in Figure 8.14. (NASA/University of Arizona)

FIGURE 8.16

Jupiter, as seen from Pioneer 11 as it approached the planet from south of the equator at a distance of about one-half a million miles. (NASA/University of Arizona)

to Section 16.2.) In spite of Jupiter's internal heat, the outer layers which we can see are quite cold, some −200 to −250°F.

The inside of Jupiter probably is mainly in the liquid state. The extremely large pressures inside the planet squeeze hydrogen into being a liquid. At the much higher temperatures inside a star, the hydrogen is gaseous, even at far higher pressures. It is possible that Jupiter possesses a small, rocky core made of heavier materials.

As seen through a telescope, Jupiter reveals a banded structure, and many detailed features have been discovered by the Pioneer 10 and 11 flybys of the planet (Figures 8.15 and 8.16). Jupiter's bands are parallel to its equator and are probably due to rising and sinking gases of different temperatures that align themselves in this way. The rapid rotation of the planet and its excess internal heat could induce this type of current in the gas. The bands slowly change shape and detailed structure, indicating that we are not seeing a solid surface. The famous red spot is a long-

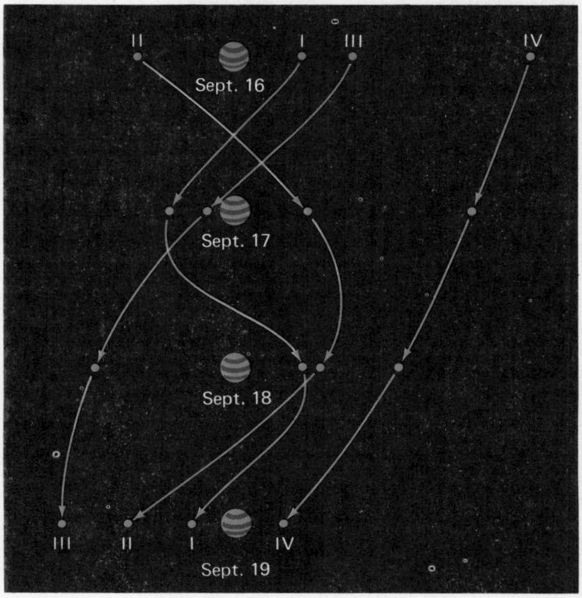

FIGURE 8.17
Jupiter and its four brightest moons. The photograph was taken on September 16, 1976, and the drawing shows the motions of the moons on successive days.
The names of the moons are Io (I), Europa (II), Ganymede (III), and Callisto (IV). These moons are quite bright and can be seen with a good pair of binoculars. They were discovered by Galileo in 1610 and are often called the Galilean satellites. The photograph was taken by Dr. E. Roemer with the 61-inch telescope at the Catalina Observatory. (Lunar and Planetary Laboratory, University of Arizona, copyright pending)

lasting feature that shows some changes in color and shape, yet it has been in existence for at least a century and a half and possibly much longer. It is a localized region of horizontal and vertical gas currents that has somehow become very stable. Local patterns of irregular gas flow continually form and dissipate in the turbulent atmosphere of Jupiter, and the larger ones can last for many years.

The space around Jupiter contains a strong magnetic field and large numbers of electrically charged particles, mainly electrons and protons. Some of these particles have very high energies, so their interaction with the magnetic field causes them to emit radio waves through the synchrotron process. (Refer to Sections 1.2 and 3.3.) Radio emission from Jupiter has been observed from the Earth for many years. The Pioneer 10 and 11 spacecraft passed through this particle and radiation zone of Jupiter without much damage, but a person on board would have received a far greater than fatal dose of radiation.

Jupiter has a fine system of 13 satellites. Astronomers discovered the thirteenth one in late 1974.[1] Jupiter with its satellites is very much like a solar system in miniature. Four of the five inner moons (see Figure 8.17) are about the

[1]A fourteenth moon was reported in 1975, but its existence remains unconfirmed.

FIGURE 8.18
The ringed planet Saturn. (Lunar and Planetary Laboratory, University of Arizona, copyright pending)

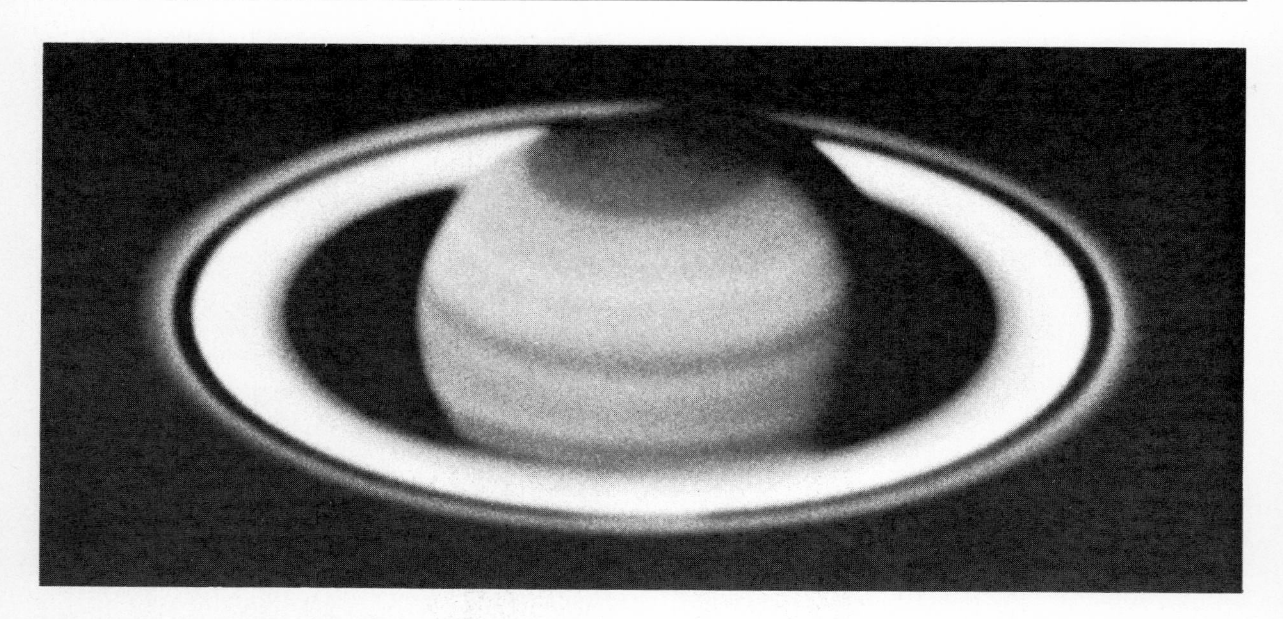

same size as or larger than our Moon, while the remainder are tiny by comparison, ranging down to a few miles in diameter. The innermost of the four large moons, named Io, has a strong influence on the radio emission of Jupiter. Io affects the charged particles and the magnetic field as it passes through the space around Jupiter, so the synchrotron emission is also affected.

SATURN

Saturn (see Figure 8.18) is the second largest planet and is the farthest away of those known in antiquity. Its properties are similar to those of Jupiter, except on a smaller scale. Saturn has the lowest average density of any planet, even lower than that of water. One way in which Saturn differs from Jupiter is that it has not been detected emitting radio waves. It also shows no evidence of having a large magnetic field.

Infrared observations from the Earth suggest that Saturn, like Jupiter, has its own supply of internal heat. This could be heat left over from the time of formation, or Saturn may be contracting and generating heat today.

FIGURE 8.19
Saturn, photographed with different types of radiation. At top violet light was used, at second from top blue light, deep red at second from bottom, and a special frequency in the infrared that is strongly absorbed by the chemical compound methane was used in the photo at the bottom. The presence of methane in the outer layers of Saturn is demonstrated by the darkness of the disk in this photograph. The methane strongly absorbs that frequency of radiation, so there is little of it left to reflect back into space. (Lunar and Planetary Laboratory, University of Arizona, copyright pending)

Figure 8.19 shows four photographs of Saturn, each taken in a different frequency of radiation. By studying the way Saturn reflects different frequencies of sunlight, we can tell something about its structure and composition.

Easily the most notable things about Saturn are its rings. The rings are composed of particles in orbit around the planet. The particles are probably made of dust, ice, and related materials, and radar studies of the way these particles reflect radiation suggest that they may be rather large, perhaps inches in diameter. The rings are very thin, so thin that they become invisible every 15 years when our view of them is edge-on. Why does Saturn have rings? Why don't other planets?[2]

We have already considered the tidal forces that one body can exert on another. The tidal force of a large planet may be capable of actually pulling apart a smaller body like a satellite if it comes too close to the planet. The near side of the satellite feels a stronger pull of gravity than the far side, and the difference between these forces is the tidal force. If the tidal force is stronger than the forces holding the satellite together, then the satellite will be pulled apart. We can calculate how close a satellite can come to a planet before the tidal force will pull the satellite apart. This limiting distance is called the *Roche* (rōsh) *limit*. All

[2]In 1977 a thin and nearly invisible series of rings was discovered around Uranus.

FIGURE 8.20

An occultation of Saturn by the Moon. This sequence of photographs was taken over an interval of about 5 minutes, and it shows Saturn emerging from behind the Moon. (New Mexico State University Observatory)

satellites are outside the Roche limits of their planets. The rings are inside the Roche limit of Saturn.

Perhaps a satellite wandered too close to Saturn and was destroyed. Perhaps the material was there originally and tried to form a satellite, but couldn't because it was inside the Roche limit. Whatever the correct explanation, the conditions must have been favorable for the formation of a ring around Saturn but not around the other planets. Somehow the solar system seems much more exciting because of Saturn's rings. How would artists depict scenes of outer space without the example of Saturn? (Figure 8.20 shows Saturn reappearing from behind our Moon.)

Saturn has a collection of 10 satellites. One of them, named Titan, is one of the largest satellites in the solar system. It is 3030 miles in diameter or about the size of Mercury. Titan was the first satellite for which an atmosphere was detected.

URANUS

In England in 1781, William Herschel was looking through his telescope and noticed an object that showed an extended disk, unlike the stars. Herschel thought that it was a comet at first, but more observations and calculations indicated that it was a large object that circled the Sun far beyond the orbit of Saturn. For the first time in history, a new planet had been discovered.

Concerning the name of the new planet, Herschel wrote[3]:

> In the fabulous ages of ancient times the appellations of Mercury, Venus, Mars, Jupiter, and Saturn, were given to the Planets, as being the names of their principal heroes and divinities. In the present more philosophical area, it would hardly be allowable to have recourse to the same method . . . if in any future age it should be asked, *when* this last-found Planet was discovered? It would be a very satisfactory answer to say, "In the Reign of King George the Third." As a philosopher then, the name of Georgium Sidus presents itself to me.

Fortunately the philosophical name "George's star" did not take hold, and eventually Uranus was adopted.

Uranus (see Figure 8.21) is about 30,000 miles in diameter. It is pretty large compared to the Earth, but it is only one-third the size of Jupiter. Uranus has a low density, about 1.6 times that of water. This is only slightly more dense than Jupiter, but it shows that Uranus has considerably less hydrogen and helium in its make-up than Jupiter does. The much smaller mass of Uranus means that it should be

[3]Quoted from *Source Book in Astronomy,* edited by H. Shapley and H. E. Howarth, McGraw-Hill, New York, 1929, p. 141.

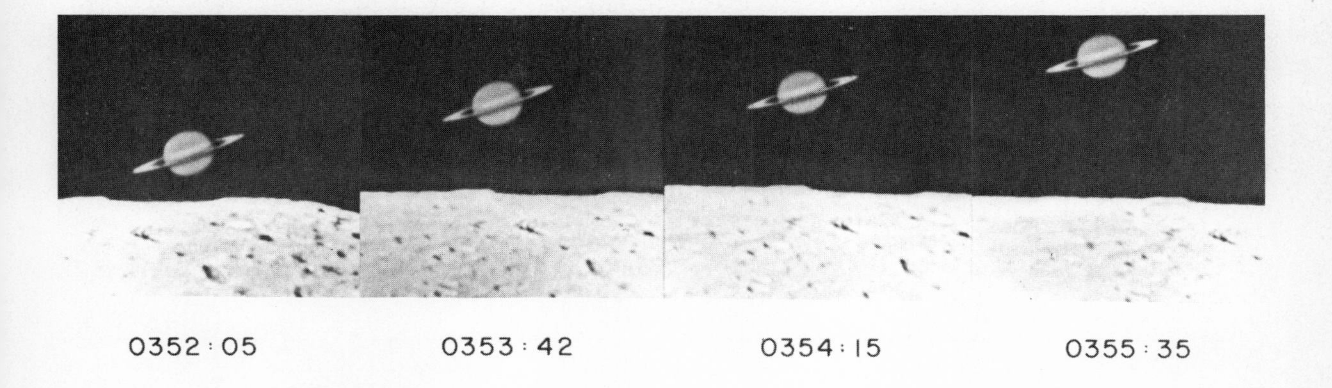

considerably less compacted than Jupiter if it were made out of the same substances. Saturn does show this effect in that it is only half as dense as Jupiter. Uranus must have a significantly higher abundance of some of the heavier elements than either Jupiter or Saturn. The most likely candidates are carbon, nitrogen, oxygen, and neon, plus a few others, for these are the most abundant elements in the Sun after hydrogen and helium. Uranus is probably composed of about 10 percent hydrogen and helium. At the low temperature of the outer layers of Uranus, various chemical compounds, such as water, ammonia, and methane, should be common.

FIGURE 8.21

Uranus and three of its satellites. (Lick Observatory photograph)

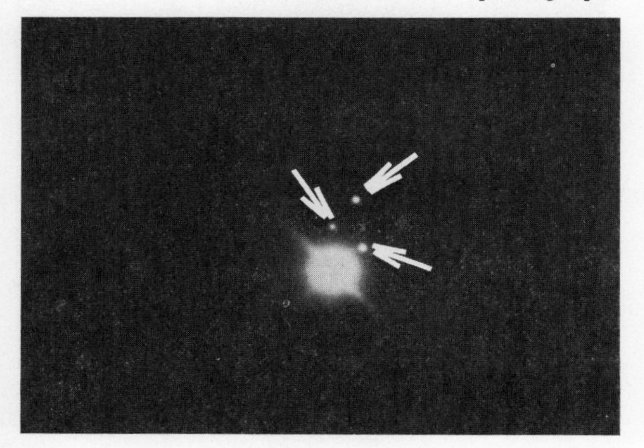

An oddity about Uranus is that its axis of rotation lies nearly in the same plane as its orbit around the Sun. Actually it is tipped by an angle of slightly more than 90 degrees, meaning that the rotation is slightly in the westward direction. Uranus has five satellites that all move around the equator of the planet, which means that they are also tipped over by this same angle with respect to the orbit around the Sun. Astronomers do not know why Uranus has this strange rotation.

NEPTUNE

As the years went by, Uranus was found to deviate slightly from the path that had been calculated for it. Its position was found to be close to, but not exactly at, the expected place, and as time went on the discrepancy became greater. What could be causing this erratic behavior? One possibility was that the laws of motion and the law of gravity as laid down by Newton were inaccurate. Another idea was that there was another planet beyond Uranus whose gravitational influence was pulling Uranus off course. The various pros and cons of these alternatives were vigorously debated during the first half of the nineteenth century.

In order to explain the observed discrepancies in the motion of Uranus, J. C. Adams of

England and U. J. J. Leverrier of France, each unknown to the other, tackled the mathematical problem of calculating what mass and position an unknown planet would have to have. Adams obtained his solution first and sent the results to the Royal Observatory in 1845. He had determined that the unknown planet, if this indeed were the cause, would have to have a certain position, and the astronomers at the observatory should look for it in that place. A combination of mistrust, procrastination, and carelessness (a hard combination to win with) caused the English observers to miss it until too late. In 1846 Leverrier sent his results, which were in good agreement with Adams's, to the Berlin Observatory. J. G. Galle found the new planet on the very first night of his search, September 23, 1846.

If the English role in the discovery of Neptune seems frustrating, consider the plight of J. J. Lelande. In 1795 he twice recorded the position of Neptune, thinking it to be a faint star. He even noticed that the two positions did not agree with each other, but he shrugged off the difference as due to an error in the earlier observation. If he had had more faith in the accuracy of his own observations, he would have discovered Neptune 50 years ahead of schedule.

The discovery of Neptune had a strong effect on the scientists of the nineteenth century. It reaffirmed their faith in the laws of physics, in particular, and in the scientific method, in general, and the discovery showed very forcefully that it is premature to assume that nature has secrets that are beyond man's ken.

Neptune (see Figure 8.22) is the twin planet to Uranus, being slightly smaller and slightly more massive. Neptune is densest of the giants and turns the most slowly on its axis; a day on Neptune is about 15 hours long. There are two satellites known to circle Neptune. One of them, Triton, is probably the most massive satellite in the solar system.

FIGURE 8.22

Neptune and its two satellites. (Lick Observatory photograph)

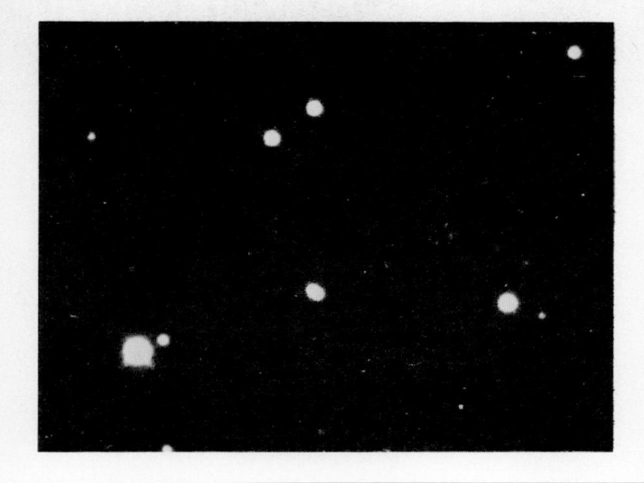

FIGURE 8.23
Pluto is the slightly oval image near the center. This is not the shape of the planet, but the image trailed somewhat during the exposure. (Lunar and Planetary Laboratory, University of Arizona, copyright pending)

PLUTO

The presence of Neptune accounted for practically all of the observed errors in the motion of Uranus, but some astronomers believed that a tiny discrepancy remained. Percival Lowell was one of these astronomers, and he applied the same type of mathematical attack that Adams and Leverrier used earlier in an attempt to find yet another new planet. Lowell died in 1916, but his search was carried on by others. On February 18, 1930, Clyde Tombaugh discovered the planet Pluto very close to the position predicted by Lowell.

Although Pluto's mass is not well known, it is certainly far less than predicted by Lowell, and it could not have produced the supposed errors in the positions of Uranus and Neptune. The errors are thought to be in the observations themselves, and the discovery of Pluto was an accident. Lowell's belief was the impetus that caused the search to be carried out—without that search, who knows when Pluto would be discovered?

Pluto (see Figure 8.23) is in many respects the most unusual planet. Unlike the other outer planets, it is small and probably has a high density. Of all the planets, it is the one whose orbit is least like a circle; as Figure 5.2 indicated, Pluto actually crosses the orbit of Neptune. It has been suggested that Pluto was not originally a planet at all but started as a satellite of Neptune. Some perturbation then caused it to escape to a separate orbit around the Sun.

Pluto is so far away from the Sun that, unless it has its own source of heat, its temperature must be around $-400°F$. The Sun as seen from Pluto appears only as a tiny dot, although a very bright one. Most of the other planets would be very difficult to see from Pluto because they would be lost in the glare of the Sun. The Earth, for example, would never get more than about $1\frac{1}{2}$ degrees from the Sun as viewed from Pluto.

IMPORTANT WORDS

Mercury
Venus
Carbon dioxide
Greenhouse effect
Mars
Jupiter
Saturn
Roche limit
Uranus
Neptune
Pluto

REVIEW QUESTIONS

1 What two planets go through phases like the Moon?

2 Why can't Mercury or Venus be seen at midnight?

3 What is unusual about Mercury's rotation?

4 Can we see the surface of Venus? How do we know it is very hot?

5 Why is the surface of Venus so hot?

6 Does Mars have large volcanoes?

7 Does Mars have much liquid water on its surface? Did it in the past?

8 What coincidence involved the moons of Mars?

9 What is the largest planet?

10 Does Jupiter have a source of heat in addition to the Sun?

11 How much more material would Jupiter need to be a star instead of a planet?

12 What are the rings of Saturn made of and how might they have formed?

13 Who discovered Uranus and how?

14 How do we know that Uranus has a different composition than Jupiter and Saturn?

15 What is unusual about the rotation of Uranus?

16 How was Neptune discovered?

17 What is unusual about the orbit of Pluto?

QUESTIONS FOR DISCUSSION

1 Study Figure 8.1 and convince yourself that when Mercury (or Venus) is in the morning sky, it is moving from the near side of the Sun to the far side, in other words, away from the Earth. When in the evening sky, it is moving toward the Earth.

2 Mercury has the same density as the Earth. Why is this evidence that Mercury and the Earth are actually different in their chemical compositions?

3 When we burn wood, coal, gasoline, and so on, oxygen is changed into carbon dioxide. Is it conceivable that, over long enough periods of time, large numbers of fires could affect the climate of the Earth as a whole? How?

4 How important is it to find out whether or not there is life on Mars? How much effort should be spent trying to find out? Suppose some discovery is made that indicates that the probability of life on Mars is now 10 times less than we had thought. Is it worth the same effort as before in trying to get to Mars and find out?

5 Suppose Jupiter had enough mass to become a regular star, though smaller and fainter than the Sun. What are some of the ways we would be affected?

6 We life inside the Roche limit of the Earth. Why aren't we torn apart by the tidal force of the Earth?

7 It has often been speculated that there could be a planet just like the Earth that exists opposite the Sun from us. This "anti-Earth" moves in the same orbit we do, always keeping the Sun between us so we cannot see it. Other than by using spaceships, how can astronomers be sure that such a planet does not exist?

8 Swift's statement about the moons of Mars and Lowell's calculations on the position of Pluto turned out to be quite accurate by coincidence, astronomers say. "Don't believe it," others say. "There is no such thing as coincidence. These are mysterious workings beyond the realm of science."

What is your opinion?

REFERENCES

Good general references on the planets of the solar system include the *Earth, Moon, and Planets*, 3rd ed., by F. L. Whipple (Harvard University Press, Cambridge, 1971), *The Solar System* by Z. Kopal (Oxford, London, 1973), and the September 1975 issue of *Scientific American* which has been reprinted as a book called *The Solar System*, (Freeman, San Francisco, 1975).

The planet Mercury is the main topic of the July 12, 1974, issue of *Science* (185). The March 29, 1974, issue of *Science* (183) devotes itself mainly to Venus. Both of these planets are well covered and illustrated in "Mariner Unveils Venus and Mercury" by K. F. Weaver [*National Geographic* 147 (June 1975), 858].

The planet Mars has become the subject of a large amount of work. Mariner 9 provided the main information for "Journey to Mars" by K. F. Weaver [*National Geographic* 143 (February 1975); 231] and "The Volcanoes of Mars" by M. H. Carr [*Scientific American* 234 (January 1976), 32]. Reports on the Viking results include the October 10, 1976, issue of *Science* (194); the 1976 Summer and Fall issues of *Sky and Telescope* (52); "Viking on Mars" by R. S. Young in the *American Scientist* [64, (November-December 1976), 620]; "Mars as Viking Sees It" [*National Geographic* 151 (January 1977), 3]; and "Sifting for Life in the Sands of Mars" by R. Gore [*National Geographic* 151 (January 1977), 9].

Jupiter also has a large volume of recent literature. Some of this is in the January 25, 1974, issue of *Science* (183); in *Pioneer Odyssey*, edited by R. O. Fimmel, W. Swindell, and E. Burgess (NASA SP-349, U.S. Government Printing Office, Washington, D.C., 1974); in "Mystery Shrouds the Biggest Planet" by K. F. Weaver [*National Geographic* 147 (February 1975), 284]; in "Jupiter 1975" by R. Smoluchowsky [*American Scientist* 63 (November-December 1975), 638]; in "The Meteorology of Jupiter" by A. P. Ingersoll [*Scientific American* 234 (March 1976), 46] and in "The Galilean Satellites of Jupiter" by D. P. Cruikshank and D. Morrison [*Scientific American* 234 (May 1976), 108].

The Discovery of Neptune by M. Grosser (Harvard University Press, Cambridge, 1962) is an interesting and detailed account of that planet. The article "Cratering in the Solar System" by W. K. Hartmann [*Scientific American* 236 (January 1977), 84] tells how studying craters on the Moon and the inner planets can give us information about the history of the solar system.

NINE
THE LEFTOVERS

In this chapter we consider the minor members of the solar system: asteroids, comets, and meteors. These objects make up a very small part of the solar system in terms of mass, but they are quite interesting in their own right. In addition, they offer important clues about the origin and early times in the solar system.

9.1 Asteroids

THE BODE-TITIUS LAW

Take the following series of numbers: 0, 3, 6, 12, 24, 48, . . . ,. After 3, each number is twice the preceding one. To each of these numbers add 4. The result is the new series of numbers 4, 7, 10, 16, 28, 52,. . . . Now let's look at the average distances of the planets to the Sun, limiting ourselves to those planets known long ago. On the scale for which the Earth's distance is 10, we have: Mercury 3.9; Venus 7.2; Earth 10; Mars 15.2; Jupiter 52.0; Saturn 95.4. We see that these two sets of numbers are nearly the same. Let's compare them side-by-side as in Table 9.1.

People had long been looking for some magic formula for reproducing the distances of the planets from the Sun. As long as exactness isn't necessary, this series works fine. J. D. Titius (tish-us) of Germany first noted it in 1772. J. E. Bode thought it was one of the greatest things ever discovered and tried to convince the astronomical world of its importance. This series of numbers is usually known as "Bode's law" or the *Bode-Titius law*.

The thing that most attracted Bode's attention was the gap between Mars and Jupiter where the series had number 28. It had long been noted that there is plenty of room for a planet in the space between these other two, but now Bode had what he considered to be hardly less than mathematical proof that another planet must exist: "Can we believe that the Creator of the world has left this space empty? Certainly not!"[1]

In 1781 Uranus was discovered. It has an average distance from the Sun of 192 on the scale, while the next number in the series after 100 (for Saturn) is 196. Excellent agreement! No doubt about it, there *had* to be a planet at distance 28 from the Sun.

Bode became more enthusiastic than ever, but he seems to have had some difficulty in spreading this excitement to others. After years of trying, he was finally able to get a group of German astronomers together in 1800 to make an organized search for the missing planet.

[1]Quoted from *Source Book in Astronomy*, edited by H. Shapley and H. E. Howarth, McGraw-Hill, New York, 1929, p. 180.

TABLE 9.1
Bode-Titius law

	Mercury	Venus	Earth	Mars		Jupiter	Saturn
Distance:	3.9	7.2	10.0	15.2		52.0	95.4
Series:	4	7	10	16	(28)	52	100

The group was just getting started on the search when word came from Italy: G. Piazzi found the new planet by accident on January 1, 1801. Oh well, Bode was still delighted with the news.

The new planet was named Ceres, and its position was almost exactly where Bode's law predicted: 27.6 as opposed to 28. At nearly the same time that Piazzi made his discovery, the famous philosopher G. Hegel published a work

> . . . showing, by the clearest light of reason, that the number of the planets could not exceed seven, and exposing the folly of certain devotees of induction who sought a new celestial body merely to fill a gap in a numerical series.[2]

So much for philosophy!

CERES AND THE REST

Ceres was very unusual among the planets in that it is so small—only about 600 miles in diameter. A year later a second tiny planet, Pallas, was discovered in very nearly the same orbit as Ceres. This was indeed strange for two planets to be in about the same orbit around the Sun. H. Olbers, the discoverer of Pallas, suggested that maybe one of the original planets exploded for some reason, and Ceres and Pallas are two of the fragments. Sure enough, two more of the tiny planets were found in the next few years. They were named Juno and Vesta. Additional ones have been found off and on ever since, and today several thousand are known (see Figure 9.1). W. Herschel gave them the term *asteroid*. "Minor planet" is another common name.

Ceres is the largest of the asteroids. There probably is no limit on how small they can be; certainly many are known that are less than a mile in diameter. The smaller asteroids do not have strong enough gravity to pull themselves into a spherical shape, and studies of their light do show that many of them are irregular in shape. Figure 9.1 shows a more recently discovered asteroid.

POSSIBLE ORIGINS

Are the asteroids really fragments of an exploded planet? What can make a planet explode? Even if they are fragments, the original planet still would have to be quite small. All of the asteroids put together don't make a respectably sized planet.

It seems more likely that instead of one exploding planet, there were originally several bodies in the asteroid belt. Maybe the nearness of the giant Jupiter kept a larger planet from forming. If these few pieces would occa-

[2]Quoted from A. M. Clerke, *History of Astronomy*, 4th ed., Black, London, 1902, p. 73.

sionally collide with each other, pieces might break off and create the asteroids as we know them today. The collisions between them should continue, and perhaps in the distant future only a fine dust will be left.

Sometimes an asteroid will be deflected from one orbit around the Sun into a new one by the gravitational disturbance of Jupiter. The new orbit might bring the asteroid into the inner part of the solar system where it could possibly collide with one of the planets or satellites there. Some of the meteors we see at night are probably tiny asteroids. The many craters appearing on the inner planets and satellites are possibly due to collisions with larger asteroids. Maybe the inner solar system was at one time filled with asteroid-type bodies. Today practically all such bodies, except those in orbit between Mars and Jupiter, have been swept up by the various planets.

Astronomers are now able to determine approximate compositions of the surfaces of asteroids through their reflection spectra. (Refer to Section 3.3.) The similarity in spectra among certain asteroids and various types of meteorites is further evidence of a connection between these objects. (Refer to Section 9.3.)

FIGURE 9.1
The tiny asteroid Väisälä, discovered in 1949 and photographed by an anonymous graduate student. The stars are trailed, and the asteroid is the faint dot near the center. (Goethe Link Observatory, Indiana University)

THE BODE-TITIUS LAW?

What are we to think of the Bode-Titius law? Is it really a scientific law, or is it just another one

Courtesy, Museum of Fine Arts, Boston. Babcock bequest

of those "coincidences" that we seem to keep running across?

The Bode-Titius law is not a scientific law in the usual sense of the term. In the first place, it is too specific. A scientific law would concern orbits in general, not just those of the specific planets that go around the Sun. Furthermore, while a scientific law does not need to be exact, it should state precisely when it is valid and when it breaks down. The Bode-Titius law makes no explanation for the near misses of most of the inner planets, nor for the breakdowns of Neptune (301 instead of the predicted 388) and Pluto (394 instead of 772).

Astronomers agree that the Bode-Titius law is far more than a coincidence. If planetary distances were completely random, it would not be difficult to find a series to fit the first six of them; but to have that series afterward fit Uranus and the asteroids is asking quite a bit. The point is that astronomers do not believe that the planetary distances are really random. Perhaps conditions at the time of formation of the solar system were such that this particular set of distances was more likely than other sets. Perhaps the orbits with these distances are more stable over long periods of time. It is possible that someday a new scientific law will be found, with the Bode-Titius law being a special case of it.

9.2 Comets

A bright *comet* is one of the most interesting sights in astronomy. Long ago the appearance of a comet was thought to be the warning of bad news, but now they are recognized as unusual objects that are subject to the same natural laws as everything else.

Comets move in elliptical paths around the Sun under the same gravitational force that controls the planets. This was first demonstrated in the case of Halley's comet, when its return was successfully predicted for the year 1758. While planetary orbits are nearly circles, most comets move in long, narrow ellipses. The orbits of many comets carry them far above and below the flat plane that contains all of the planetary orbits.

The main part of a comet is a very small nucleus, perhaps only a few miles across, that contains almost all of the mass. When it is far from the Sun, a comet is faint and difficult or impossible to see from the Earth since we see it by reflected sunlight. If it comes close to the Sun, it warms up, and some gas and dust are evaporated from it. This gas and dust are driven away by the radiation and particles emitted by the Sun, causing a tail to stream out from the nucleus. The closer to the Sun it comes, the more material is evaporated, and

FIGURE 9.2
Comet Ikeya-Seki in 1965. This was a very spectacular naked-eye comet. (Lick Observatory photograph)

FIGURE 9.3
Comet Kohoutek, photographed by S. Kutoroff at the Catalina Observatory, in 1974. An artificial satellite crossed the field of view during the exposure. (Lunar and Planetary Laboratory, University of Arizona, copyright pending)

FIGURE 9.4
Comet Alcock, photographed in 1959 by E. Roemer at the Flagstaff Station, U. S. Naval Observatory. The telescope was moved to follow the comet, so the images of the stars are trailed. (Official U. S. Navy photograph)

the larger and more conspicuous is the tail. It is the long tail that gives a comet its distinguished appearance, but the tail disappears when the comet moves back far away from the Sun (see Figures 9.2 through 9.5).

A comet loses some of its tail-producing material every time it comes near the Sun. Eventually it must run out of material, leaving only a core that cannot be easily vaporized and, most likely, cannot be seen from the Earth. A number of comets have actually been observed to disintegrate, and others have not shown up on schedule. The destruction of comets is not a rare event. If comets are so short-lived, why haven't they all disappeared by now? Where is the new supply coming from?

One theory is that a very large number of cometary nuclei move in more or less circular

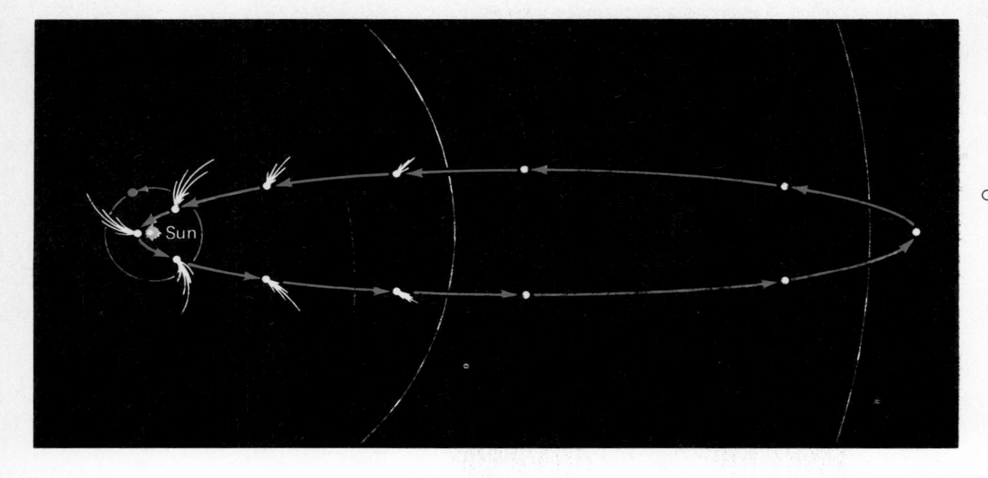

FIGURE 9.5
A comet orbit. The dotted curves represent the nearly circular orbits of the planets. The comet tail points away from the Sun and becomes less conspicuous as the comet moves further away from the Sun.

orbits on the edge of the solar system, far beyond the orbit of Pluto. They are material left over from the formation of the solar system. As long as the comets stay far out in the cold, dark reaches of space, they are safe from the destructive power of the Sun and from the prying eyes of astronomers. On occasions, the feeble gravitational force of a nearby star may disturb this cloud of frozen comets. Perhaps a few will be drawn free of the solar system, while others will be driven into long, narrow ellipses that bring them into the inner parts of the solar system. It is only these latter ones that we can see and that have their lives endangered by passing close to the Sun. Further disturbances by the planets, particularly the giant Jupiter, could make some of the comets permanent residents of the inner solar system. If comets did form far away from the heat of the Sun, they could offer important clues about the composition of the original material from which the solar system formed. They could also be a link with the tenuous matter that exists in the space between the stars. (Also see Chapter 13.)

9.3 Meteors

The shooting stars or *meteors* that we can see on any clear night have long been believed to be events in the upper air of the Earth. This is another case in which the ancient belief turned out to be correct. In fact, the word "meteor" in its older sense meant anything having to do with the air. Meteorology, for example, is the study of the atmosphere of the Earth.

PARTICLES IN SPACE

The space around the Sun has material of all sizes in it—from the giant planets through specks of dust down to individual atoms, nuclei, and electrons. On the average the density of this material is extremely low, but nevertheless the space is not quite empty. All of these objects feel the gravity of the Sun and tend to travel in paths around the Sun. The smallest particles feel forces, such as the pressure of radiation from the Sun, that are negligible for the planets, and their paths may be more complicated than ellipses or hyperbolas; still, they move in orbits around the Sun.

The Earth is continually colliding with small particles in space, sweeping up the volume around it. When a particle of the size of a grain of sand meets the Earth at a speed of many miles per second, the friction of its motion through the upper air causes the particle to vaporize in a second or two. Light and heat are given off, and a meteor can be seen from the ground if it is a clear night. The much smaller particles do not have enough energy to make a visible splash when they strike the air. Much larger particles will not be completely

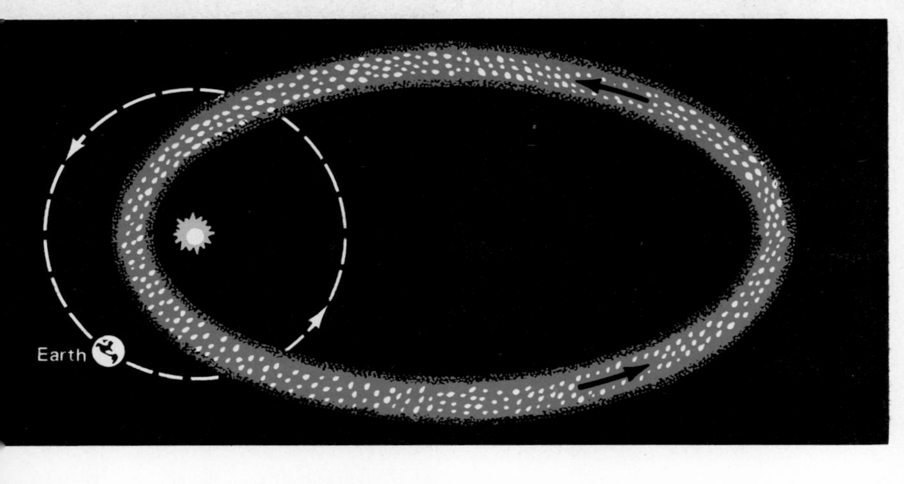

FIGURE 9.6
A meteor stream. The meteorites in a stream all have very nearly the same orbit around the Sun. They are the debris from a comet that might have broken up by the present time.

vaporized by the high-speed trip through the atmosphere. They survive the fall to the ground and provide us with samples of matter from outer space. It is common to distinguish between the flash of light that we see in the sky—the *meteor*—and the particle that causes it—the *meteorite*. Astronomers also like to distinguish between particles that have landed on the Earth—meteorites—and those still in orbit around the Sun—meteoroids. The latter distinction is not important for our purposes, so I shall use the term "meteorite" for the particle, whether or not it has encountered the Earth.

SPORADIC AND STREAM METEORS

Some meteors are random or *sporadic*, meaning that their appearance is not related to other meteors. Under conditions good for observation, a person can generally see several sporadic meteors per hour. More meteors are visible after midnight than before. (See number 4 in the Questions for Discussion at the end of this chapter.)

Sometimes many meteorites are moving in the same orbit around the Sun. This group is called a *meteor stream* (see Figure 9.6). If the stream happens to cross the Earth's orbit, there will be a larger than usual display of meteors whenever the Earth is at the crossing. This display is known as a *meteor shower*, although the name is misleading in most cases. Usually a meteor shower only adds a few meteors per hour to the number of sporadic meteors that can normally be seen. For a few days around August 11 each year, the Earth crosses a major stream of meteorites. This meteor shower is well worth the loss of the few hours of sleep that it takes to see it.

Meteor streams often have the same orbit around the Sun as a known comet, and the stream particles are obviously material from the comet. Occasionally meteor streams produce spectacular showers. For a short time on November 17, 1966, many meteors were visible each second as the Earth passed through a swarm of particles that was the remains of a former comet. Another display occurred on October 9, 1946, when the Earth passed through the tail of a comet named Giacobini-Zinner.

Sporadic meteors, on the other hand, are probably tiny asteroids that have wandered from their usual place beyond Mars. Two meteorites found on Earth were observed well enough during their fall to have accurate orbits calculated for them. In both cases, the orbits extended well out into the asteroid belt between Mars and Jupiter.

TYPES OF METEORITES

Meteorites are divided into three classes according to their composition: *iron*, *stony-iron*, and *stony*. The iron meteorites are almost pure metallic iron and nickel, the stony ones are composed mostly of the kinds of materials found in various types of Earth rocks, and the stony-irons are in between. These meteorites remind us very much of the core and the mantle or crust of the Earth: The meteorites seem to be pieces from a much larger body or bodies that were once molten and had a large enough gravity to separate into core and mantle. The iron meteorites were once part of the core of such a body, and the stony ones might have come from the mantle.[3] Many of the asteroids are large enough to satisfy these conditions, so we have direct evidence of the break-up of the asteroids into smaller fragments, probably through collisions with each other.

RADIOACTIVE AGES

By studying the tiny amount of radioactive material contained in meteorites, scientists are able to determine how long ago the meteorites formed. These are the same radioactive substances used to date Earth rocks that are listed in Table 6.2. Scientists have found meteorites that are up to 4.6 billion years old. This is older than any rocks found on the Earth or Moon, and we can understand why. The surfaces of the planets and satellites were subjected to many forces that could destroy early records of radioactivity. On the much smaller asteroids, from which the meteorites apparently came, these forces were much smaller, although collisions between asteroids could cause considerable heating. The ages of the meteorites should give us a fairly accurate date for the formation of the asteroids. This agrees very well with the date of formation of the Sun. (Refer to Section 15.2.) It is generally believed that the entire solar system took shape during a rather short interval about 4.6 billion years ago.

It has long been known that space contains some extremely rapidly moving particles known as *cosmic rays*. Most of the particles are protons, and their origin is not known for certain. The cosmic rays constantly strike anything that exists in space, including the meteorites. When cosmic rays hit a meteorite in space, they cause certain kinds of nuclear reactions to occur; in other words, they break apart certain atomic nuclei and form others. Some of these newly formed nuclei are radioactive and can be used to find ages, as we learned in Chapter 6. In the present case, however, the ages scientists find do not tell when the material became solid, but how long it has been exposed to the cosmic rays.

Scientists have found that most iron meteorites have been exposed to cosmic rays in space for lengths of time ranging from about 20

[3]See, however, the discussion that follows on exposure ages of the stony meteorites.

million to 1 billion years. These times are much shorter than the 4.6 billion years since the meteorites solidified. What does this mean? Apparently the meteorites spent most of their existence inside larger bodies, where they were shielded from the cosmic rays. Only after these bodies were broken up were the meteorites exposed to the cosmic rays. By studying the two kinds of radioactivity in a meteorite—the original radioactivity represented by uranium, thorium, and argon, plus the radioactivity induced by cosmic rays—we can tell when the parent body was formed and when it broke up. A number of iron meteorites have cosmic-ray exposure ages that cluster around 600 million years, and a second group has ages that cluster around 900 million years. It would seem that these are the dates of two major asteroid collisions.

Stony meteorites present a problem, as they show exposure ages that are much shorter than those of the irons: from less than 1 million years to nearly 100 million years. The asteroid collisions that produced the iron meteorites were apparently quite separate from the ones that produced the stony meteorites. The very short exposure ages of the stony meteorites are a puzzle. Why are none found that even approach 1 billion years? Some astronomers consider this evidence that the stony meteorites, unlike the irons, did not originate in the asteroid belt. Perhaps they came from asteroids whose orbits brought them closer than usual to the Earth. This would make collisions with the Earth more likely and could explain the young cosmic-ray exposure ages; however, there may not be enough such asteroids to account for all of the known stony meteorites. Some astronomers believe that the stony meteorites may come from comets instead of asteroids.

There is a certain rare type of stony meteorite called the *carbonaceous chondrite* (car-bon-A-shus KON-drite). These meteorites have interested scientists very much because, unlike other kinds of meteorites, they do not seem to have been subjected to very high temperatures or pressures in the past, and they do not appear to have been melted. High temperatures can drive out the more volatile substances, and melting in a region of moderate gravity can cause a separation of substances according to density. The carbonaceous chondrites may be examples of the primitive presolar system material (minus the hydrogen and helium) before it had a chance to be processed by high temperatures and pressures.

IMPORTANT WORDS

Bode-Titius law | **Comet**
Asteroid | **Meteor**

Meteorite
Sporadic meteor
Meteor stream
Meteor shower
Iron meteorite
Stony meteorite
Cosmic rays
Carbonaceous chondrite

REVIEW QUESTIONS

1 What is the Bode-Titius law?

2 Did Uranus fit the Bode-Titius law?

3 Who found the new planet to fill the gap in the law?

4 What are asteroids, and how do they differ from the major planets?

5 What evidence is there that a connection might exist between asteroids and meteorites?

6 Why is the Bode-Titius law not a true scientific law?

7 How do cometary orbits differ from planetary orbits?

8 What causes a comet to have a tail?

9 If a comet comes very close to the Sun, can it last for a very long time?

10 Where does the new supply of comets come from?

11 Is a shooting star really a star?

12 Are meteors in our atmosphere?

13 What causes a meteor?

14 What is the difference between a meteor and a meteorite?

15 What is the difference between a sporadic meteor and a stream or shower meteor?

16 Where do meteorites come from?

17 Why do we believe that meteorites are fragments from larger bodies that broke up sometime in the past?

QUESTIONS FOR DISCUSSION

1 What part did the Bode-Titius law play in the discovery of the asteroids?

2 On Ceres, the largest of the asteroids, gravity is so weak that a person on its surface could easily jump up 50 feet. Do you think such a person would probably be hurt by falling 50 feet on Ceres?

3 A bright comet appears in 1978, and someone says that it is the same comet that was seen in 1900. Since a comet does not have any permanent markings to identify it, how can we check to see if the two comets really are the same one?

4 Why are there more meteors after midnight than before? This question is closely related to the question of why the front window of a car gets more bug splatters than the back window.

5 What if cosmic rays were either more or less abundant in the past than they are today? Would our estimates of the lengths of time that the meteorites have been exposed in space be affected? Can you think of a way in which a variable amount of cosmic rays could account for at least some of the

puzzling aspects of the very short exposure ages of stony meteorites?

REFERENCES

General references are: "The Smaller Bodies of the Solar System" by W. K. Hartmann [*Scientific American* 233 (September 1975), 142] and Chapters 8, 9, and 10 of *The Solar System* by Z. Kopal (Oxford, London, 1973).

Asteroids are discussed by C. R. Chapman in "The Nature of Asteroids" [*Scientific American* 232 (January 1975), 24]; and by C. R. Chapman and D. Morrison in "The Minor Planets: Sizes and Mineralogy" [*Sky and Telescope* 47 (February 1974), 92].

Comets are the topic covered by F. L. Whipple in "The Nature of Comets" [*Scientific American* 230 (February 1974), 48] and K. F. Weaver in "What You Didn't See in Kohoutek" [*National Geographic* 146 (August 1974), 241].

The carbonaceous chondrites are the subject of "The Most Primitive Objects in the Solar System" by L. Grossman [*Scientific American* 232 (February 1975), 30].

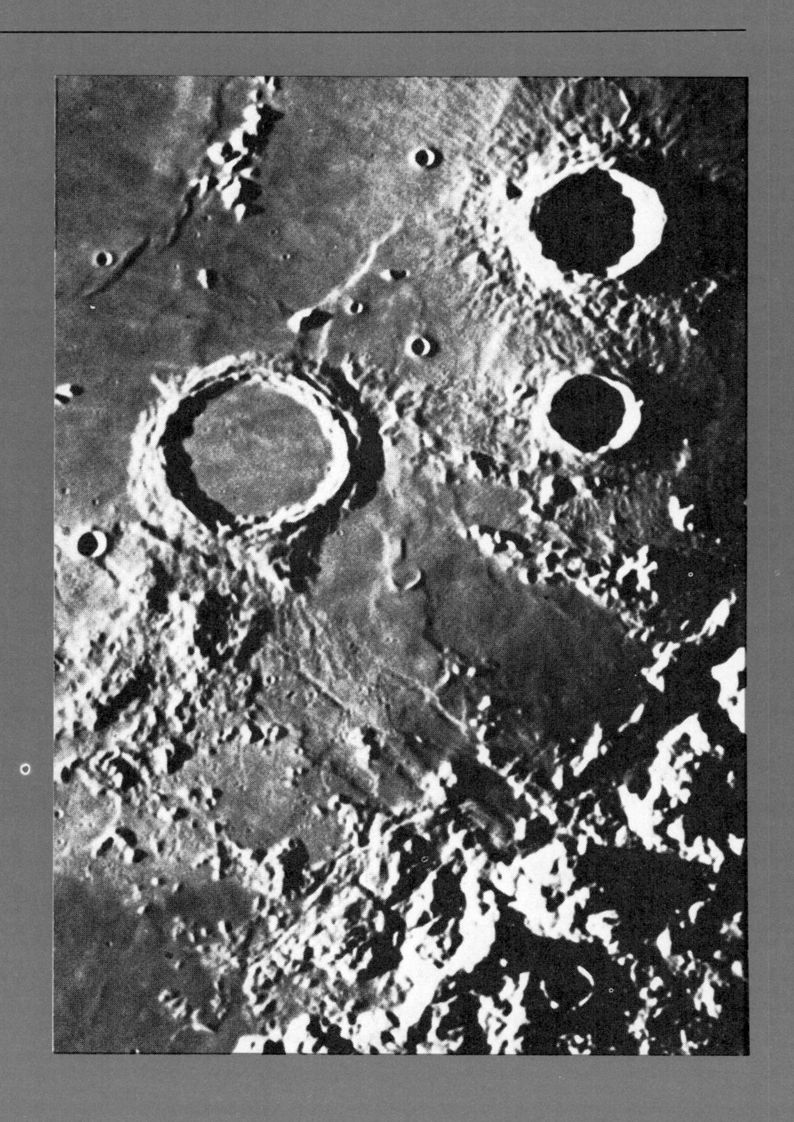

TEN
ORIGINS

Here we consider the general process by which the solar system might have been created. Since the formation of planetary systems is very closely related to the interesting question of life elsewhere in the Universe, we also look at the scientific basis for trying to answer this question.

10.1 Formation of the solar system

Much is known about how the solar system could *not* have been made, but relatively little about how it *was* made is known for certain. There are some general ideas that have a consensus approval which are probably correct, but when it comes to details, disagreement among the experts is the rule.

GENERAL CONSIDERATIONS

Today it is almost universally agreed that the planets were made in the same process and at essentially the same time as the Sun. Having the Sun capture the planets after they were already formed or having the planets somehow created from a collision between the Sun and another star are believed to be nearly impossible.

As mentioned in Section 8.2, Jupiter and the Sun would be a *double star*, instead of a planet and a star, if Jupiter were sufficiently massive. Scientists have observed many of these double and even multiple stars; in fact, only about half of the stars are single like the Sun. This suggests that the formation of a star with planets may be a variation of the common theme of the creation of two or more stars together.

The solar system is very flat (except for many of the comets), and practically all of the matter is moving in orbits and rotating in the same general eastward direction. These facts indicate that the material from which the solar system evolved had an eastward rotation and that this rotation had an influence on how the system formed.

COLLAPSE OF THE SOLAR NEBULA

In Section 15.3 it is explained that stars form when a cloud of the extremely tenuous gas and dust that exist in the space between the stars collapses under its own gravity. As the cloud or nebula contracts to a smaller size, any rotation it possesses will cause it to become flat. (Refer to Section 18.1.) In our solar system, the bulk of the matter fell into the center to form the Sun, while the peripheral material somehow condensed into the planets.

The *solar nebula*, that is, the cloud of

material from which the solar system was made, was predominantly hydrogen and helium. Many of the atoms of the heavier elements formed molecules that tended to stick together and make tiny "dust" particles. As we will note in Chapter 13, the dust particles are an important part of the interstellar matter. As the solar nebula shrank in size, the dust particles often stuck together, making particles of larger size.

The composition of the particles depended on their position in the nebula. Near the center where the Sun itself was forming and heating up, the matter was hot, and the particles were made of material that could survive moderately high temperatures. The more *volatile* substances, that is, the more easily evaporated material, could not readily condense on the particles. Farther away from the center, the material was cooler, and volatile substances had a better chance to condense on the particles. Thus particles of high density were created close to the Sun, while lighter particles tended to form farther away from the Sun.

All of these particles were moving in orbits around the rapidly forming Sun at the center, and there were many collisions among the particles. The more violent collisions tended to break the particles apart, but most collisions were mild enough that the tendency was to stick together. Larger and larger bodies were built up, the larger ones swallowing the smaller ones, until planet-size objects finally came into being. These had enough gravity to attract other bodies to them, and they grew in size even faster. Finally, nearly all of the particles were swept up by one or another of the larger bodies, except possibly for the asteroid belt, and the planets ceased growing. The end of the planets' growing era can possibly be identified as the time of formation of many of the craters visible on Mercury, the Moon, and Mars.

Near the Sun temperatures were high, and the atoms of hydrogen and helium were moving very rapidly. The inner planets—Mercury, Venus, the Earth, and Mars—probably never did become massive enough to capture the hydrogen and helium gases through their gravities. They remained made out of rocky and metallic substances. Farther away from the Sun, temperatures were much lower, and the hydrogen and helium were easier to capture. Thus, more massive bodies formed in these regions, and these bodies contained large amounts of hydrogen and helium from the solar nebula. This is how the very low density giant planets came into being.

There is evidence that near the end of its process of formation, the Sun was much brighter than it is today and emitted large numbers of energetic particles. The large emission of radiation and particles from the Sun blew the excess gas out of the solar system, and it also blew off any *primitive atmospheres* that the inner planets might have had. The present atmospheres of Venus, the Earth, and Mars were released from the interiors of these planets over the past several billion years.

This is at least a plausible outline of the formation of the solar system. There are many unsolved problems remaining. One of these is the question of why the Sun rotates so slowly. The solar system has a certain amount of rotational motion or *angular momentum*, and practically all of it is concentrated in the giant planets. It is a puzzle how so much of the angular momentum was transferred to such a small part of the total mass of the solar system.

OTHER PLANETARY SYSTEMS

Was the formation of the solar system a rare or common event? When we see other single stars like the Sun, can we expect them to have a family of planets? Or were the events that led to the creation of the planets so unusual and so improbable that we can expect our solar system to be unique? Unfortunately, we don't know enough about its origins to answer these questions, but we can get some hints from other considerations.

A planetary system around another star could not be seen directly, but it could perhaps be detected indirectly. The faint star called Barnard's star has been closely observed for many years by P. van de Kemp, and he detects a small wiggle in its motion. This has been the subject of some controversy, but it has a good chance of being real. What can cause a star to wiggle like that? A small star or a large planet moving in an orbit around it. Van de Kemp calculates that the invisible body has a mass only a little bit larger than that of Jupiter. If this turns out to be correct, Barnard's star does possess a planetary system. There are other reported cases of nearby stars showing similar wiggles in their motions, and investigations in this area are continuing.

Observational evidence is not yet conclusive, but it is pointing toward the conclusion that planetary systems are quite plentiful. This is not really surprising, for if something happens once, why can't it happen again? This is not to mean that it *must* happen again; the assumption that it cannot, however, is much more special and arbitrary than the assumption that it can.

10.2 Life in the Universe

Observational evidence of life elsewhere in the solar system is quite meager. Most experts agree that there *is* a small chance of finding living organisms on Mars, although the Viking spaceships were not successful in this, and Jupiter is sometimes mentioned in a very speculative way. Elsewhere in the solar system the chances seem quite unlikely; beyond the solar system, there is no evidence at all.

Can we then speak intelligently and scientifically about the likelihood of life existing in the Universe? Can we do other than guess?

SCIENTIFIC BASIS FOR LIFE?

In Section 1.1 we saw that the scientific point of view assumes that a given situation results automatically if the conditions are right. Science does not prove this; it is only an assumption, and conceivably there may be instances when it is wrong. It is worth noting, however, that the assumption has led to many impressive accomplishments.

What are we to think of life? Are we to believe that the laws of science can account for the properties of inanimate matter and radiation, but that they break down when living matter enters the scene? This might very well be correct. Maybe life is not understandable by mankind. But to assume this before all the evidence is in is to set up artificial barriers limiting what areas science can investigate. Why not admit that we don't know, and wait to see what investigations bring out?

Of course we all have philosophical prejudices of various kinds. We often feel ahead of time that the results ought to turn out in a certain way, and our feelings are not always based on scientific fact. This is fine as long as we admit that our feelings are biased; the problems arise when we try to peddle our prejudices as scientific facts. Scientists are as guilty of this as anyone. Many scientists have claimed that the existence of life beyond the Earth is essentially a proven fact. This statement has no more scientific basis than does the denial that life can exist elsewhere. Maybe in the near future it will become a scientific fact.

THE CONDITIONS FOR LIFE

Let us pretend that life and its origins are subject to the laws of science and see what happens. We do not understand all of these laws, but we are assuming that they are in principle understandable. Life exists on the Earth today, so it must have originated sometime in the past. If you want to believe that life came to Earth from somewhere else, then you are simply transferring to another place the origin of life. Sometime and somewhere conditions must have been just right for life to develop out of nonlife.

Living matter is composed of extremely complex molecules. From our present perspective, it is reasonable to suppose that given the proper environment and a long enough period of time, some of these life molecules would eventually form through natural chemical reactions. This supposition is strengthened by the discovery in interstellar space of certain organic molecules that could be among the very early steps in the evolution of life. (Refer to Section 13.1.)

What conditions are needed for life to get started? The answer is not known, but it certainly is plausible to suppose that the necessary conditions were met long ago on the surface of the Earth. The presence of a planet at the proper distance from a star appears to be one of the requirements. We saw in Section 10.1 that other planetary systems may be plentiful. With many billions of stars in our galaxy and with many billions of galaxies existing in the Universe, it is reasonable to expect that there are plenty of places where conditions are or have been about right. The probability of finding life elsewhere in the Universe thus revolves around the question, "If conditions are about right, how likely is it that life will form spontaneously?"

Is it conceivable that life requires such special conditions for its evolution from nonlife that the Earth is the only place where it formed? Most scientists would consider this possible but very unlikely, as it requires an extremely restrictive assumption about the origin of life. If there is a scientific basis for life, then it probably is common in the Universe. Of course there may not be a scientific basis for the origin and evolution of life, but we have not yet received proof of this.

If there are other advanced civilizations, can we hope to contact them? Astronomical distances are so huge that unless our knowledge of physics is considerably in error, traveling to other planetary systems and looking for life is nearly impossible. Communicating by radio signals is conceivable, although years would be required to receive an answer from even the nearest star. Nevertheless, means of communicating with other civilizations in our galaxy are the object of serious scientific study today.

IMPORTANT WORDS

Double star
Solar nebula
Volatile
Primitive atmosphere
Angular momentum

REVIEW QUESTIONS

1 Were the planets probably made at the same time as the Sun?

2 Why are the inner planets made of heavier material than planets farther away from the Sun?

3 Why do the giant planets have much hydrogen and helium, while the smaller planets do not?

4 Why do scientists wonder about the slow rotation of the Sun?

5 Are other planetary systems known?

6 Is there any evidence about life beyond the solar system?

7 Why doesn't the assumption that life came to Earth from another place solve the question of the origin of life?

8 What does the discovery of certain organic molecules in space have to do with the question of life in the Universe?

9 Are there probably many or few places in the Universe where conditions are about right for the evolution of life?

QUESTIONS FOR DISCUSSION

1 Can you think of any reasons why it might be unlikely that the Sun captured the planets?

2 Recall the meaning of the velocity of escape from Section 5.3. What does this have to do with whether a planet can hold onto gases for its atmosphere? How is the temperature of the planet involved?

3 Suppose the Moon was formed in another part of the solar system and was later captured by the Earth. Look at the densities of the planets listed in Table 5.1 or in Appendix 4. Can you see any arguments favoring the formation of the Moon at the orbit of Mars or beyond?

4 Speaking of the study of life beyond the Earth, G. Simson said, "Exobiology is still a 'science' without *any* data, therefore no science." What does he mean? Do you think this is an important point?

5 C. Sagan says[1]:

> In conclusion, let us reflect on man's uniqueness in the universe. Tortuous paths have led, over some five billion years of evolution, to a technical society on this planet—an evolution marked by so many branch points that we can have no hope that a similar path has been followed elsewhere. There can be no humans beyond the earth.

In other words, Sagan believes that while there are numerous examples of intelligent life throughout the Universe, none of them can have evolved similarly enough to Earth life that they would be humans. Is his statement more of a scientific fact or a philosophical prejudice? Do you agree with him?

6 The number of advanced civilizations in our galaxy is often expressed as a mathematical equation involving the number of habitable planets, the probability that such a planet will have life originate on it, the probability that this life will evolve to an advanced form, and the probable lifetime of an advanced civilization. What do you think would be

[1]Quoted from *Interstellar Communication*, edited by C. Ponnamperuma and A. G. W. Cameron, Houghton Mifflin, Boston, 1974, p. 23.

the average lifetime of an advanced, technological civilization? 100 years? 1000 years? One billion years?

REFERENCES

Certain aspects of the origin and development of the solar system are contained in: "The Origin and Evolution of the Solar System" by A. G. W. Cameron [*Scientific American* 233 (September 1975), 32]; "The Chemistry of the Solar System" by J. S. Lewis [*Scientific American* 230 (March 1974), 50]; Chapters 11 and 12 of *The Solar System* by Z. Kopal (Oxford, London, 1973); and Chapter 14 of *Earth, Moon, and Planets*, 3rd ed., by F. L. Whipple (Harvard University Press, Cambridge, 1974). S. S. Kumar discusses the possibilities of other planetary systems in "Planetary Systems" in *The Emerging Universe*, edited by W. C. Saslaw and K. C. Jacobs (University of Virginia Press, Charlottesville, 1972).

The following is a sample of the many articles and books on life in the Universe: *Intelligent Life in the Universe* by I. S. Shklovskii and C. Sagan (Dell, New York, 1966); *Interstellar Communication*, edited by C. Ponnamperuma and A. G. W. Cameron (Houghton Mifflin, Boston, 1974); *Communication with Extraterrestrial Intelligence*, edited by C. Sagan (MIT Press, Cambridge, 1973); and "The Search for Extraterrestrial Intelligence" by C. Sagan and F. Drake [*Scientific American* 232 (May 1975), 80].

A discussion of the possible evolution of life is given by M. Calvin in "Chemical Evolution" [*American Scientist* 63 (March-April 1975), 169].

PART THREE

THE STARS

Here we expand our attention from the solar system to the stellar system. The stars are the heart and soul of astronomy; the term "astronomy" comes from the Greek word *aster*, meaning "star."

In Chapter 11 we study the Sun in a close-up view of a star. Chapter 12 describes some of the more important properties of the stars and how they are determined, and in Chapter 13 we find out about the tenuous interstellar matter that exists in the space between the stars. The final three chapters of Part Three are based primarily on a theoretical approach to understanding a star: Chapter 14 indicates how the structure of a star can be determined by applying the known laws of physics, while Chapters 15 and 16 show what we can find out about the life histories of the stars.

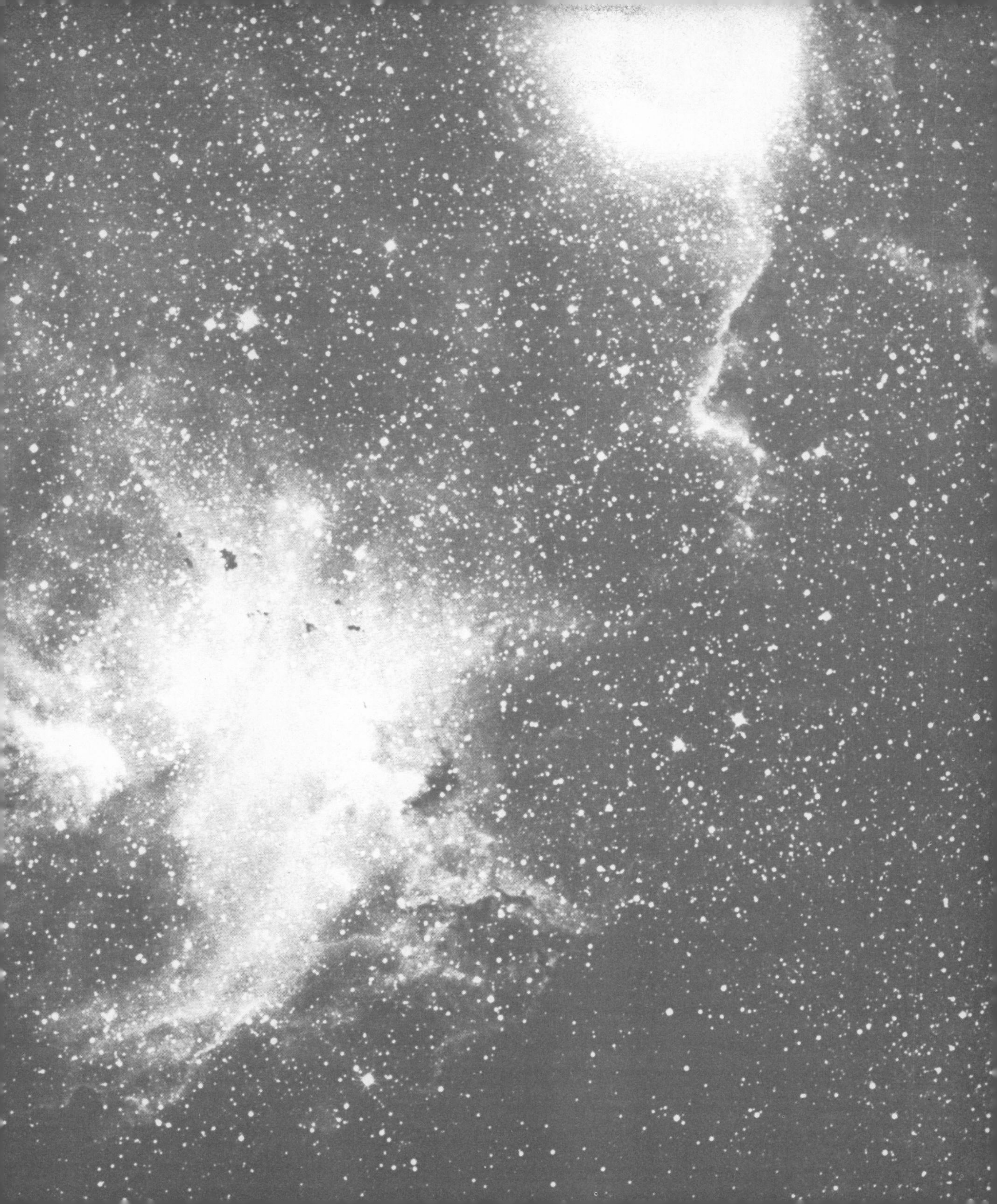

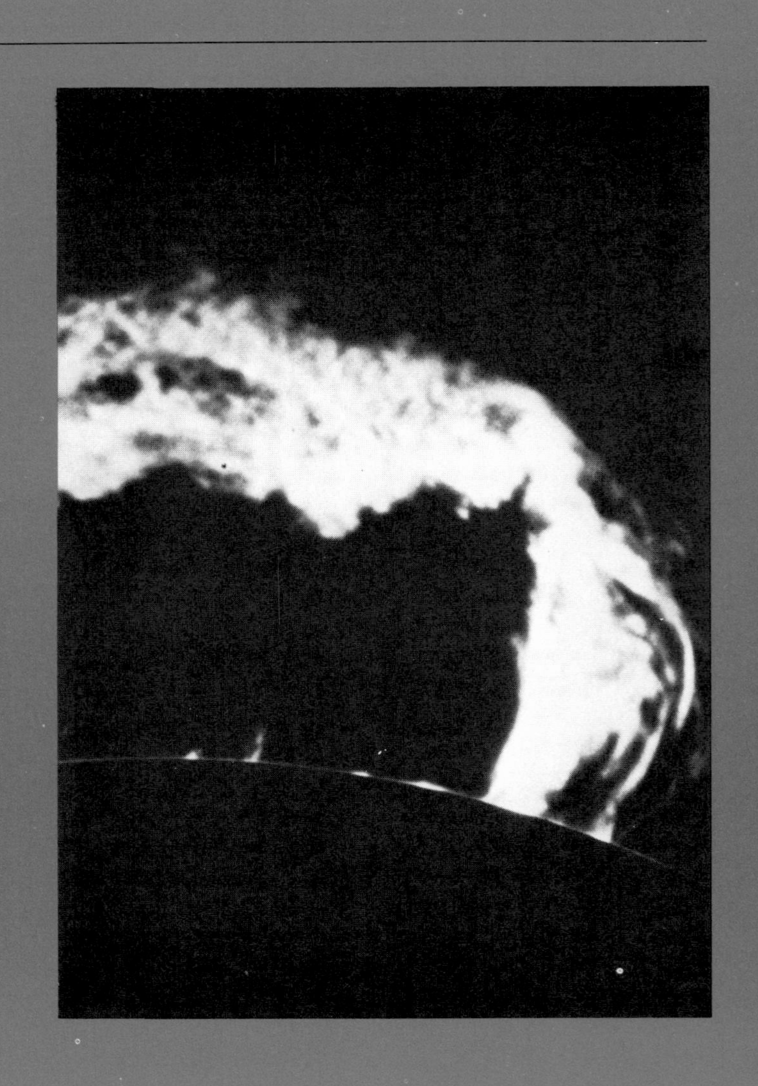

ELEVEN
THE SUN

The Sun has much more in common with other stars than it does with the other members of our solar system, so it is appropriate to study it in Part Three. There are certain properties of the Sun which it undoubtedly shares with other stars, but which are not observable for other stars because of their great distances. These are the properties which are emphasized in the present chapter.

11.1 The outer atmosphere

The very brightness of the Sun not only makes it difficult to observe but dangerous to observe, because eye damage will result if proper precautions are not taken. By using appropriate equipment, however, we can look at it directly or with a telescope.

The Sun appears to have a well-defined surface (see Figure 11.1). We might guess from this that its surface is liquid or solid like that of the Earth, but this is not true. The Sun is gaseous throughout, and if we looked at it very closely under high magnification, its edge would appear somewhat blurred, like the edge of a cloud.

CORONA AND CHROMOSPHERE

Above the surface of the Sun we normally do not see anything. When this blindingly bright surface is covered by the Moon in an eclipse, however, then we can see a faint white halo around the Sun that extends far above the surface (see Figure 11.2). This halo is called the *corona*. At the base of the corona, just above the visible surface, is a narrow region known as the *chromosphere*. The corona and the chromosphere make up the outer atmosphere of the Sun. (The atmosphere of a star, as distinguished from its outer atmosphere, is the surface itself and will be studied in Section 14.1.)

The soft light of the Sun's outer atmosphere is usually lost in the glare of the solar disk. Instruments have been devised so that the outer atmosphere can be studied without waiting for an eclipse (see Figure 11.3), but eclipses still offer the best viewing conditions. All total eclipses of the Sun attract large numbers of astronomers to study some aspect of the outer solar atmosphere.

The chromosphere consists of hot gases, and its spectrum shows the emission lines we would expect in such situations. (Refer to Section 3.3.) The corona has a continuous spectrum with emission lines superposed on it. The continuous spectrum is radiation from the surface that is scattered in all directions by electrons in the corona. The emission lines arise directly in the corona.

Radiation from the surface of the Sun shows that it has a temperature of about 10,000°F. We would expect the inside of the

FIGURE 11.1
The disk of the Sun, photographed at Kitt Peak National Observatory. Some sunspots are visible. (The Kitt Peak National Observatory)

FIGURE 11.2
The total eclipse of the Sun on March 7, 1970. The Moon is blocking off the disk of the Sun, thereby allowing the corona to be seen. (The Kitt Peak National Observatory)

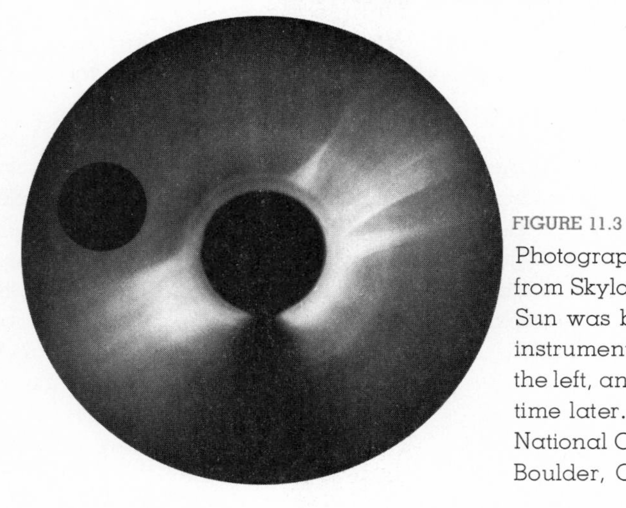

FIGURE 11.3
Photograph of the corona of the Sun taken from Skylab on June 30, 1973. The disk of the Sun was blocked off artificially by the instrument. The disk of the Moon is seen to the left, and a solar eclipse occurred a short time later. (High Altitude Observatory, National Center for Atmospheric Research, Boulder, Colorado)

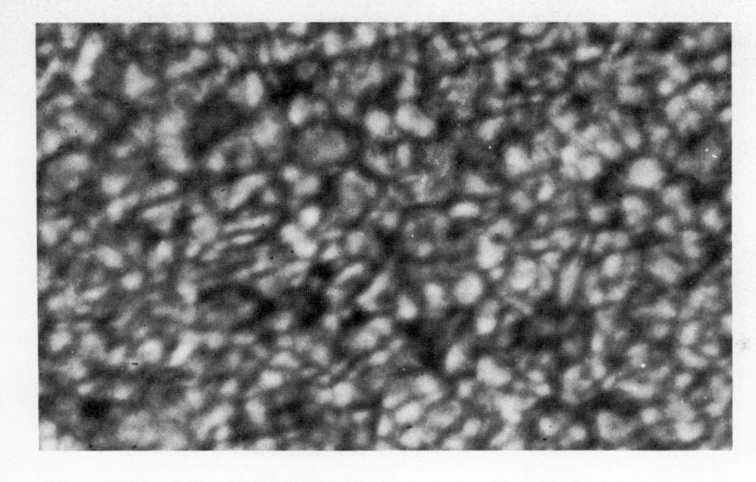

FIGURE 11.4
A close-up photograph of the surface of the Sun. The churning motions of the gas cause the observed patterns. The light areas are rising, hot gases, and the dark areas are cooler gases that are sinking down into the Sun. (The Kitt Peak National Observatory)

Sun to be much hotter than this (and in Section 14.2 we shall see that this is correct). Likewise we also expect the regions above the surface to be cooler, but in this expectation we are wrong. The emission lines of the corona are the kind that can be emitted only by an extremely hot gas, and the corona appears to have a temperature of several million degrees Fahrenheit. If the corona is so hot, why doesn't it heat the surface to a much higher temperature? It does try to, but there is so little material in the corona that it cannot succeed. It is like trying to heat a large house with one small candle.

WHY SO HOT?

What makes the corona so much hotter than the surface? For many years astronomers have puzzled over this question, and even today there are aspects that are not completely understood. The overwhelming part of the energy emitted by the Sun leaves the surface and passes through the corona without affecting it in any way. There must be another supply of energy that somehow is dumped into the corona and makes it hot; an object needs to actually absorb energy in order to be heated by it.

The material just below the surface has a sort of churning, up-and-down motion called *convection* that gives the surface a mottled or granular appearance (see Figure 11.4). The convective motions produce a noise that sends sound waves out in all directions through the outer parts of the Sun. The sound waves are not a form of radiation but consist of wave motions of the atoms in the gas. These waves do carry a certain amount of energy, and they can pass through the surface and the chromosphere without much hindrance. When the waves reach the corona, however, some difficulties arise. The reason is that the density of the material in the corona is extremely low, so there are fewer particles that can share or take part in the wave motions. The wave has the same energy as it did when it was down at the surface of the Sun or in the chromosphere, so the particles in the corona must wave much more violently. The sound wave changes into a shock wave and deposits most of its energy in the corona, thereby heating it to a very high temperature. The energy deposited in the corona in this way is very small compared to the energy radiated by the surface. The amount of matter in the corona is very small, so it doesn't take much energy to heat it to a very high temperature.

The corona is not placid, but it acts much like a pan of boiling water. The high temperature causes matter to be continuously boiled off and evaporated into space. This produces an outflow of particles from the Sun known as the *solar wind*. The Sun loses 100,000 tons of material each second through the solar wind, but

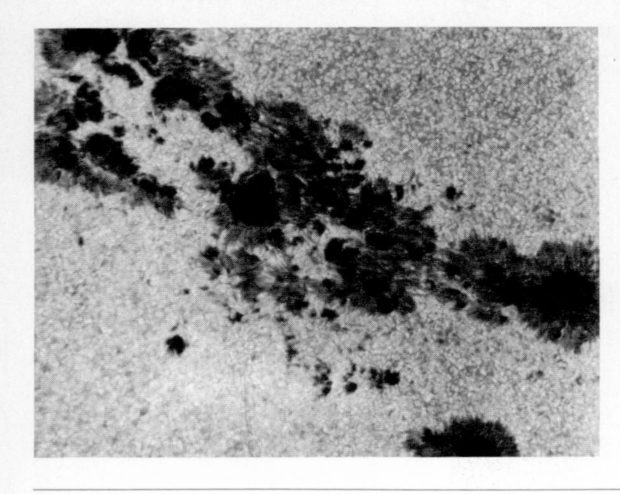

FIGURE 11.5
A large sunspot group on the surface of the Sun. The granulation is also visible. (Sacramento Peak Observatory, Association of Universities for Research in Astronomy, Inc.)

even over its lifetime of some 5 billion years this is a negligible fraction of the mass of the Sun. The solar wind is not a uniform flow but is subject to rather large fluctuations.

11.2 Solar activity

SPOTS, PROMINENCES, AND FLARES

In Figure 11.1 you can see several dark spots on the surface of the Sun; Figures 11.5 and 11.6 show close-ups of groups of such spots. The spots look dark because they are some 2500°F cooler than the surrounding areas and so emit less light. Spots are not permanent features; on the average they come and go in a few weeks. They usually occur in groups rather than alone. Scientists have observed that sunspots have very strong magnetic fields.

If we plot the number of sunspots visible at different times, we find that the numbers fluctuate from one month to another. Over longer periods of time, however, a systematic trend becomes noticeable. On the average the number of sunspots builds up to a maximum every 11 years, with smaller numbers occurring in the intervening years. There is likewise a minimum number occurring every 11 years between the maxima. This 11-year period is the *sunspot cycle*. The cause of the cycle is not well understood.

Bright clouds of gas are often observed above groups of sunspots. These are the *prominences* (see Figures 11.7 and 11.8). With proper equipment, astronomers can easily detect prominences beyond the edge of the Sun. Sometimes material in a prominence appears to be moving up into the corona, sometimes it moves down toward the surface, and sometimes no motion is apparent. Prominences give

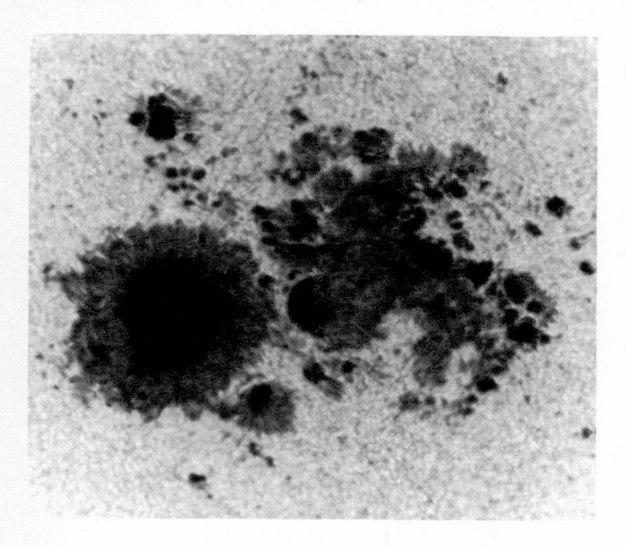

FIGURE 11.6
A detailed photograph of sunspots. (Sacramento Peak Observatory, Association of Universities for Research in Astronomy, Inc.)

FIGURE 11.7
A detail of the edge of the Sun, showing the hot, churning gases near the surface, the ragged edge, and the prominences above the surface. (The Kitt Peak National Observatory)

the impression of matter that is stretched out along strong magnetic fields. Some prominences stay quiet for months, while others expand explosively.

Occasionally a region near a large sunspot group will be the site of a major explosion known as a *flare*. In a flare, large amounts of radiation of all frequencies are emitted, and clouds of gas are also expelled from the Sun into space. These clouds can have significant effects when they strike the Earth.

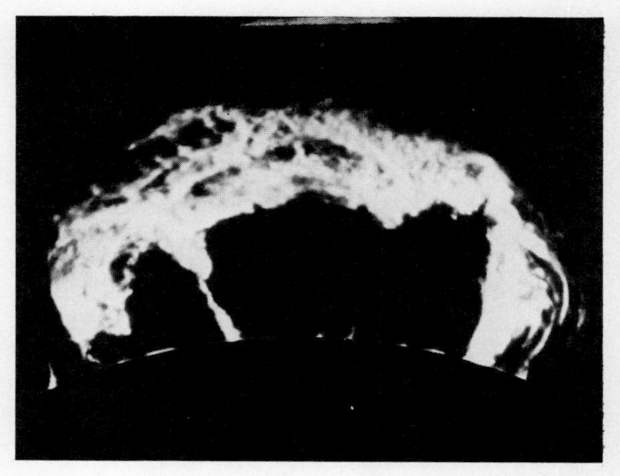

FIGURE 11.8
A prominence extending far above the surface of the Sun. The solar disk is artificially blocked off by the instrument. (Sacramento Peak Observatory, Association of Universities for Research in Astronomy, Inc.)

EFFECTS ON THE EARTH

When clouds of particles ejected by solar flares strike the upper part of the Earth's atmosphere, they can produce *auroras* (northern and southern lights) (see Figure 11.9). The auroras are usually seen at high latitudes because the Earth's magnetic field allows the electrically charged particles from the Sun to reach the Earth most easily in the regions near the poles. Particles from solar flares can also disrupt radio communications and temporarily modify the Earth's magnetic field. There is also concern for the safety of astronauts who happen to be in space during strong flares. Flares are quite common—several occur every day during sunspot maxima—but the very strong flares that appreciably affect the Earth are much more rare.

Since solar flares are correlated with the sunspot cycle, so, too, does the amount of particles and x rays that bombard the Earth vary with the 11-year cycle. There is evidence that large-scale weather patterns are influenced by these particles and by high-energy radiation. When averaged over large enough areas on the Earth over long enough periods of time, weather conditions do seem to vary in accordance with the sunspot cycle. This does not

FIGURE 11.9
A photograph of the aurora or the northern lights, taken from a U.S. Air Force Weather Service satellite. The aurora is over Canada, and the lights of cities in the north central United States are visible below. Chicago and Lake Michigan are easily identified. The aurora is caused by particles from the Sun striking the upper atmosphere of the Earth. (U.S. Air Force Meteorological Satellite photograph, courtesy of E. H. Rogers and D. F. Nelson of the Aerospace Corporation)

mean that we can look at the Sun and decide what the weather will be next Tuesday in Elkhart. It could possibly mean, though, that knowing what part of the sunspot cycle we are in would enable us to guess what the average rainfall, temperatures, and so on, would be for the whole country, averaged over a year.

The energy released in a strong flare is only a very tiny fluctuation on top of the steady output of energy from the surface of the Sun. Spots, prominences, flares, and other signs of solar activity would be invisible if we did not have such a close position to the Sun. Other stars no doubt have similar properties, but they are apparent only in a few stars that have developed them to an extreme degree.

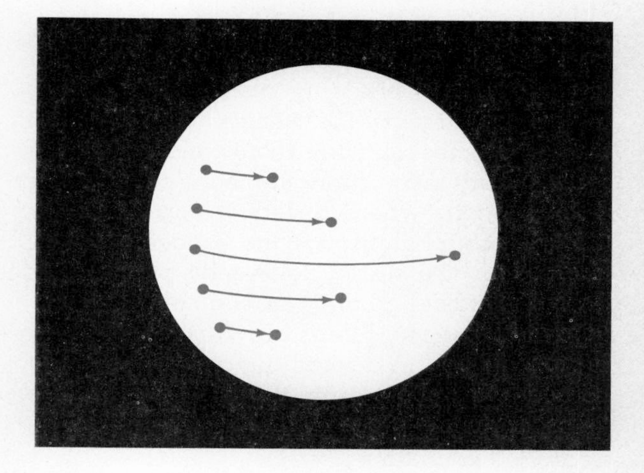

FIGURE 11.10
Solar rotation. Points near the equator of the Sun rotate more rapidly than those closer to the poles.

MAGNETIC FIELDS

What causes all of this activity on or above the surface of the Sun? Studying this question is one of the very active fields of research in astronomy today. Magnetic fields are obviously involved. When a magnetic field occurs in a gas in which some of the particles are electrically charged, the gas and the field exert forces on each other. If the gas wants to move in a certain direction, it must drag the field along with it, and this requires energy. The magnetic field acts something like a rubber band that can be stretched, squeezed, and tangled up by the gas motions; if it becomes distorted too much, it will try to snap back the way it was, pushing back on the gas.

The Sun spins around on its axis about once a month. It does not rotate at a uniform rate, but its equator turns slightly faster than the poles. This effect is known as *differential rotation* (see Figure 11.10). A point on the equator of the Sun will take about 25 days to be

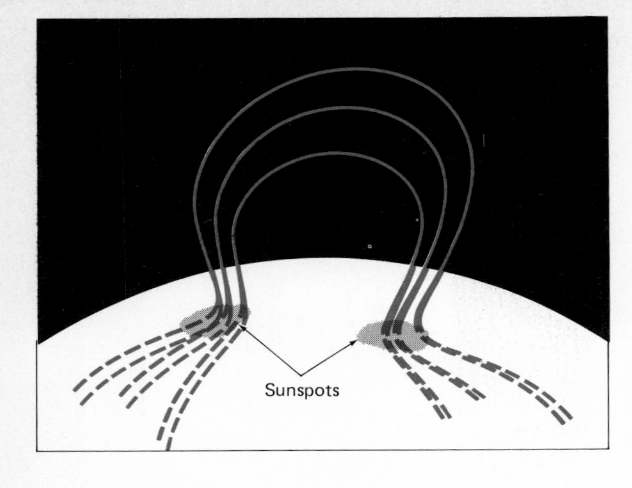

FIGURE 11.11
A sunspot model. The spots are pictured as regions where the magnetic field breaks through the surface of the Sun.

carried once around; for a point near the poles, this time is nearly 37 days. Differential rotation causes a slow, shearing motion in the solar gases, and below the surface the magnetic field becomes stretched out and, in some places, very much strengthened. Spots appear to be regions where a very strong magnetic field has erupted through the surface (see Figure 11.11). The strong fields may push rather dense clouds of gas up into the corona, causing prominences. Flares may result when the magnetic field in a small region that has built up to a very large strength by much twisting and squeezing suddenly snaps back to a less contorted condition.

The 11-year cycle of sunspots and other solar activity is not well understood. The differential rotation tends to stretch out the magnetic field in an ordered way along the direction of rotation, but random gas motions try to disrupt this alignment. There is a struggle between these opposing effects and, alternately, one and then the other gains the upper hand. It is possible that 11 years is the length of time that it takes for each of these effects to lose and then regain its supremacy.

IMPORTANT WORDS

Corona | **Convection**
Chromosphere | **Solar wind**
Sunspot | **Aurora**
Sunspot cycle | **Differential rotation**
Prominence
Flare

REVIEW QUESTIONS

1 Does the Sun have a solid or liquid surface?

2 What are the two parts of the Sun's outer atmosphere?

3 Why is the outer atmosphere best seen during a total eclipse of the Sun?

4 What is the approximate temperature of the corona?

5 Why is the corona so hot?

6 What is the solar wind?

7 What are sunspots?

8 Does the number of visible sunspots change in a regular fashion with time?

9 What are prominences?

10 What are flares?

11 Are flares more likely to occur at certain times than at others?

12 Is there evidence that solar activity might affect our weather?

13 Are magnetic fields an important part of solar activity?

QUESTIONS FOR DISCUSSION

1 If the corona keeps losing matter into space, why doesn't the Sun eventually lose its corona completely? Where do you think the top of the Sun's corona is?

2 How can we know that the Sun is rotating, since it has no solid surface?

3 Sunspot maxima occurred in 1958 and again in 1969. If you would like to see some brilliant displays of northern lights, in what year should you plan a trip to Alaska or another northern land?

4 Make a series of diagrams showing how differential rotation of the Sun stretches the magnetic field out in the direction of the rotation.

REFERENCES

A good general reference is *The Quiet Sun* by E. G. Gibson (NASA SP-303, U.S. Government Printing Office, Washington, D.C., 1973). Although parts of this text are advanced, most of it can be understood by the layman.

Additional readings related to the material we covered in this chapter are: "The Solar Corona" by J. M. Pasachoff [*Scientific American* 229, (October 1973), 68]; "The Rotation of the Sun" by R. Howard [*Scientific American* 232, (April 1975), 106]; "The Sun as Never Seen Before" by E. G. Gibson [*National Geographic* 146, (October 1974), 494]; "The Sun" by E. N. Parker [*Scientific American* 233, (September 1975), 42]; and "The Turbulent Sun" S. Lindsay, ed., [*Natural History* 85, (November 1976), 54].

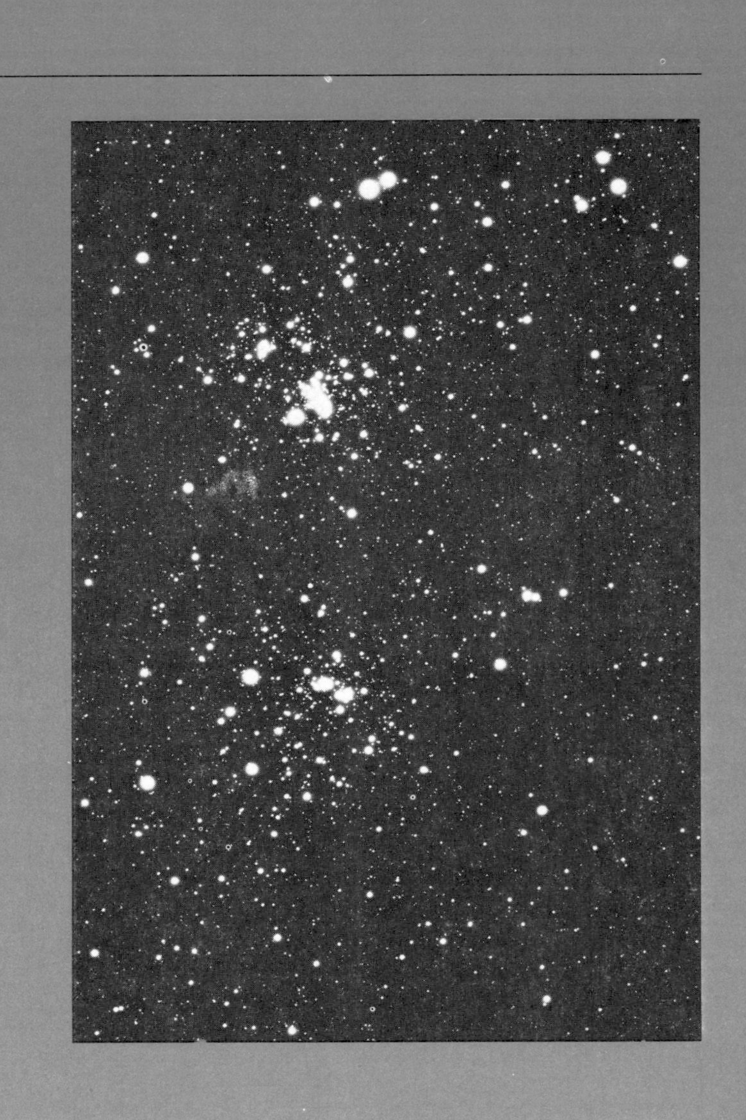

TWELVE
PROPERTIES OF THE STARS

The study of anything begins with the collection of data. In this chapter we consider what data concerning the stars the astronomers would like to have and how they obtain those data. We will also see how they organize their data in order to provide a better basis for understanding the stars.

12.1 Distance and luminosity

PARALLAX

The discussion of radiation in Chapter 3 suggested that astronomers might be able to obtain much information from the spectra of stars, that is, from knowing how much energy is received from the stars at different frequencies of the radiation. This suggestion is true, and we will consider the details later in the book. It may be surprising to learn that astronomers can also get important information simply by measuring the directions to the stars.

Directions to stars are most accurately measured by taking photographs of them with large telescopes. The telescopes are usually fixed to the Earth, so wherever the Earth moves, the telescopes are carried along, too. But when we view an object from different positions, we see it in different directions; this change in direction is known as *parallax*. Since the Earth has several motions, the direction to any given star will change with time in a rather complicated way.

The turning of the Earth on its axis causes the stars to appear to move across the sky each day, but it does not change the relative positions of the stars. The patterns or constellations outlined by the stars remain the same day after day. The daily motions of the stars, therefore, do not tell us much about the stars themselves.

The Earth moves around the Sun in a nearly circular path that is about 186 million miles across. At different times of the year we are in different parts of the orbit, and parallax should cause a yearly shift in the direction of each star (see Figure 12.1).

In part (a) of Figure 12.1, we see the orbit of the Earth around the Sun. Points A and B are positions of the Earth six months apart. W is a nearby star, while X, Y, and Z are more distant stars. Note how the direction to W changes through the angle k as the Earth moves around the Sun. Part (b) of the figure shows what a photograph of the stars would look like if it was taken while the Earth was at position A, and part (c) represents a picture taken from position B in the Earth's orbit. By comparing these two pictures, we see directly that the image of star W has shifted in position. Stars X, Y, and Z will

FIGURE 12.1
Parallax of a star. (a) The Earth's orbit around the Sun. (b) What the stars look like from position A. (c) What the stars look like from position B.

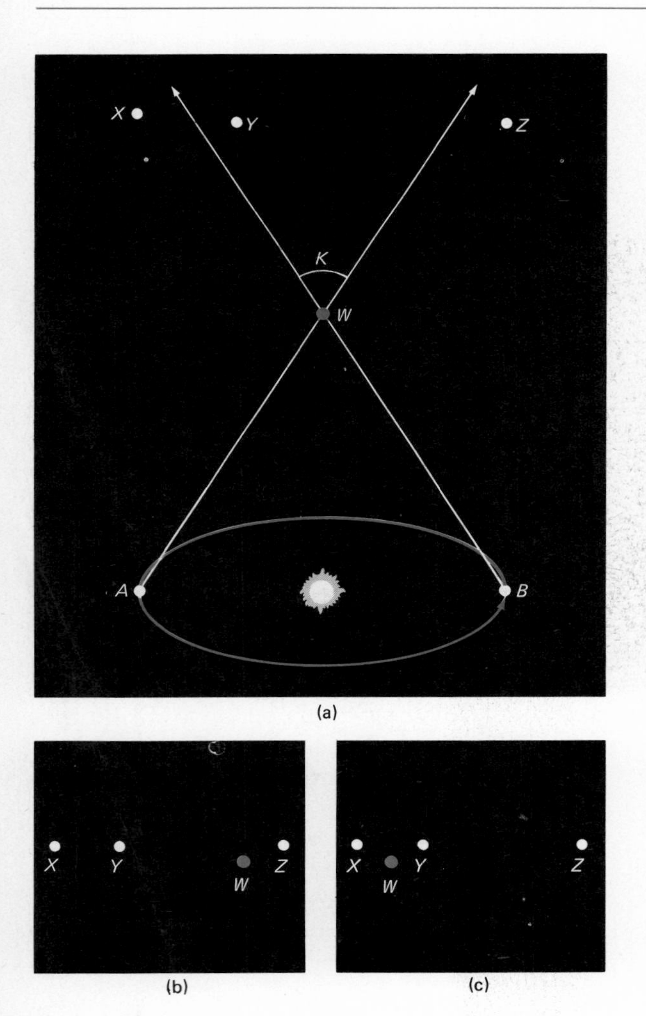

also have a change in direction, but the amount will be smaller because they are farther away. By measuring the positions of the image of star W on photographs taken six months apart, astronomers can find the size of angle k, which is the change in the direction of the star. The distance between the two positions A and B is known to be about 186 million miles; a simple trigonometric formula can then be used to find the distance to the star.

Applying this method is in practice quite difficult, because the stars are so far away that angle k is always extremely small. Measuring the shift in direction of even the closest star is like trying to measure the width of a penny from nearly 2 miles away: It can be done, but it isn't easy. Most stars are so far away that the parallax shift is too small to be measured. For these stars other methods must be used to find their distances.

THE NEARBY STARS

The star nearest to the solar system is named Alpha Centauri, and it is about 25 trillion miles away. This is typical of the distances between stars in the space around us. The stars are so far apart that it is difficult to keep an accurate perspective. The following review might help:

Light travels very fast, 186,000 miles in 1 second. It would take less than 1/7 second for light to travel completely around the Earth.

PLATE 1
A Chinese armillary sphere. The armillary sphere, not used much in modern times, illustrates the relationships among the more important astronomical circles in the sky. (Norton Scientific, Tucson, Arizona)

PLATE 2
The McMath Solar Telescope at Kitt Peak National Observatory. The vertical tower on this huge telescope is about 100 feet high. The telescope is not moved, but a movable mirror on top reflects sunlight down the slanting tower to a working area underground.

PLATE 3
The 1000-foot radio telescope at Arecibo, Puerto Rico. The Arecibo Observatory is part of the National Astronomy and Ionosphere Center which is operated by Cornell University under contract with the National Science Foundation.

PLATE 4
The Hawaiian volcano Halemaumau erupting in 1968. (Robert G. Strom)

PLATE 5
(Left) A spectacular lightning display on Kitt Peak, near Tucson, Arizona. Telescopes of Kitt Peak National Observatory and of Steward Observatory, University of Arizona, are visible. (© Gary Ladd 1972)

PLATE 6
(Right) Star trails recorded during a time-exposure photograph taken on Kitt Peak. The trails are all small parts of circles with the north celestial pole at the center. The north star Polaris is the bright star slightly to the left of the pole. Each trail covers about 1/48 of a full circle, which indicates that the exposure lasted for about 1/48 of a day, or about 30 minutes. (Steven A. Grandi)

PLATE 7

(Left) Twilight, the crescent Moon, and the planets Venus (top) and Mercury. Note that the dark side of the Moon is faintly visible from sunlight that was reflected off the daylight side of the Earth. The photograph was taken from Mt. Lemmon, near Tucson, Arizona. (Massino Tarenghi)

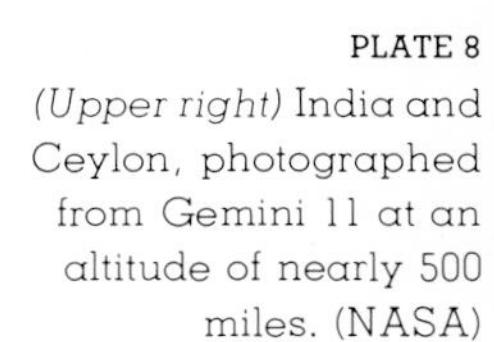

PLATE 8

(Upper right) India and Ceylon, photographed from Gemini 11 at an altitude of nearly 500 miles. (NASA)

PLATE 9

(Lower right) The Earth, photographed by Apollo 17. Africa is the main land mass visible. (NASA)

PLATE 10

(Above) A spiral pattern of clouds photographed over the Pacific Ocean by Apollo 9. (NASA)

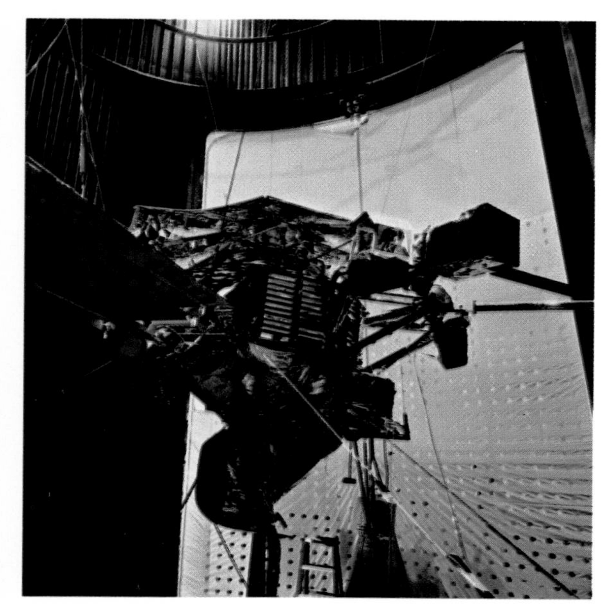

PLATE 11

The Mariner 10 spacecraft prior to its launching on November 3, 1973, for close encounters with Venus and Mercury. (NASA/JPL)

PLATE 12

(Upper right) The corona of the Sun, photographed during the total eclipse of the Sun on March 7, 1970. The photograph was taken with infrared radiation, which is invisible to the eye, and the colors are not realistic. (Norton Scientific, Tucson, Arizona)

PLATE 13

(Lower right) The Moon coming out of the shadow of the Earth during the eclipse of the Moon on January 30, 1972. A small amount of sunlight, primarily red light, survives passage through the atmosphere of the Earth and is bent into the shadow, thus illuminating the eclipsed Moon. The faint stars visible in the photograph are in the constellation Cancer. (Norton Scientific, Tucson, Arizona)

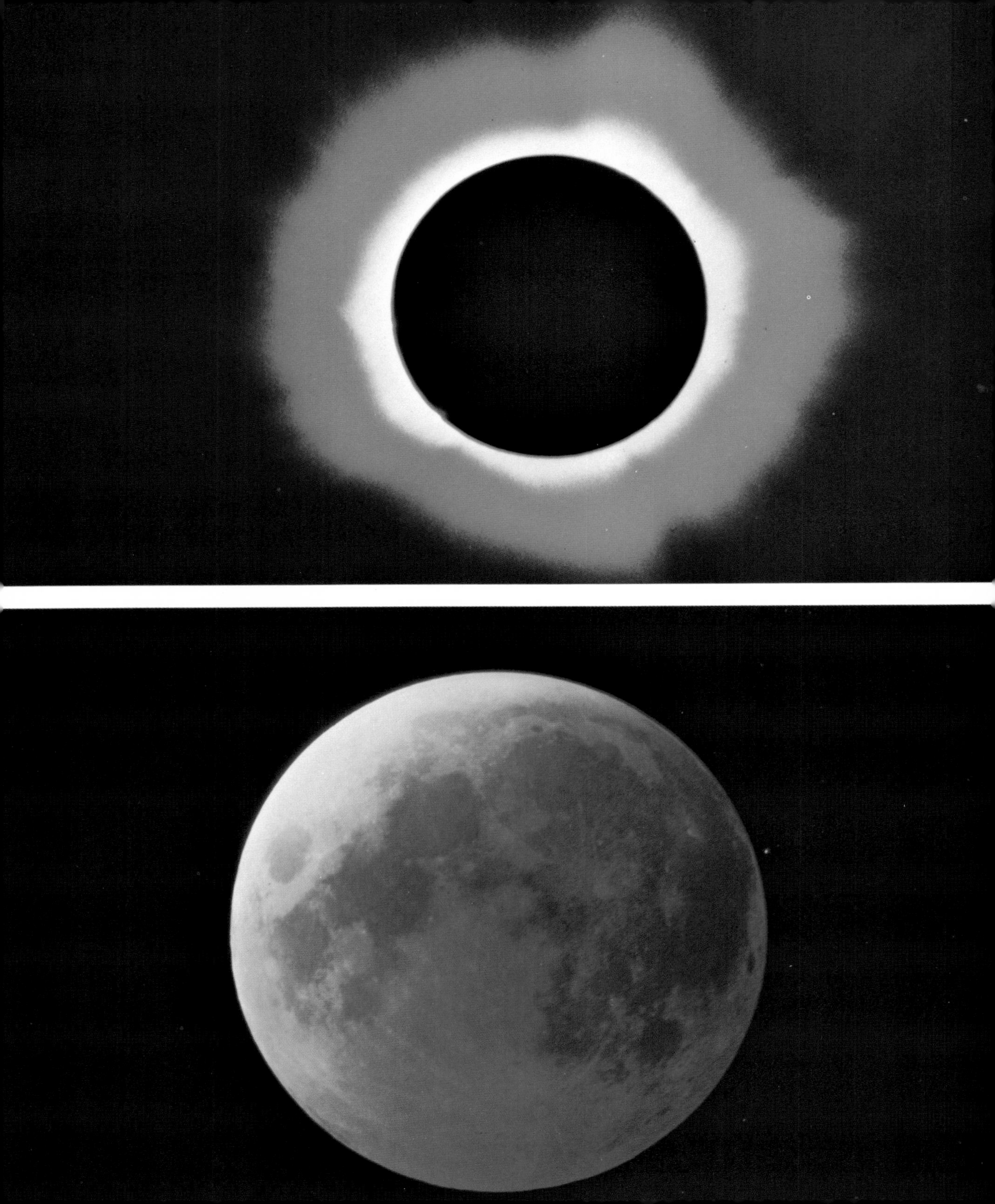

PLATE 14

A close-up view of the red sands and some small rocks on the surface of Mars. Taken by Viking 1 in 1976. (NASA)

PLATE 16

(Below) The planet Venus, photographed by Mariner 10 from a distance of about one-half million miles on February 6, 1974. The planet looks like it has the white clouds and blue oceans of the Earth, but this is deceiving. The photograph was made with invisible ultraviolet radiation, and the colors are not realistic. Both the white and the blue refer to regions in the atmosphere, and the surface of Venus cannot be seen. (NASA/JPL)

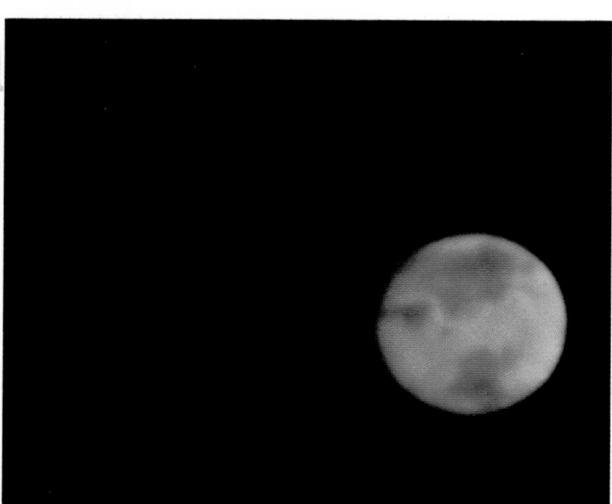

PLATE 15

(Above) Mars, photographed with the 61-inch telescope at Mt. Lemmon Observatory on May 24, 1969. This was about as good a view of Mars as was possible prior to the space age. (Lunar and Planetary Laboratory, University of Arizona, copyright pending)

PLATE 17

(Below) Sunset on Mars as seen by Viking 1 in 1976. It must be a lonely feeling to know that the nearest people are many millions of miles away. (NASA)

PLATE 18

(Above) Jupiter with its mysterious red spot. Photographed with the 61-inch telescope of the Mt. Lemmon Observatory. (Lunar and Planetary Laboratory, University of Arizona, copyright pending)

PLATE 19

(Below) Jupiter, photographed by Pioneer 11 in 1974. The photograph was computer-enhanced at the University of Arizona. (NASA/University of Arizona)

PLATE 20

(Overleaf, left) The Orion Nebula, photographed with the 200-inch telescope on Mt. Palomar. This is a large emission nebula with many hot, young stars in the vicinity. (Copyright by the California Institute of Technology and Carnegie Institution of Washington. Reproduced by permission from the Hale Observatories)

PLATE 21

(Overleaf, right) A region of thick interstellar gas and dust and a nearby star cluster in the constellation Serpens. The red color is due to a strong emission line of hydrogen in the interstellar gas. (Copyright by the California Institute of Technology and Carnegie Institution of Washington. Reproduced by permission from the Hale Observatories.)

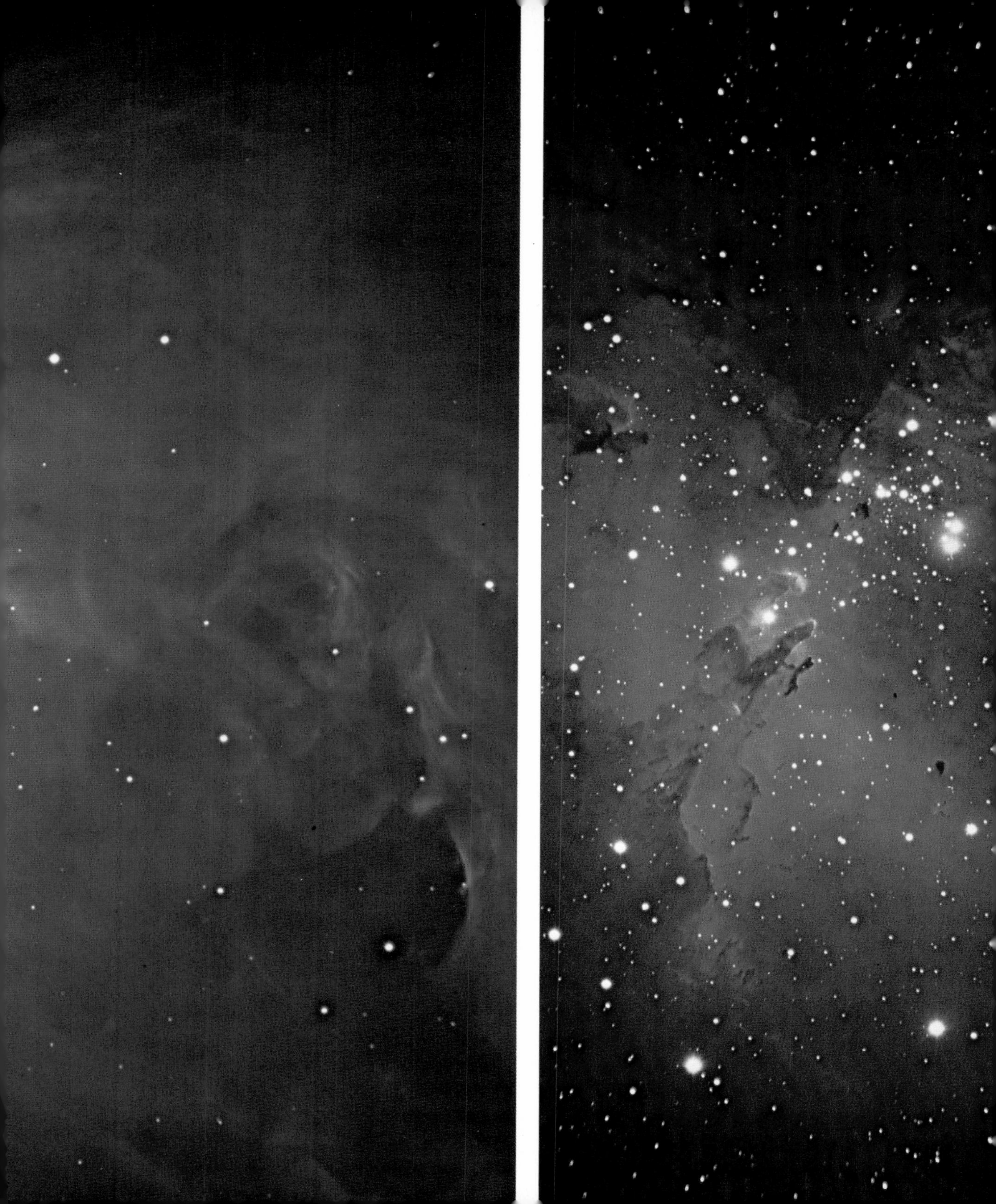

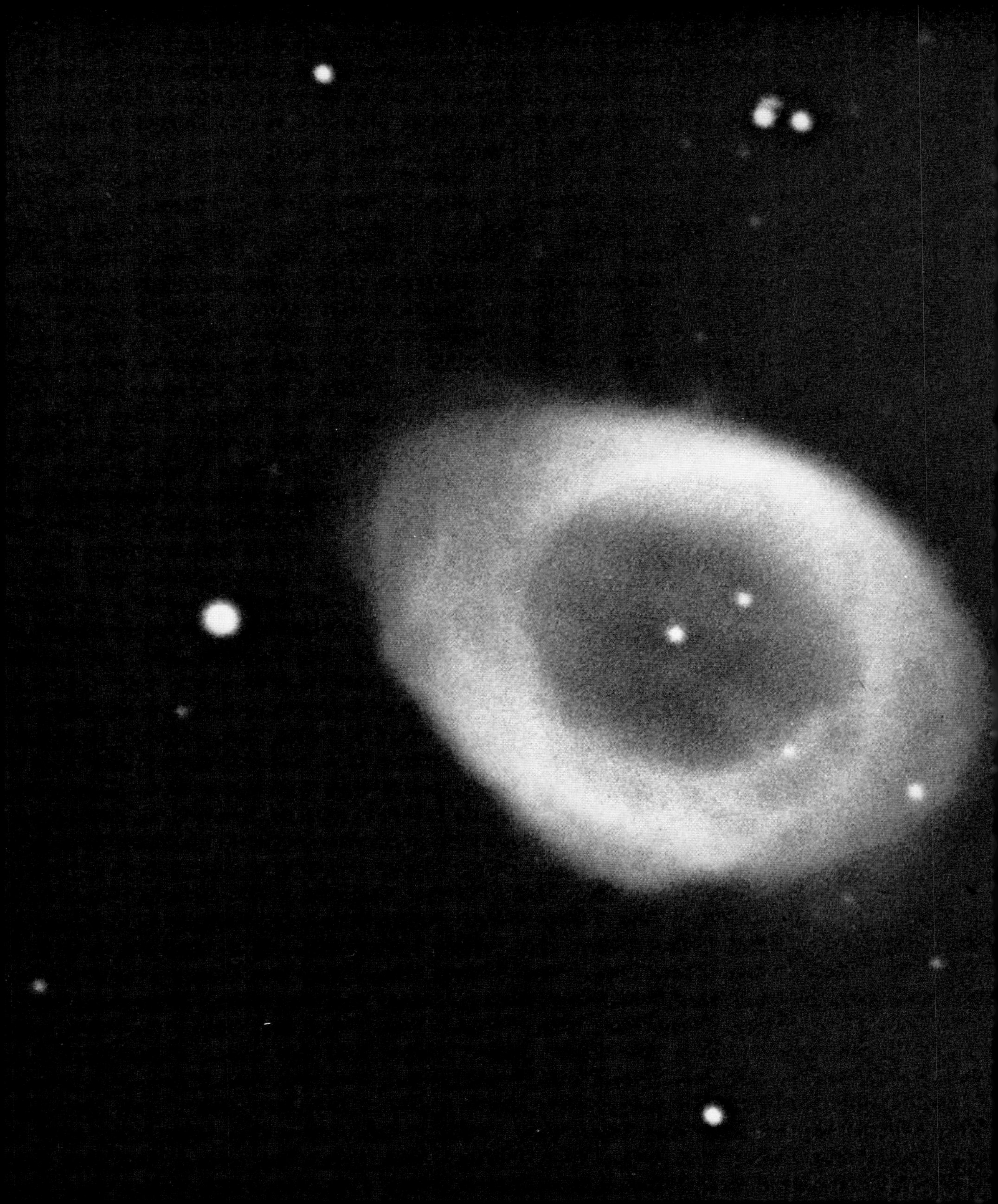

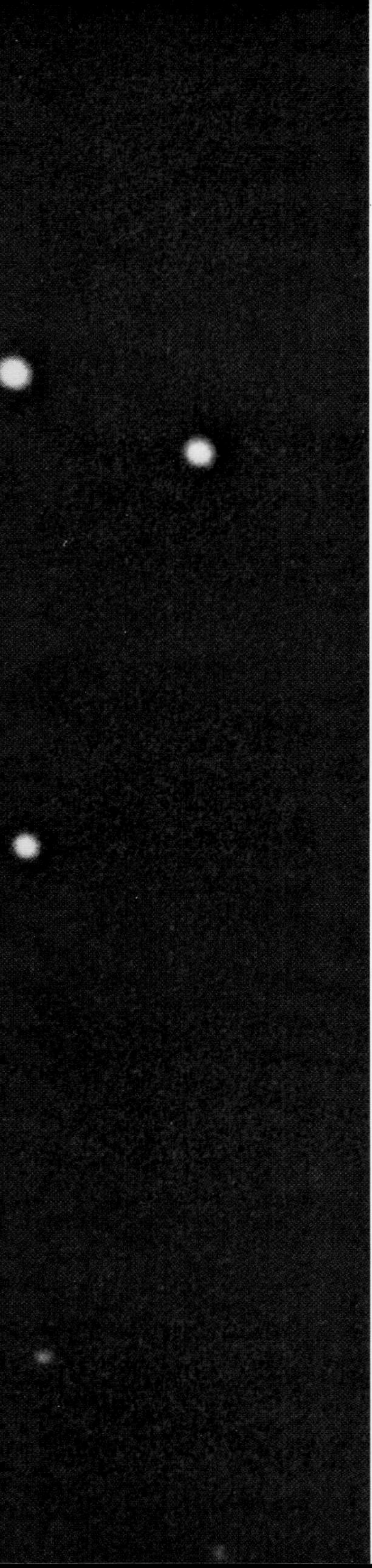

PLATE 23

Saturn, photographed by S. Larson at Mt. Lemmon Observatory in 1974. (Lunar and Planetary Laboratory, University of Arizona, copyright pending)

PLATE 22

The Ring Nebula in Lyra, photographed with the 200-inch reflecting telescope on Mt. Palomar. The yellow color is due to strong emission lines of oxygen in the gaseous shell around the hot, central star. (See also Figures 14.7 and 16.3.) (Copyright by the California Institute of Technology and Carnegie Institution of Washington. Reproduced by permission from the Hale Observatories)

PLATE 24
The Pleiades, a star cluster in the constellation Taurus. This rather young group of stars is still surrounded by the nebulosity out of which it formed. The blue light is starlight reflected by the dust grains around the stars, just as the blue daylight sky is sunlight scattered by particles and molecules in the atmosphere of the Earth. (Copyright by the California Institute of Technology and Carnegie Institution of Washington. Reproduced by permission from the Hale Observatories)

PLATE 25
(Overleaf) Comet Ikeya-Seki, a very spectacular comet visible in the morning hours during the fall of 1965. This photo was taken with infrared film, and the blue color of the sky is not realistic. (Norton Scientific, Tucson, Arizona)

Light crosses the distance between the Earth and the Moon in slightly more than 1 second, while the Sun is far enough away that it takes 8 minutes for light to cover that distance. Pluto, out at the edge of the solar system, must wait 5½ hours for sunlight to reach it. When we look at Alpha Centauri, we see light that was emitted by that star 4⅓ years ago—and that is only the *nearest* of the stars.

Light travels nearly 6 trillion miles in 1 year, and that distance is called a *light year*. Note that this is a distance, not a length of time: Alpha Centauri is then 4⅓ light years distant from us.

On a dark night, away from city lights and smog, the sky appears to be crowded with stars. There are so many stars that they seem to blend together, and you would think that they would be continually bumping into each other. But once again appearances are wrong. If someone were to hollow out the entire Earth and toss in a couple of ping pong balls, then the Earth would be filled with ping pong balls in about the same way that space is filled with stars. "Empty" is a pretty good way to describe space.

In Section 1.3 we estimated the distance to the star Sirius by assuming that Sirius is identical to the Sun. The result was about 100,000 times the distance to the Sun, which is the same as 1.6 light years. Modern measurements of the parallax angle of Sirius put its distance at 8.6 light years, about five times farther away than our estimate. Even so, there are only seven stars, including the Sun, that are closer to us than Sirius.

LUMINOSITY

These data about Sirius can be used to tell us how much energy Sirius radiates into space. How bright a star appears to us depends both on how much radiation it emits each second and on how far it is from us. The amount of energy a star gives off each second is known as the *luminosity* of the star. Luminosity is one of the very important properties of a star, and it does not depend on how far away the star is. A star that appears faint may be nearby and have a low luminosity, or it may have a large luminosity and be very far away.

In Section 1.3 we also noted that Sirius appears to us to be about 10 billion times fainter than the Sun, meaning that we receive about 10 billion times less energy per second from Sirius. How much of this apparent faintness of Sirius is due to its distance, and how much is due to its luminosity?

Sirius is 550,000 times further away from us than is the Sun. If it had the same luminosity as the Sun, therefore, it would appear 550,000 × 550,000 or 300 billion times fainter than the Sun does. (Remember that we must multiply the

FIGURE 12.2
Two photographs of the same region of the sky, the left being taken with red light and the right with blue light. Stars that are red in color will show a stronger image on the red photograph than on the blue one, while blue stars will show the opposite. Examine the two pictures and pick out stars you think are very cool and some you think are very hot. (Copyright by the National Geographic Society–Palomar Observatory Sky Survey. Reproduced by permission from the Hale Observatories)

distance by itself in order to find its effect upon the apparent brightness of an object.) But Sirius does not appear 300 billion times fainter than the Sun, only 10 billion times fainter. We conclude that Sirius is about 300 billion/10 billion or 30 times more luminous than the Sun. In this way we have calculated the energy output of Sirius.

We can find the luminosity of any star in exactly the same way if we know its distance and if we can measure how bright it appears in the sky. The apparent brightness of a star is not always easy to measure, for it must include radiation of all frequencies, not visible light alone. With the proper instruments and equipment, this measurement can be made for most stars with a high degree of accuracy.

12.2 Temperature, size, and mass

Most stars show continuous spectra with absorption lines and often emission lines superposed (see again Figures 3.8 or 4.9.) Often hundreds or thousands of absorption lines are observed. The discussion in Section 3.3 gave the background for understanding how we can obtain information from stellar spectra. Later we will consider the spectral lines; for now we will concentrate on the continuous background part of the spectra.

TEMPERATURE

Stars give off radiation because they are hot. The continuous part of the spectrum of a star does not arise from a solid or liquid but from a very thick gas. The frequency at which most of the energy comes out depends on how hot the star is. The atoms of a very hot star are moving around very fast, bumping into each other with much kinetic energy and producing many high-energy photons. A cooler star has atoms that move more slowly, the atomic collisions are less violent, fewer photons are emitted, and the photons have less average energy. If an astronomer can accurately measure what frequencies most of the photons have, he can tell how hot the star is. This really measures how hot the outer layers are, for the interior of a star is much hotter still.

If an object is heated to a temperature of 20,000°F and emits a continuous spectrum, then it will appear white to our eyes. It doesn't matter what the object is made out of, it will always be white in color. Such an object is emitting radiation in all frequencies, and the mixture of red, blue, green, and other colors of light is just right to appear white to us. The object also emits invisible radiation, such as infrared, x rays, and so on; but of course these do not affect the color we see.

Suppose the object is heated to the higher temperature of 50,000°F. Then the photons

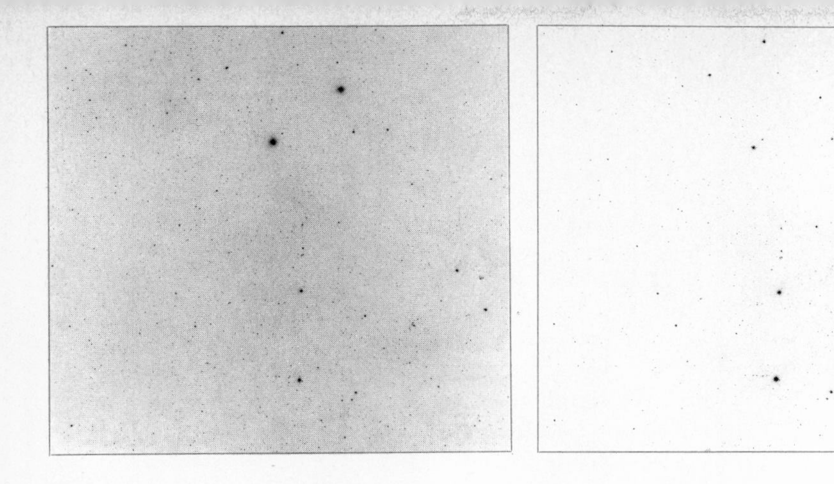

emitted will be, on the average, even more energetic and have higher frequencies than before. (Refer again to Figure 3.6.) Since blue light has more energy and a higher frequency than red light, the object will now radiate a greater fraction of its energy in blue light and a lesser fraction in red light. As a result, the object will now appear slightly bluish.

Now, let the same object cool down to 10,000°F, the temperature of the Sun. At this lower temperature, a smaller fraction of the photons are in high-energy blue light, and a larger fraction are in the lower frequency red and orange colors. The object will then have a slight yellowish color. (Sunlight is often said to be white in color but is more accurately somewhat yellow.) If the object is further cooled to 5000°F, it will have very few of its photons with energies as great as that of blue light. Most of the emitted energy will be in the invisible infrared region, while the rather small amount of visible light will be concentrated in the low-energy part, that is, red and orange. The object will appear deep red to the eye.

These arguments about color apply to any object with a continuous spectrum whose energy comes from internal heat; they do not apply, for example, to a thin gas whose spectrum consists only of emission lines. Neither do they apply to an object whose emission is due to the synchrotron process.

Astronomers have instruments which they can attach to telescopes that can be used to measure the colors of stars with very high accuracy. This is one way in which the temperatures of stars are found. There can be complications (as we will note in Section 13.1), but the process is usually straightforward. The important point is to understand why the coolest stars appear red, and the hottest seem slightly bluish, while those of intermediate temperature have yellow to white colors. Figure 12.2 is two photographs of the same region of the sky, one taken with red light and the other with blue light. A red-colored star will appear brighter in the red-light photo, while a bluish star will have a stronger image in the blue-light photo. Examine the photos and pick out some stars you think are very cool and some you think are quite hot.

SIZE

Consider a grain of sand and a mountain. If these two objects have the same temperature, then each molecule of the grain of sand has the same kinetic energy, on the average, as each molecule that makes up the mountain. The heat energy of a body is the total kinetic energy of all the molecules and atoms of which it is composed; the much larger mass of the mountain means that it has much more total heat energy than the grain of sand. Both of these objects are emitting radiation due to their temperature (most of this radiation will be in radio and infrared frequencies unless they are quite hot, of course), and the much larger mountain will emit more energy than the grain of sand; the mountain will have a larger luminosity. Because they have the same temperature, however, they will both have the same "color," that is, they will have the same fraction of their total luminosity at the various frequencies. In other words, if the mountain has one part in a thousand of its energy in red light, the grain of sand will, too.

Now let's heat the grain until it is much hotter than the mountain. The average molecule in the sand now has more kinetic energy than one in the mountain, and it will produce more photons and higher energy photons than a molecule in the mountain. If we measured the spectra of the two bodies, we would find that the grain would be bluer than the mountain.[1] The size of the mountain, however, will more than make up for its lower temperature, so the mountain will still have the greater luminosity in that it will radiate more total energy per second.

[1]When I used the words "redder" and "bluer" in a talk one time, a purist informed me that there are no such words. "Something is red or is not red, so one object cannot be redder than another." Astronomers, however, measure the color of a star and assign a number to it. It is accepted terminology to say that one star is redder than another, indicating a lower temperature, if the numerical value of its color is greater; the other star is said to be bluer than the first.

The temperature of a mountain or a grain of sand or a star fixes the relative amounts of high- and low-frequency radiation that it emits, and the temperature also fixes the total amount of energy emitted by each square inch of the surface. The higher the temperature, the more total energy emitted through each square inch of surface area. The luminosity, therefore, depends both on temperature and on how much surface area there is. This information is used by astronomers to find the sizes of the stars.

Let's examine the star Sirius again. Its distance has been measured as 8.6 light years. We used this distance and the known apparent brightness to find that Sirius is about 30 times more luminous than the Sun. The color of Sirius is almost pure white, corresponding to a temperature of 18,000°F. Any star (or any other object for that matter) of this temperature radiates about 10 times as much energy per second through each square inch of its surface area as does the Sun. Luminosity is the energy radiated through the whole surface each second. If Sirius and the Sun had the same amount of surface area, then Sirius would have 10 times the luminosity; instead it has 30 times the luminosity. It follows that Sirius must have three times the surface area of the Sun. A

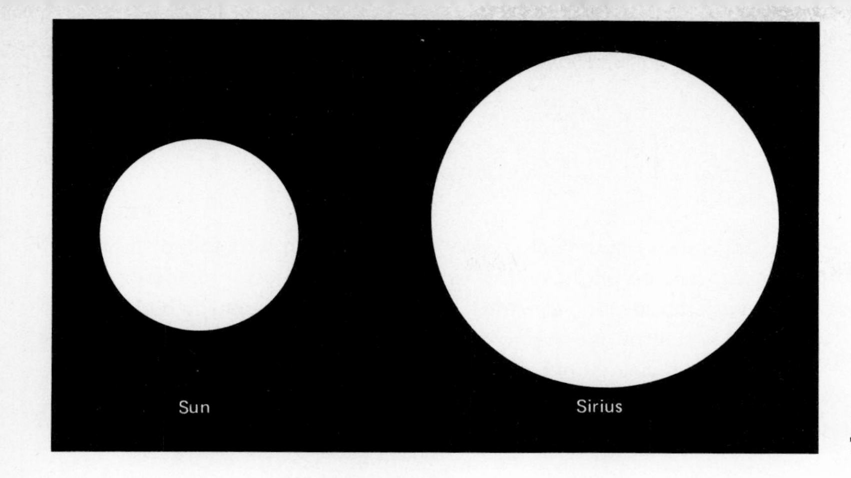

FIGURE 12.3
The Sun and Sirius.

ball having three times the area of the Sun must have 1.7 times its diameter. Since the diameter of the Sun is known to be about 860,000 miles, we find that Sirius must have a diameter of 1.7 × 860,000 or 1,500,000 miles. It is thus about 1½ million miles from one side of Sirius to the other (see Figure 12.3).

Using this idea we can make an analogy to a farmer planting a crop of corn. The "temperature" of his field tells us how many bushels of corn he gets to the acre, while the "luminosity" of the field is the total yield. If we know both the yield per acre and the total yield, we can easily figure out how large a field the farmer has. If we know the temperature and the luminosity of any star, we can easily figure out how big the star must be in order to produce that much luminosity.

COMPOSITION

What are the stars made of? Already we have seen how to answer this, at least partly. The absorption and emission lines that are superposed on the continuous spectra of stars are the fingerprints of the atoms and molecules that make up the stars. Certain lines occur at frequencies that hydrogen atoms strongly absorb and emit. If these lines are observed in the spectrum of a star, we know that the star must contain hydrogen. The same applies to the other atoms and molecules in the star. (We will take up the question of the chemical composition of the stars in greater detail in Section 14.1.)

MASS

Probably the most important property of a star is its mass. The mass of an object tells how much matter it contains. If an object is on the surface of the Earth, we can find its mass by measuring how much it weighs. Since we cannot bring a star to the Earth and weigh it, we must find another way to determine its mass.

The planets and other members of the solar system are held together by the gravitational force due to the mass of the Sun. Astronomers know many examples of two or more stars that are held together in much the same way by their mutual gravity (see Figure 12.4). Two stars that move in orbits around each other, like the orbits of the planets around the Sun, are called *binary stars*. The gravitational forces that binary stars exert on each other depend on their masses. By carefully measuring the orbits in which the binary stars move, astronomers can calculate how much mass the stars must have for their gravities to hold them in orbit. In this way they have determined the masses of a number of stars. Many other stars are single or else have binary orbits for which the required measurements cannot be accurately made. For these, the masses cannot be

FIGURE 12.4

The eclipsing binary star WW Cygni. A photograph was taken, then the telescope was moved slightly, and another photograph was taken. Each star shows two images side-by-side. The star near the center, which is WW Cygni, has a stronger image on the left than on the right. This "star" actually consists of two stars that are too close together to be seen separately. As they move in their orbits around each other, one star periodically passes in front of the other, blocking off its light and causing the stars to appear dimmer. (The Hale Observatories)

directly determined, and astronomers must resort to less direct means. (We will consider these means in Section 12.3.)

SOME INTERESTING STARS

When the properties of a large number of stars are measured, we find that there are all kinds: large ones and small ones, hot ones and cool ones, massive ones and light ones, dense ones and tenuous ones, bright ones and faint ones. The largest stars are at least several hundred times bigger than the Sun, while the smallest are only a few miles in diameter. Stars with temperatures of several hundred thousand degrees are known, while there are also very cool infrared stars with temperatures under 1000°F. The most luminous stars are the *supergiants* that emit energy at a rate up to a million times greater than the Sun. There is probably no limit on how faint a star can be, but astronomers have measured some that have luminosities a million times fainter than that of the Sun. The most massive stars have perhaps 60 or 70 times as much matter as the Sun, while the least massive have about 1/12 the mass of the Sun.

Betelgeuse (BEETLE-jooz) is a bright red star in the constellation Orion, and Antares is the name of another red star in the constellation Scorpius. (See the star maps in Appendix 7.) These two well-known stars are very red, indicating that they have low temperatures. They are both very luminous supergiants, however, so they emit large amounts of energy each second despite their very low temperatures. The only way they can do this is to have extremely large surface areas. These two stars are not only much larger than the Sun, they are *even* larger than the orbit of Jupiter around the Sun. Actually these huge supergiants are nearly vacuums; they are over 1000 times less dense on the average than the air around us at sea level. Of course, near their centers the matter is much more closely packed than this (see Figure 12.5).

The star Sirius is again worth considering. It is not alone in space but part of a binary star system. A small, faint star named Sirius B moves in an orbit around Sirius. Sirius B has a luminosity 20 times fainter than the Sun, but it is far hotter than the Sun.[2] The only way such a low luminosity can come from such a hot star is for the star to be very small; Sirius B is only about 7000 miles in diameter, about the size of the Earth. Hot and very small stars like Sirius B have been called *white dwarfs* (although today the term has a somewhat different

[2]Data based on J. L. Greenstein, J. B. Oke, and H. L. Shipman, *Astrophysical Journal* 169 (1971), 563.

FIGURE 12.5
The sizes of some stars. Note that Aldebaran is the largest star in the first part of the figure and the smallest star in the second part. (a) Small to moderately large stars. (b) Some very large stars.

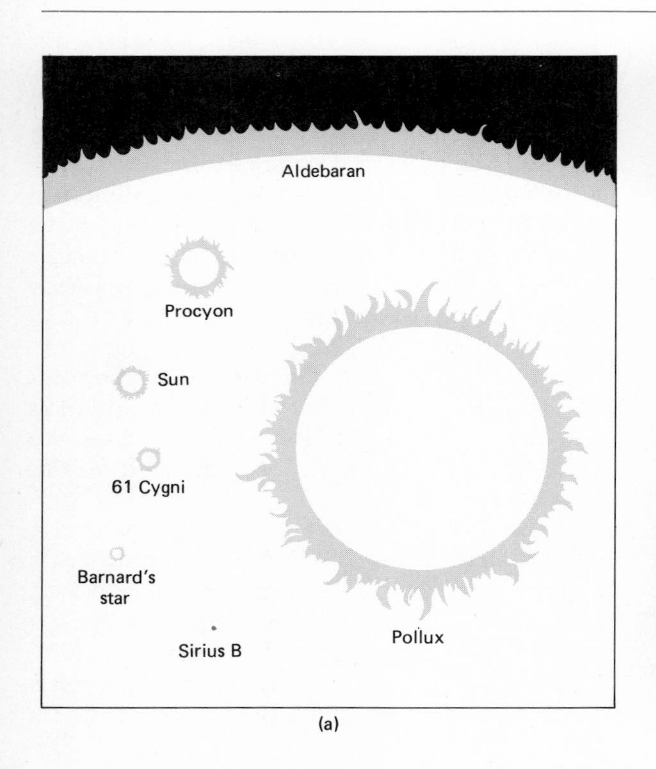

(a)

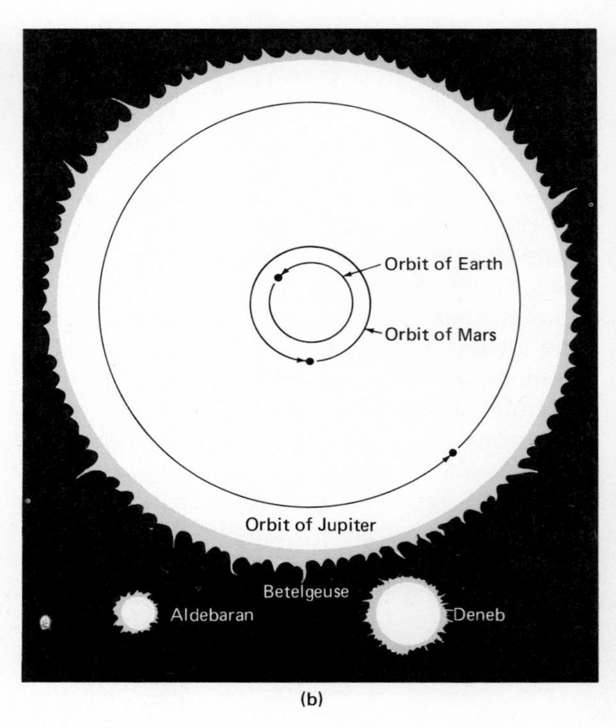

(b)

meaning as we will see in Chapter 16). Even more interesting is that Sirius B has about the same mass as the Sun. Since that mass is squeezed into a sphere only 7000 miles in diameter (remember that the Sun is 860,000 miles in diameter), the matter in Sirius B is compacted to an extremely high density. A teaspoonful of matter from Sirius B has a mass of about 10 tons.

But even this is mild in comparison with the strange stars discovered in 1967 called *neutron stars*. Neutron stars also have about the same mass as the Sun, but they are only a few miles in diameter. A teaspoon of this kind of

star would have a mass of some 10 *billion* tons—the mass of a respectable mountain.

These numbers and others that could be stated are so fantastic that maybe you wonder whether astronomers really are making them up. Perhaps astronomers live in a theoretical dream world—a world in which stars can be larger than the orbit of Jupiter and can have a mountainful of material compacted down to a teaspoonful in size but a world that has nothing to do with the reality we live in. Yet it is not by dreaming that astronomers have come up with numbers like these, it is by applying the same principles of observation, experimentation, and reasoning that have been so successful in expanding science here on Earth. Perhaps the dreamers are those who believe that reality is limited to the narrow range of conditions that we find in our everyday lives, while the whole Universe lies unseen and unknown around them.

12.3 Hertzsprung-Russell diagrams

STELLAR DATA

The supergiants Betelgeuse and Antares are extremely unusual types of stars. Most stars are more modest in their properties, the Sun being closer to what might be called a typical star. But even the Sun has some pretty impressive statistics: nearly a million miles across, a mass equivalent to 2 billion billion billion tons (2×10^{27} tons), and the energy which the Sun radiates in just a few hours is enough to vaporize the entire Earth.

Many people find facts like the above interesting, but not everybody. Walt Whitman wrote:

When I heard the learn'd astronomer,
When the proofs, the figures, were ranged in columns before me,
When I was shown the charts and diagrams, to add, divide, and measure them,
When I sitting heard the astronomer where he lectured with much applause in the lecture-room,
How soon unaccountable I became tired and sick,
Till rising and gliding out I wander'd off by myself,
In the mystical moist night-air, and from time to time,
Look'd up in perfect silence at the stars.

We all have heard dull lectures, but one that is so bad as to make the audience sick is indeed worthy of note. Whitman makes an important point: There is (or should be) a lot more to astronomy than lists of facts and figures.

Scientists are not satisfied with only knowing many facts. What they want to do is study the facts to see what underlying patterns they reveal. Then they want to be able to see how those patterns could be explained as logical consequences of some basic physical principles. This is the general way in which scientists try to come to an understanding of their subject.

There are many ways in which the data about stars could be studied. The most common way is through an *H-R diagram*. The letters stand for E. Hertzsprung and H. N. Russell, two astronomers who first used a diagram of this type. In a Hertzsprung-Russell or H-R diagram, each star is represented by a point, and the position of the point depends on the temperature and luminosity of the star (Figure 12.6).

FIGURE 12.6
The Hertzsprung-Russell diagram.

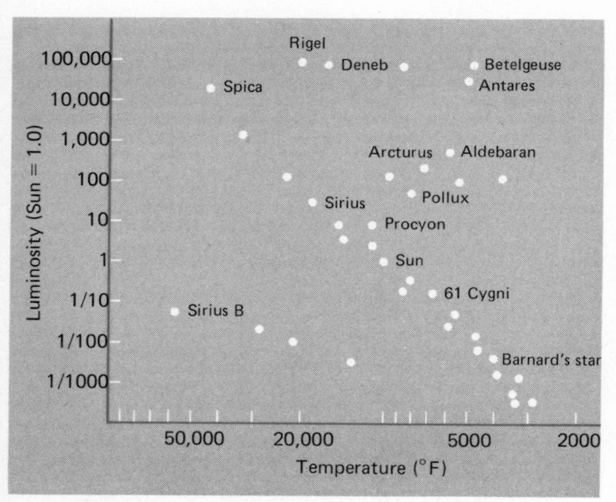

In the Figure 12.6 each dot represents a star. The temperature corresponding to any dot is found on the scale along the bottom: The dots further to the left are for hotter stars, and those to the right are cooler. The luminosity is read from the scale on the left; the higher up in the diagram a dot is, the greater the luminosity of the corresponding star. Note that in this figure the luminosity is given on the scale such that the luminosity for the Sun is 1. Thus a star whose point in the H-R diagram is level with the number 100 has 100 times the luminosity of the Sun.

HEIGHT-WEIGHT DIAGRAM

An H-R diagram for stars is somewhat similar to Figure 12.7, which is a height-weight diagram for people. In this diagram each dot represents a person, and the position of a dot in the diagram depends on the height and weight of the person it stands for. The taller persons have dots farther to the left in the diagram, while heavier people are represented higher up in the diagram. If you see a dot in the upper middle of this diagram, you know the person it

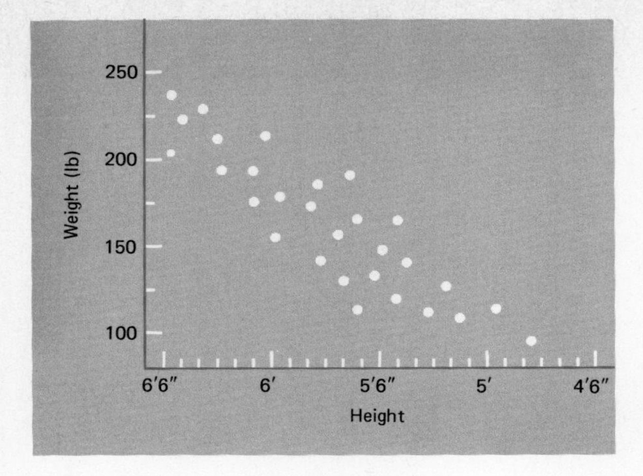

FIGURE 12.7
Height-weight diagram.

represents is of medium height and quite heavy. In the same way, a point in the upper middle part of the H-R diagram stands for a star of high luminosity and medium temperature. If you know two things about a person, namely, the height and weight, then you can plot that person in the diagram of Figure 12.7. If you know the luminosity and temperature of a star, you can put a point for that star in the H-R diagram.

It appears that Figure 12.7 simply gives the heights and weights of some people. Actually, it is considerably more informative than that. The points in the figure form a pattern that runs from the lower right side of the diagram to the upper left. This pattern suggests that a very tall person is likely to also be fairly heavy, while a short person probably weighs less than average. A 200-pound man is much more likely to be 6 feet tall than 4 feet tall. It is easy to say that this is obvious, that you don't need Figure 12.7 to tell you this. But to make the analogy to the H-R diagram, we must think of what information the figure would convey to an intelligent creature who had never seen a human being. The approximate relation between height and weight shown by the figure would be an important clue to such a creature about how people are put together and, perhaps, how they age. Can we get similar information about the stars from the H-R diagram?

DIFFERENT TYPES OF STARS

The H-R diagram shows that most stars have points that are in a band that runs from the lower right to the upper left side of the diagram. This band is called the *main sequence*. As the figure shows, the Sun and Sirius are both main sequence stars. Why do most stars fall on the main sequence? Are stars that are not on the main sequence peculiar in some way? (These are important questions relating to the structure and aging of stars that will be considered in some detail in Chapters 15 and 16.)

The stars that have points at the very top of the H-R diagram are called supergiants. They are the most luminous of all stars. We have already met two supergiants—Betelgeuse and Antares. Figure 12.6 showed two more: Rigel (RYE-jel), a hot star in the constellation of Orion that is near in the sky to Betelgeuse, and Deneb, a bright, moderate-temperature star in Cygnus. Stars somewhat less luminous than the supergiants but still above the main sequence in the H-R diagram are called *giants*. Examples of giants are Pollux, one of the bright stars in Gemini the Twins, and Arcturus, the bright orange star in Boötes. Arcturus can be located in the sky by a backward extension from the handle of the Big Dipper. Giants are rather rare stars, and supergiants

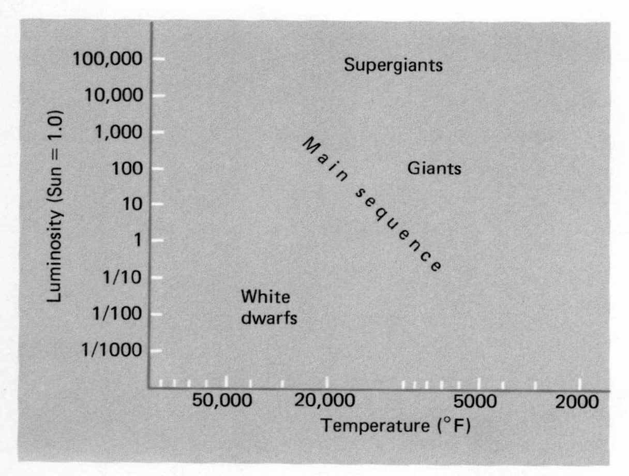

FIGURE 12.8

Schematic H-R diagram.

are extremely rare. These stars are so luminous that they easily stick out in a crowd, while the fainter but much more common main sequence stars are easily overlooked unless they happen to be quite close to us. It is also observed that the cooler, fainter main sequence stars are much more abundant in space than the hotter, more luminous main sequence stars.

The stars in the lower left center of the H-R diagram, below the main sequence, are of low luminosity and are moderately hot. They are the white dwarfs. There are many other special types of stars which have points that represent them in special parts of the H-R diagram. Most of these are very rare stars that will not concern us further (although some of them will be mentioned at an appropriate time later in the book). Figure 12.8 shows the names given to some of the more important types of stars and where they occur in the H-R diagram.

USING THE H-R DIAGRAM

If we already know something about people, then we can use height and weight to guess some other characteristics of a person. For example, the sex and overall appearance of a person are suggested by height and weight, and we can sometimes infer something about the strength and general physical condition from these statistics. Our guesses may be wrong in individual cases, of course, but they are more reliable than guesses without basis.

If we know where a star belongs in the H-R diagram, we can also tell other things about the star. Some of these things we don't need to guess about. We saw in the last section, for example, that the size of a star can be calculated if we know its luminosity and temperature; therefore the position of a star in the H-R diagram fixes its diameter as well. It is also observed that the mass of a star is closely related to its position in the H-R diagram. (There are some complicating features to this relation but we will discuss them later.)

We can't tell everything about a person from height and weight alone, and we can't tell everything about a star from its luminosity and temperature alone. Specifically, we cannot tell what materials the star is made of from its position in the H-R diagram. Locating a star in the H-R diagram does not answer all questions about that star, but it is a good start.

Suppose we see a star at night and decide to find out where in the H-R diagram it belongs. The direct way to do this is to measure its temperature and luminosity. It is usually not very difficult to find the temperature of a star since all we need to know is its color; few measurements are as simple as getting the color of a star, so this part is usually not hard. The luminosity, however, can be a much greater problem. The difficulty here is being able to find the distance accurately. Measuring stellar parallaxes (as mentioned in Section 12.1) requires much work, and it is good only for the stars that are nearby. Thus it is not easy to make the direct measurements that are needed in order to know where a star occurs in the H-R diagram. Fortunately it is usually not necessary.

The spectral lines greatly simplify this problem. Most stars have hundreds or thousands of lines in their spectra. Which specific lines show up and how strong they appear depend on the composition of the star, of course, and also on its temperature and luminosity. With some practice, then, an astronomer is able to predict the temperature and luminosity of a star from the general appearance of its spectral lines. This is the easy way to find out in which part of the H-R diagram a star belongs.

IMPORTANT WORDS

Parallax	**White dwarf**
Light year	**Neutron star**
Luminosity	**H-R diagram**
Binary star	**Main sequence**
Supergiant	**Giant**

REVIEW QUESTIONS

1 What is parallax?

2 Why are parallax angles of stars so difficult to measure?

3 How far away is the star nearest to the solar system?

4 Are stars really close together in space?

5 What is the luminosity of a star?

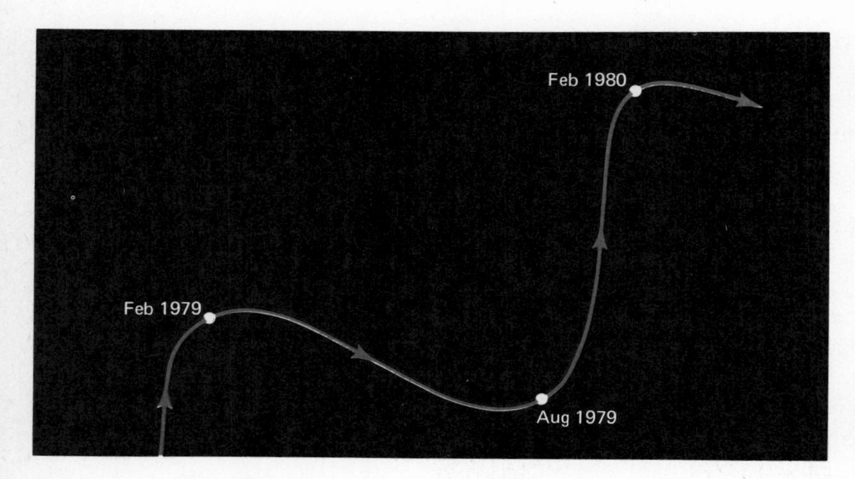

FIGURE 12.9
A star's motion on the sky.

6 What measurements do we need in order to find the luminosity of a star?

7 Why do stars give off radiation?

8 How can we tell the temperature of a star from its color?

9 If a star is deep red in color, is it probably very hot or very cool?

10 How can astronomers calculate the size of a star if they know its temperature and luminosity?

11 For what types of stars can astronomers determine the mass?

12 How big are the largest stars?

13 What is most unusual about white dwarfs and neutron stars?

14 What do scientists want in addition to mere lists of facts?

15 What is an H-R diagram?

16 What is the main sequence?

17 What are giant stars?

18 Are giants and supergiants common types of stars?

QUESTIONS FOR DISCUSSION

1 Suppose that we could set up telescopes on the planet Pluto. How would this help us to measure more accurately the parallaxes of stars? What problems would this introduce in addition to the obvious one of how to get to Pluto and back?

2 If we very accurately measured the position of a star in the sky, we would find that it follows a wavy path, as in Figure 12.9. Can you explain why the path is in this form?

3 What is meant by the statement that the Sun is 8 light minutes from the Earth?

4 A welding torch has a much higher temperature than a forest fire, yet the heat from the forest fire can be felt much farther away. Why? Compare the spectra of the torch and the fire.

5 Use the same reasoning that you just did in question 4 to compare a white dwarf with a red supergiant. Which star has the higher temperature? Which star can be seen from farther away? Why?

6 An astronomer photographs the spectrum of a star, examines it for a while, and then says, "Aha! I recognize this spectrum as the kind produced by a type of star 10 times more luminous than the Sun." Can you explain how this information might be useful in trying to find out the distance to the star? (A measurement of how bright the star appears would also be useful for this.)

REFERENCES

The material in this chapter is covered in detail in almost any popular book on astronomy. Encyclopedia articles are another important source of information.

THIRTEEN
BETWEEN THE STARS

The space between the stars is much more nearly empty than any vacuum we can create on Earth, yet the matter that does exist there is quite important in astronomy. Some of the reasons for this are discussed in this chapter, and the subject will remain an important part of our discussions throughout the remainder of the book.

13.1 Gas and dust

INTERSTELLAR MATTER

A number of years ago an observatory invited the public to come and look through one of its telescopes. When a certain woman had her turn at the telescope, she reportedly told the astronomer that she would like to see infinite space. The astronomer is supposed to have replied, "Simply look between the stars, madam."

The sight between the stars might or might not be a good view of "infinite space" (maybe closing your eyes would give an even better view), but the astronomer's statement could have been prompted partly by the belief that the space between the stars is completely empty. Today astronomers know that there is a significant amount of matter between the stars known as *interstellar matter*.

The interstellar matter averages a few atoms per cubic inch of space. The air we breathe is like a stone wall by comparison: There are about 100,000,000,000,000,000,000 or 10^{20} atoms and molecules in each cubic inch of air. The interstellar matter does seem almost nonexistent by comparison.

Interstellar matter is not spread evenly between the stars but tends to clump together into clouds. These clouds are called "nebulae." Figures 13.1 through 13.4 show a few of the many known nebulae. One of the best known is the Orion Nebula of Figure 13.1. This is a hazy patch that can be seen with the naked eye on a dark night. (Also see the star maps in Appendix 7.)

There are two kinds of interstellar matter: *gas* and *dust*. The gas consists of individual atoms or molecules of various substances, primarily hydrogen and helium. The dust is made of very tiny particles or grains, an individual grain having a size that is typically $1/10,000$ to $1/100,000$ inch. As small as these dust grains are, each is still large enough to contain billions of atoms. The total mass of the gas in space is about 100 times the mass of the dust, which shows that there must be an immense number of gas atoms for each grain of dust.

TYPES OF NEBULAE

When a cloud of interstellar gas occurs near a very hot star, the radiation from the star will heat the gas and make it glow. This type of

FIGURE 13.1

The Orion Nebula. This is a mass of glowing gas, heated by the very hot stars embedded in it. (Lick Observatory photograph)

FIGURE 13.2
A complex mixture of interstellar dust, gas, and surrounding stars in the constellation Serpens. Slightly to the upper left of center is the star cluster Messier 16. (Lick Observatory photograph)

FIGURE 13.3
(Upper photo) The nebula IC 2944, photographed at Cerro Tololo, Chile, by B. J. Bok. (The Cerro Tololo Inter-American Observatory)

FIGURE 13.4
(Lower photo) The Trifid Nebula in the constellation Sagittarius, photographed by B. J. Bok. (Steward Observatory, University of Arizona)

FIGURE 13.5
A reflection nebula caused by a thick dust cloud surrounding the star Merope in the star cluster Pleiades. (Also see Plate 24.) (The Hale Observatories)

FIGURE 13.6
Dark nebulae in the constellation Sagittarius. The apparent holes in the star background are due to clouds of dust blocking off the starlight. Photographed by B. J. Bok with the 90-inch reflector. (Steward Observatory, University of Arizona)

cloud is known as an *emission nebula;* the Orion Nebula of Figure 13.1 is an example. Inside the nebula is a very hot star that is causing the gas to shine. The spectrum of an emission nebula shows the emission lines from the gas in the nebula. By identifying the lines which are observed, astronomers can determine the composition of the gas in much the same way as is done for stars from their spectra. If the nearby stars are not hot enough, the gas will not be made to glow, and there will be no emission nebula. Only stars with temperatures above about 40,000°F are hot enough to cause emission lines in the nebula. If part of the gas is directly in the line of sight to a star, it will produce absorption lines in the spectrum of the star. These lines are in addition to the absorption lines produced by the star itself.

Whenever the gas is clumped together into a relatively high density cloud, it is likely that the dust is clumped together also. Thus a nebula will usually contain a concentration of both gas and dust. While the gas produces absorption lines in the spectra of background stars and is made to glow if there are very hot stars around it, the dust is very good at simply reflecting or scattering photons from the stars in all directions. If a star is in or in front of a nebula and is too cool to produce an emission nebula in the gas, it still might be bright enough to light up the dust around it. This produces a *reflection nebula,* so called because the dust grains are rendered visible by their reflecting some of the starlight into our direction. Figure 13.5 is a photograph of a reflection nebula. The nearby stars are not hot enough to cause the gas to shine, but they are luminous enough to illuminate the dust around them. The spectrum of a reflection nebula is the same as that of the star that causes it, modified by the reflection properties of the dust particles.

Since the dust particles very efficiently scatter starlight that shines on them, they make it difficult for light to continue on its way through the dust. If there is very much dust between us and a star, it can block off essentially all of the visible light and many of the other frequencies, making the star nearly invisible. Figure 13.6 shows a *dark nebula,* which is a region where the dust is thick enough to cut off the light from the more distant stars. Such "holes" look like regions where stars do not exist, but they are actually concentrations of the dust grains. Figure 13.7 shows how the dust can form either a bright reflection nebula or a dark nebula, depending on the positions of the stars around or behind it.

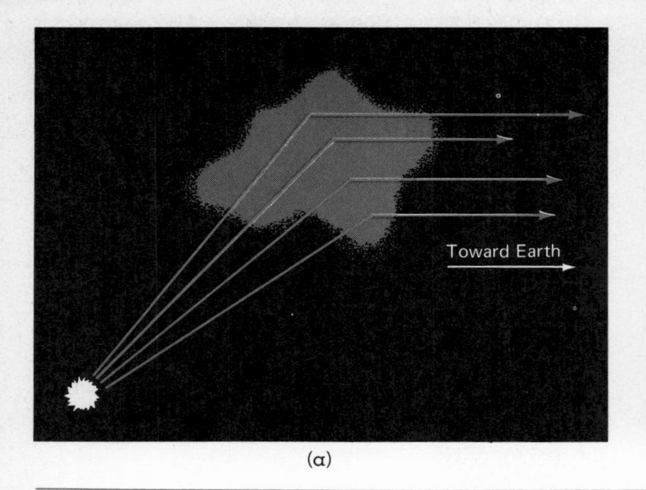

(a)

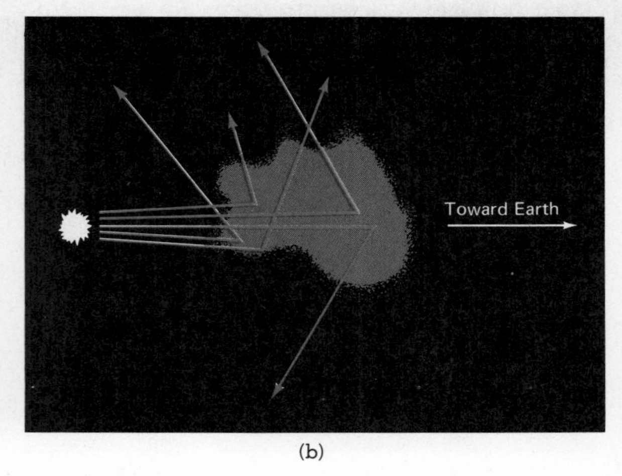

(b)

FIGURE 13.7

Reflection and dark nebulae. (a) A dust cloud scatters some starlight toward the earth. (b) A dust cloud scatters some starlight away from the earth.

OBSCURING POWER OF THE DUST

It is difficult to believe that the interstellar dust, thin as it is, is capable of dimming and even cutting off completely the light from distant stars. In space there is an average of only one tiny dust grain for every 100 million cubic feet; yet when a photon travels a sufficiently great distance, the amount of dust it must traverse is enough to seriously impede its progress.

Consider the following example: A thin sheet of paper is able to block off most of the light that tries to pass through it. Now suppose that we cut this sheet of paper into many tiny pieces and spread them out in a long column of the same area that will stretch from here all the way to the nearest star, Alpha Centauri (see Figure 13.8). (We will ignore the engineering problems involved in getting this done!) The paper is now spread extremely thin, but it is still as capable as it was before of blocking off light that tries to pass through it. If space were filled with paper to this extent, we would have trouble seeing even this nearest star. Fortunately the dust in space is millions of times more tenuous than the paper in this example, but it still manages to make it difficult to detect the more distant stars in our galaxy. A few grains of sand, ground into a fine powder and spread over a volume the size of the Earth, would illustrate how tenuous the dust really is.

INTERSTELLAR REDDENING

Interstellar dust has another important effect on the radiation of stars. The dust scatters high-frequency photons more readily than low-frequency ones. Blue light has a higher frequency than red light, so it is more easily scattered by the dust. Radiation that passes through a great amount of dust loses much of its blue light, while a smaller amount of its red light is scattered; thus the radiation becomes redder in color. This is the same type of process that causes the light from the Sun and Moon to be reddened by the scattering in the Earth's atmosphere. The air molecules scatter blue light better than red light. Light coming from most directions in the sky is scattered light, predominantly blue; light coming directly from the Sun or Moon is light that has survived the trip through the atmosphere without being scattered, and is predominantly red. The effect is strongest when the Sun or Moon is near the horizon since a greater amount of the Earth's atmosphere is then involved.

The reddening of starlight by interstellar

FIGURE 13.8
If an ordinary sheet of paper were cut into tiny pieces and spread in a long column of the same area, stretching all the way to a bright star, that star would be very effectively blocked from view. This shows how empty space must be for us to be able to see the stars.

dust can cause some confusion for astronomers. As we have seen, the color of a star is an indication of its temperature. If the color of a star can also be changed by the dust between us and the star, the effects of the dust must be taken into account if astronomers are to find an accurate temperature. There are stars that appear very red in color, and without further information, we would suppose that they were very cool stars, yet their spectra show lines that indicate very high temperatures. These stars are hot, but their radiation has been reddened very much by the dust in space.

The dimming power of dust on stellar radiation must also be taken into account when astronomers try to find other properties of the stars. For example, it was stated in Section 12.1 that the apparent brightness of a star depends on both its luminosity and its distance. We must add that it also depends on how much dust is between us and the star. The interstellar dust can be quite a nuisance to astronomers who are trying to determine the properties of stars.

RADIO WAVES

While regions where dust is abundant can block off higher frequency radiation, the low-frequency radio waves are not much affected by the dust. Fortunately the interstellar gas produces a large number of spectral lines in radio frequencies, so the gas can be studied without interference from the dust. The most important radio line occurs at a wavelength of 21 centimeters (frequency $= 1.4 \times 10^9$ cycles per second) and arises from cold hydrogen gas. Since most of the interstellar gas is cold hydrogen and does not produce any radiation in visible light, radio telescopes that can detect the 21-centimeter line of hydrogen are essential for our finding the properties of most of the interstellar gas.

INTERSTELLAR MOLECULES

A large number of interstellar molecules have been discovered in recent years, and new ones are being added at a rapid rate. In most cases the discoveries have come from the identification of absorption or emission lines in the ultraviolet and radio regions of the spectrum. The most common molecule in space is the hydrogen molecule consisting of two hydrogen atoms. Some of the molecules in the interstellar gas are moderately complicated organic molecules, such as formaldehyde and methyl alcohol, and their existence in space caught astronomers by surprise.

A molecule is formed when two or more atoms become bound together by the electric forces that act between them. The correct kinds of atoms must have a chance collision or must somehow find each other before the molecule can be formed; interstellar space has such low densities that collisions between atoms are extremely rare, especially collisions involving the less common atoms, such as carbon, oxygen, and so on. (Remember that the gas is almost all hydrogen and helium.) Furthermore, it seems that the high-energy ultraviolet radiation from hot stars would be sufficient to destroy most molecules even if they could be formed; despite these arguments, astronomers have observed that the molecules do exist in space.

The preceding arguments could also be applied to the interstellar dust. An extremely large number of very rare atoms must come together to form a single dust grain, and why shouldn't the ultraviolet rays also destroy the grains? How the grains and the molecules are formed is not known for certain. The rather new science of space chemistry has come into being to attack these interesting and puzzling problems.

Astronomers have suggested that the dust particles are formed in the outer parts of very cool stars, where matter is much more dense than in space. The dust is then somehow blown off the stars and into space. Many cool stars are actually observed to be losing mass into space. The dust grains are probably not easily destroyed by the small amount of ultraviolet radiation in interstellar space; only when it is very close to hot stars is the dust vulnerable to being torn apart by the high-energy photons. It is interesting that those molecules that are most sensitive to being destroyed by high-frequency radiation seem to occur only within very dense dust clouds, where they are well shielded from that radiation.

It is sometimes suggested that the organic molecules observed in space indicate the possible presence of some form of life in space. This conclusion is certainly not warranted by the available data. Although nothing is known for certain about the formation of life, we already noted in Section 10.2 that it is scientifically reasonable to assume that given the proper ingredients, the right physical conditions, and a long enough period of time, life will develop spontaneously through natural chemical processes. It is also reasonable to suppose that some of the organic molecules observed in space are in the important early steps of this chemical evolution of life. If part of

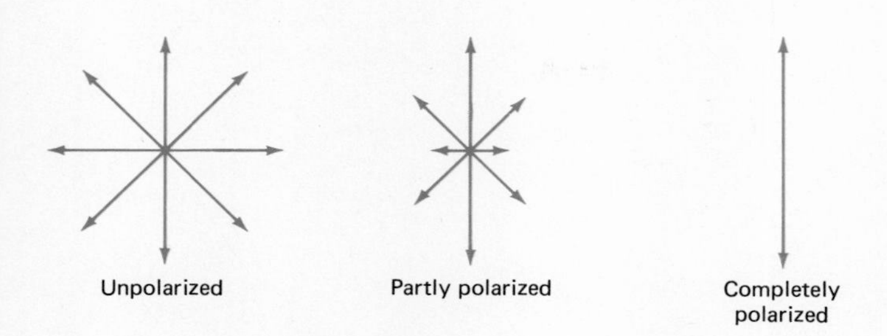

FIGURE 13.9
Polarized radiation. The arrows represent the vibrations of an electromagnetic wave traveling into or out of the page.

the complicated chain of life can be forged in the hostile environment of interstellar space, it is not difficult to believe that the complete chain can be formed under the much more favorable conditions that might prevail on the surfaces of some planets.

13.2 Magnetic fields

Starlight that passes through very much interstellar dust is observed to be *polarized*. Radiation is polarized if the electric field vibrations are greater in one back-and-forth direction than in others (see Figure 13.9). When radiation is polarized, it means that the direction of polarization, that is, the direction of maximum vibration, has something special about it that other directions do not have.

How can the dust polarize the radiation passing through them? We have already seen that dust scatters high-energy radiation more easily than low-energy radiation. In order to explain the polarization observations, we must assume that the dust can scatter radiation that is vibrating in a certain direction more readily than that vibrating in other directions. In other words, the dust grains must be able to distinguish different directions. A sphere looks the same from any direction, therefore a spherical grain could not scatter differently according to the direction of vibration of the radiation. Thus, we conclude that many of the dust grains cannot be perfect spheres. A long, narrow grain does have one direction that is special for it: the direction in which it is long. Such an elongated grain can scatter radiation in a way that can produce the observed polarization.

Having elongated grains is not enough. If the grains are pointed in different directions, each will try to polarize the radiation in its own way. Many grains with random orientations will cancel out the polarization, leaving unpolarized radiation. The dust particles, therefore, must have a tendency to line up with similar orientations. The only mechanism known that could cause this is a magnetic field. Thus we are led to the conclusion that much of interstellar space contains a significant magnetic field.

A magnetic field can exert a force on magnetic materials as well as on any moving electric charge. A magnetic field is a form of energy, and the exchanges of energy between it and matter are in some cases very important. The interstellar fields are very weak; the magnetic field at the surface of the Earth is about a million times stronger. Yet the enormous volume covered by the interstellar field indicates that a significant amount of energy is involved.

A magnetic field plays an essential part in the production of synchrotron radiation. As we noted earlier, this kind of radiation is emitted by electrically charged particles having very large energies and moving in a magnetic field. A few nebulae, as well as many galaxies and quasars, are strong sources of synchrotron radiation. This shows that magnetic fields and large supplies of energetic particles are common throughout the Universe. It has been known for many years that very energetic particles called "cosmic rays" are bombarding the Earth from space. In Chapter 16 we will discuss where these particles might get their high energies.

IMPORTANT WORDS

Interstellar matter
Interstellar gas
Interstellar dust
Emission nebula
Reflection nebula
Dark nebula
Polarized radiation

REVIEW QUESTIONS

1 How does the density of the interstellar matter compare with that of the air we breathe?

2 What is a nebula?

3 What are the two kinds of interstellar matter?

4 What type of star is needed to produce an emission nebula?

5 What is a reflection nebula?

6 How does the dust sometimes produce a reflection nebula and sometimes a dark nebula?

7 How does the dust cause starlight to be reddened?

8 What is the most common molecule in space?

9 Where are the dust particles probably formed?

10 Is there life in space?

11 What is polarized radiation?

12 Why do we believe the dust grains have irregular rather than spherical shapes?

13 What part do magnetic fields play in polarizing starlight?

14 What part do magnetic fields play in the production of synchrotron radiation?

QUESTIONS FOR DISCUSSION

1 Compare the color of the radiation that comes to us from a reflection nebula and the color of the star causing the nebula.

2 Measurements show that two stars are at the same distance from us and that they have the same brightness as seen from the Earth. There is a dust cloud between us and the first star that blocks off one-half the light passing through it, while the second star is in a clear region of space. Which star has the greater luminosity, and by how much?

3 It is possible that when the Galaxy formed, the stars were made first; later the interstellar matter could have been provided by mass that was lost by the stars. Another possibility is that the interstellar

matter came first, and the stars formed from it afterward. Finally, it is possible that the stars and interstellar matter were formed together at the beginning. Can you think of any "scientific" reasons for believing that one of these three possibilities might be more or less likely than the others?

4 Refer to question 6 in the Questions for Discussion at the end of Chapter 12. Suppose that unknown to the astronomer, there is a dust cloud dimming the starlight by a significant amount. How might this affect the astronomer's idea of the distance to the star?

REFERENCES

The interstellar gas is described by J. S. Miller in "The Structure of Emission Nebulas" [*Scientific American* 231 (October 1974), 34]. A particular gaseous nebula is the subject of "The Gum Nebula," by S. P. Maran [*Scientific American* 225 (December 1971), 20]. The dust is covered by G. H. Herbig in "Interstellar Smog" [*American Scientist* 62 (March–April 1974), 200].

The molecules in space are the topics of: "Chemistry between the Stars" by L. E. Snyder, in *The Emerging Universe*, edited by W. C. Saslaw and K. C. Jacobs (University of Virginia Press, Charlottesville, 1972); "Interstellar Molecules" by P. M. Solomon [*Physics Today* 26 (March 1973), 32]; "Interstellar Molecules" by B. E. Turner [*Scientific American* 228 (March 1973), 50]; and "Molecules Between the Stars" by R. McCray [*Natural History* 83 (December 1974), 72].

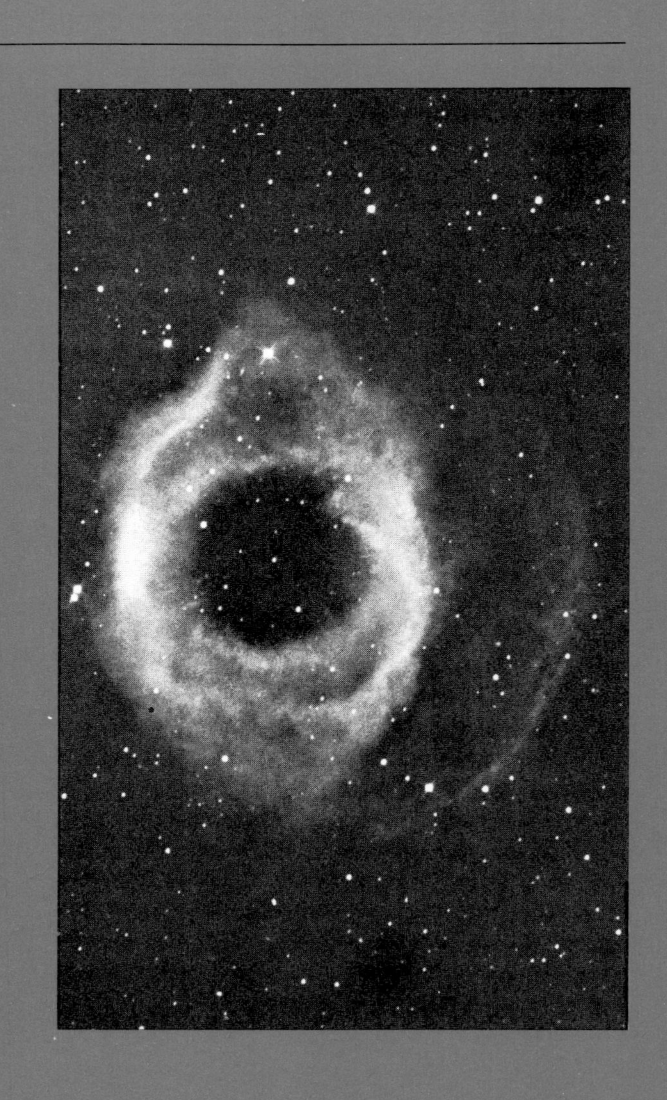

FOURTEEN
HOW STARS ARE PUT TOGETHER

Here we change our approach to studying the stars from one based mainly on observational work to one that is primarily theoretical in nature. Trying to understand the structure of a star takes us to many areas that cannot be easily checked by direct observation, areas where we must extrapolate our knowledge beyond what can be duplicated by experiments in the laboratory. This does not mean that we are moving into the field of speculation, but we should be even more cautious than usual about accepting results as final. Some ways to check our theories with observational data do exist, of course, and they are an extremely important part of our study.

14.1 The atmosphere

RADIATION PASSING THROUGH A STAR

Most stars are hot enough at their centers to cause nuclear reactions to occur there. These reactions emit gamma rays in large numbers, and through them large quantities of energy are liberated in the cores of stars. The stars must pass on this liberated energy through their interiors to their surfaces, from which it is radiated into space. If this were not the case, the energy would keep piling up at the center, causing the temperature and pressure to build up indefinitely. The very thin outer shell of a star from which radiation is emitted into space is called the *stellar atmosphere* (see Figure 14.1).

A gamma ray emitted at the center of a star will travel only a very short distance before it is absorbed by an atom. Very quickly this atom will release its newly gained energy by emitting one or more photons in random directions. This process of absorption of photons followed by the emission of other photons is repeated a very large number of times until the energy finally makes its way to the surface and leaks out. The radiation escapes through the surface at the same rate that new gamma rays are produced at the center, so the star as a whole keeps an accurate balance between gains and losses of energy. If a photon could come straight out from the center without being stopped by absorption, it would only take a few seconds more or less, depending on the size of the star, for it to escape into space. However, the absorptions and emissions cause the successive photons to zigzag back and forth many times before the energy finally reaches the atmosphere and escapes, and in this way it takes millions of years for the radiation to find its way out of the star. The sunlight we see today was released in the form of gamma rays at the Sun's center millions of years ago.

Each absorption and emission tends to

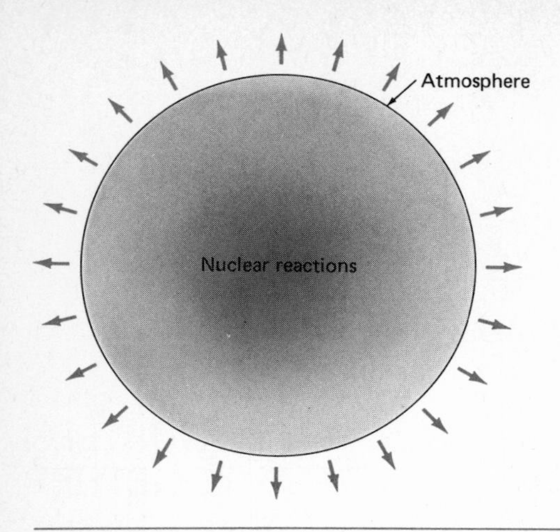

FIGURE 14.1
The atmospheres of most stars are very thin outer shells.

FIGURE 14.2
Spectra of four different stars. The temperatures given to the left are in degrees absolute; convert to degrees Fahrenheit by multiplying by 9/5. The symbols identifying the lines are H (hydrogen), He (helium), Ca (calcium), and Fe (iron). (The Kitt Peak National Observatory)

change one high-energy photon into a number of photons with less energy, keeping the total energy fixed. The reason for this is that as we move away from the core of a star, the temperature drops. We have already seen that the lower the temperature of matter, the more its emitted radiation is concentrated in lower energy photons. As the radiation gradually diffuses away from the center of a star, absorptions and emissions slowly change it into a larger number of photons with lower average energy. This is the way that the Sun breaks up the relatively small number of gamma rays and x rays in its core into a very large number of visible-light photons, plus a smaller number of ultraviolet, infrared, and radio photons that escape from its surface.

The atmosphere of the Earth is composed of the gases that exist above the solid and liquid surface. This definition is not useful for a star, since it is gaseous throughout. The atmosphere of a star is the name given to the outer regions emitting the radiation that escapes into space. The atmosphere is a very thin outer shell containing a negligible fraction of the matter in the star. Yet the atmosphere is a very important part of the star, as all radiation emitted by a star must be screened through that atmosphere. Stellar radiation carries with it the fingerprints of the atmosphere.

STELLAR SPECTRA

When we discussed the temperature of a star in Chapter 12, we were talking about the temperature of its atmosphere. We also saw in Chapter 12 how the spectral lines of a star are the source of a tremendous amount of information. These lines are formed in the atmospheres of stars, so they are really giving us information about the outer layers.

Figure 14.2 shows photographs of the spectra of a number of stars. Some of the spectral lines are identified by the symbols shown. Note that lines due to the element helium show up only in very hot stars, hydrogen lines are most prominent in fairly hot stars, and the molecule titanium oxide is strong only in very cool stars.

Astronomers once thought that these differences in spectra of stars are caused by differences in what makes up the stars. They believed that certain stars were composed mainly of iron and other metals, other stars were almost pure hydrogen, and so on. In the 1920s M. N. Saha showed that differences in temperature and, to a lesser extent, pressure in the atmospheres could account for most of the observed differences in spectra.

Let's see how this happens. Helium atoms do not absorb and emit visible-light photons unless they are made very hot. Because of this,

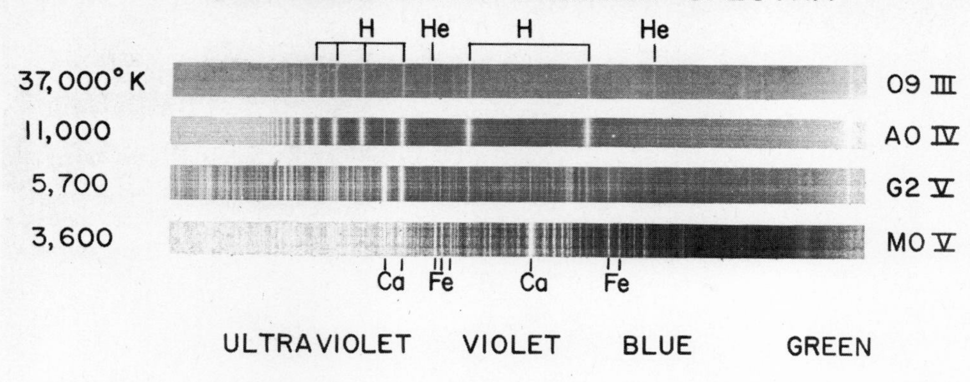

a cool star will not show helium lines in its visible-light spectrum even though there is helium in the star. Titanium oxide is a molecule that is formed when atoms of titanium and oxygen come together; this molecule will break apart into atoms of titanium and oxygen if the temperature becomes very high, so only very cool stars have spectral lines of this molecule. Hydrogen atoms do not absorb or emit visible light unless the temperature is fairly high; however, if the temperature becomes too high, the hydrogen atoms are torn apart or ionized, and the ionized hydrogen atoms can no longer absorb or emit spectral lines. Hydrogen lines are strongest in stars having a temperature of about 20,000°F. In much hotter stars, the hydrogen is mostly ionized, and the lines are weaker. In much cooler stars, the collisions between atoms are not violent enough for hydrogen to absorb or emit well in the visible region, and again the lines are weaker.

CALCULATING THE PROPERTIES

We should not conclude that there are no composition differences among stars; there are some important variations in the chemical make-up of stars. If we want to find the composition of the atmosphere of a star, the temperature and pressure effects must be taken into account. When a spectral line of a certain element is very weak in a star, that element might have a low abundance in that star; it might also happen that the temperature and pressure in the atmosphere are such that most of the element cannot absorb or emit that line. It is up to the astronomer to find out which is the correct explanation.

The basic problem of stellar atmospheres is as follows: Can we find a temperature, a pressure,[1] and an abundance for each chemical element such that each spectral line will have the same predicted strength as is actually observed in a star and such that the correct color of the star is also predicted? Suppose that we guess a certain temperature and pressure for the atmosphere of a star. Then if we further guess how much of a certain element is in the star, we can use theory to calculate how strong a spectral line of that element should be. This calculation of course requires some rather complicated physics and mathematics which we will not worry about here, but the important point is that one can determine theoretically how strong a line should be. If the theoretical line strength does not agree with observations, we can change the assumed abundance of the

[1] More accurately, we must find a series of temperatures and pressures, since these quantities vary from the top of the atmosphere to the bottom.

element and carry out the calculations again. In this way we can find an abundance for which the theory and observations agree. Figure 14.3 illustrates what a highly magnified spectral line might look like for three different assumed abundances of the element producing it.

How can we be sure that we started with the correct temperature and pressure and that the calculations were carried out properly? There are a number of ways to check this; the most important is the agreement of the theoretical calculations with the observations for every line, not just one or two. If we can find one temperature, one pressure, and one iron abundance that give good results for all iron lines, for example, we can feel fairly sure about it. If the same temperature and pressure will also work for all sodium lines, for a certain assumed abundance of sodium, for all calcium lines, and so on, then we gain confidence that our determinations are highly accurate. This is how the physical conditions and the chemical composition of the atmosphere of a star are found.

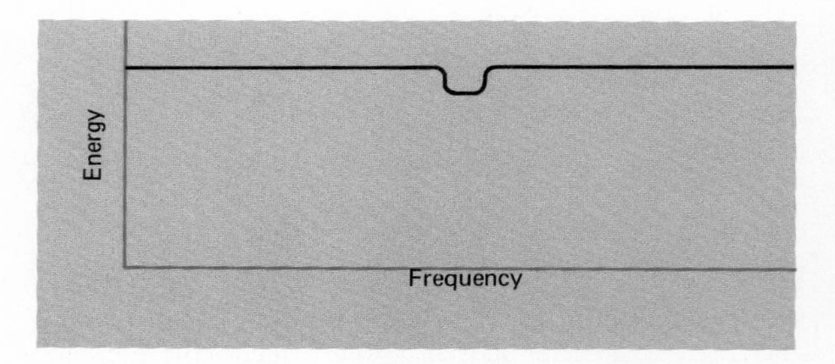

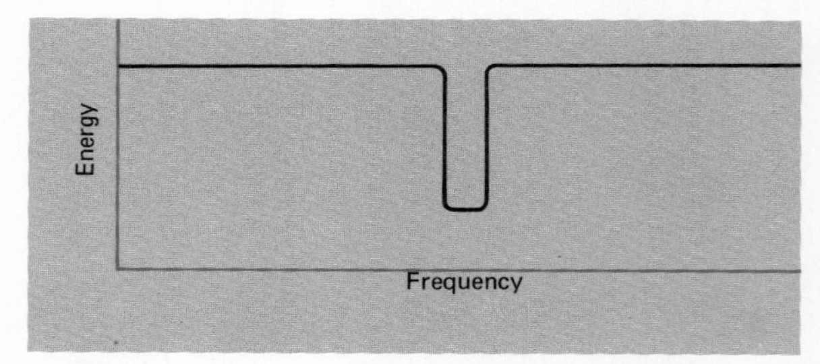

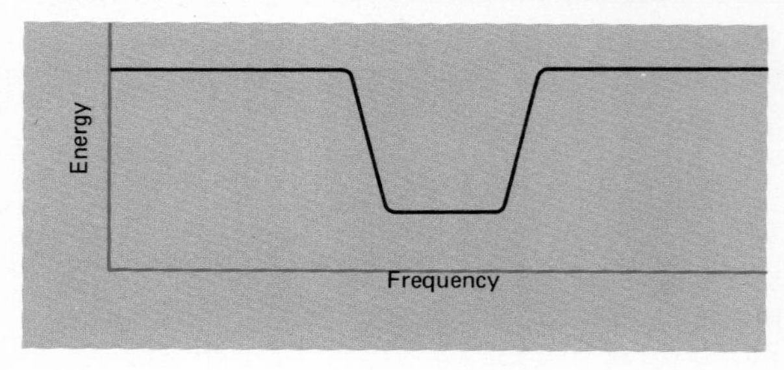

An absorption line: weak (top); intermediate (middle); strong (bottom).

ABUNDANCES OF THE ELEMENTS

Table 14.1 shows the approximate composition found for the Sun and most of the stars around us. The table gives the number of atoms of an element per million hydrogen atoms and the

TABLE 14.1
Abundances in the solar atmosphere

Element	*Atoms per million atoms of hydrogen*	*Percent of all atoms*
Hydrogen	1,000,000	90.7
Helium	100,000	9.1
Oxygen	1000	0.09
Carbon	500	0.05
Neon	500	0.05
Nitrogen	100	0.01
Silicon	30	—
Magnesium	25	—
Sulfur	20	—
Iron	4	—
Sodium	2	—
Aluminum	2	—
Calcium	1	—
Nickel	1	—
All other elements together	less than 1	—

percent abundance of the element. Over 90 percent of all atoms are hydrogen, and practically all of the remainder are helium. Only about one atom in 500 is anything other than hydrogen and helium. Studies of the interstellar material and of other galaxies indicate that a similar composition holds for most matter throughout the Universe.

Not all stars have compositions identical to those in Table 14.1. While there are a number of variations observed in individual stars, the most common difference occurs in stars in which the heavy elements—those other than hydrogen and helium—are even less abundant than they are in the Sun. (We will discuss this point further in Chapter 17.) It should again be emphasized that these abundances hold for the atmosphere. As we will see, there are good reasons for believing that, in some cases, the composition in the deep interior might be quite different.

The interior properties of a star fix its luminosity, that is, the total amount of energy the star radiates into space each second. The atmosphere of the star determines how that luminosity is distributed among the different frequencies of the radiation and what spectral lines will be formed. Both the atmosphere and the interior must be studied if astronomers hope to understand as much as possible about stars.

APPROACHING THE INTERIOR

How can we learn anything about the interior of a star? The atmosphere can at least send us

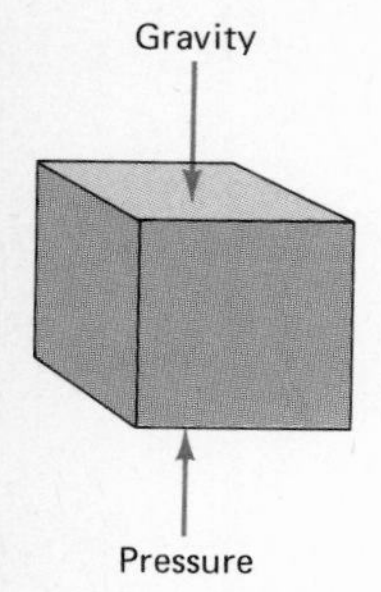

FIGURE 14.4
Gravity versus pressure.

radiation directly, but we will have to be satisfied with only indirect evidence about what the interior is like. It is again a question of whether we understand the properties of matter and radiation well enough to be able to calculate what the conditions are on the inside of a star. The principle is no different from any other area of interest since scientists always want to be able to calculate numerical details. In stellar interiors there is the practical difference that no direct observational checks can be made,[2] so we could stray a good way from the truth without necessarily being aware of it. It should be emphasized, however, that the theories that form the basis for interior calculations have been tested and checked under a large variety of circumstances (but *not* inside stars).

The method of investigating the internal properties of a star is based on the assumption that a star does not have a will of its own. It cannot decide how bright it wants to be, how hot, and so forth; these things are decided for it by the laws of physics. If we know these laws of physics, we should be able to calculate how bright the star *has* to be. The belief that such laws of physics exist and that they are capable of being discovered by rational inquiry is basic to all science.

[2] Except for the detection of neutrinos. (See the end of Section 14.3.)

A typical star like the Sun has settled down to a condition in which the various forces and processes that take place inside have come into a balance with each other. The star is stable and does not explode, collapse, or otherwise change its properties, at least over the time interval we have observed it. If we can discover what the important forces and processes are, we will be able to determine what conditions will bring about this balance or *equilibrium*, as it is often called. In this way we should be able to calculate the internal properties.

FORCE BALANCE

First let us consider the different forces important in a star. Remember that a force is a push or a pull that tries to move matter around. It is apparent that the gravity of a star is a force that tries to pull all matter, including the matter in the star itself, into the center of the star. If there were no force to oppose gravity, a star would collapse to a point. The fact that all stars have not collapsed to a point indicates that there is a force that opposes gravity. The opposition force is provided by the *pressure*. Pressure is a force per square inch exerted by matter (and also by radiation). Figure 14.4 shows a small piece of matter somewhere inside a star. Gravity pulls down on the piece, giving a weight to it. In order to fall into the center of the star, the piece

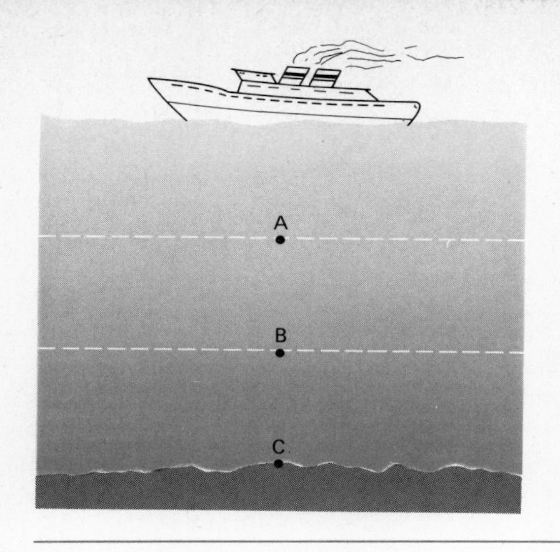

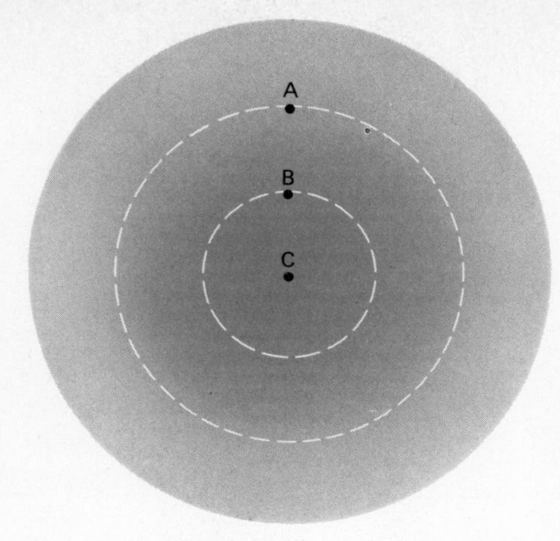

FIGURE 14.5

Pressure balances gravity at all points in the ocean and in a star.

would have to compress the matter below it or push it out of the way. The matter below resists being squeezed or pushed out of the way by exerting a pressure. If the pressure is too small, gravity will win the struggle, and the piece of matter will fall toward the center, much the way a stone sinks in water. If the pressure is too great, it will overcome gravity, and the piece will be pushed upward, the way an air bubble rises through water. If the pressure and gravity exactly balance each other, the piece of matter will feel no net force; it will remain motionless, like a submarine that is holding still under water.

The Sun and most, but not all, stars are observed to be in force balance; they are neither expanding nor collapsing, so the pressure and gravity forces must exactly balance each other at all points throughout these stars. Is this a coincidence? Or is there a reason to expect this? How did the pressure and gravity forces happen to become *exactly* equal at *all* points in a star?

Suppose that at some point in a star the pressure force is stronger than gravity. Then the material at that point would be forced to move away from the center, causing an expansion of that part of the star. Now this expansion will cause the force of gravity to become smaller because it moves the different parts of the star farther apart; however, it decreases the pressure by an even greater amount, since the thinner material can provide less support for the matter above it. As the expansion continues, therefore, the pressure will keep dropping until it no longer produces a force stronger than gravity. When this condition is reached, there is no longer any net force trying to push the material outward, and the expansion will stop.

Likewise, if gravity is stronger than the pressure, the matter will sink toward the center, causing a contraction of the star. Contraction brings the different parts of the star closer together, so it increases the force of gravity; however, contraction also makes the material more dense and more capable of supporting other matter above it, which means that it increases the pressure. The increase in pressure more than compensates for the increase in gravity, so the star will contract until the pressure has built up enough to balance gravity. When this condition holds, there is no longer a force causing the star to shrink, so the contraction halts. We see that the forces will push a star in or out until they reach a balance. A star will naturally adjust itself to the condition in which pressure and gravity forces exactly cancel each other at all points in the star. Force balance is not a coincidence but the natural condition of a stable star.

Figure 14.5 illustrates force balance in the

FIGURE 14.6

A pulsating star. At its smallest size pressure overbalances gravity and pushes the star out to a larger size. Instead of stopping at the equilibrium size, it overextends itself and gravity then becomes too strong for the pressure. This causes the star to contract to a smaller size, and the process then repeats. In most cases the change in size is much less than indicated by the figure.

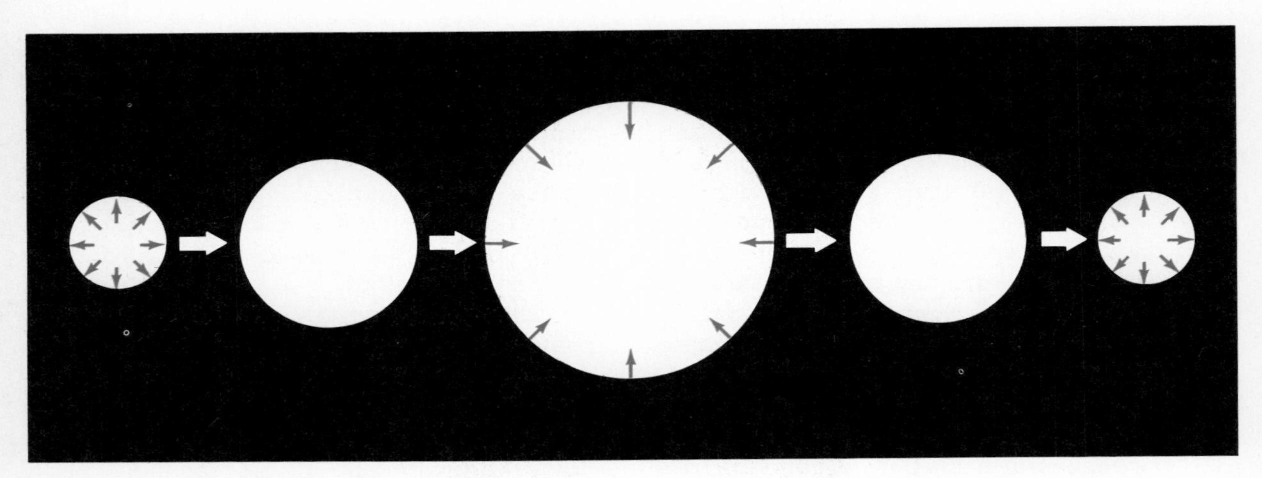

ocean and in a star. The ocean is divided into horizontal slabs, while the star is divided into spherical shells. In both ocean and star, point *A* (and all points on the lower surface of the top layer) has a pressure that is just enough to support the weight of the upper layer; this is another way of saying that the pressure pushing up at point *A* must equal the force of gravity pulling down at the same point. Point *B* must have a larger pressure since it is at a depth from which the weights of the top two layers must be supported. Likewise at point *C*, the bottom of the ocean or the center of the star, the pressure is greatest because all layers in the ocean or in the star are bearing down on it. Finding the pressure at any point is simply a matter of adding up the weights per square inch of the overlying layers. In water this is easy to do; in a star it is more complicated because of compression effects, but the principle is the same.

UNSTABLE STARS

Not all stars are in perfect force balance. There are a number of *pulsating stars* known that alternately expand and contract. In a pulsating star, the pressure and gravity come close to balancing, but they periodically get out of balance and cause the star to be alternately pushed out and pulled in. Pulsating stars change not only their size periodically but also their temperature and luminosity. They are

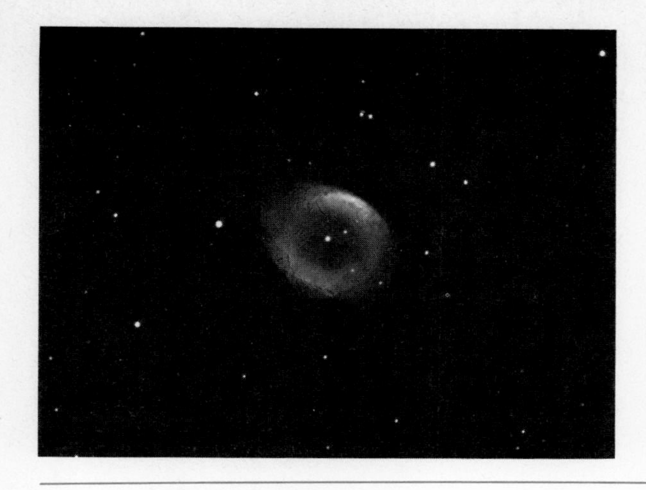

FIGURE 14.7

The Ring Nebula in the constellation Lyra. The ring consists of gases ejected by the star that is in the center when it became unstable several thousand years ago. (The Kitt Peak National Observatory)

also called *variable stars* since at certain times they appear brighter than they do at other times (see Figure 14.6).

Pulsating stars are on the verge of being unstable. Although the star tries to find a position in which gravity and pressure are in balance, the energy passing through the star will not let it completely succeed in this. Fortunately for us, the Sun and most stars do not have this energy problem. There are yet other stars, however, that have such severe energy problems that they become completely unstable and throw off part of their material. Figure 14.7 is a photograph of the Ring Nebula. Several thousand years ago, the star in the center of the ring became unstable and threw off a shell of gas that we now see as the ring. Shedding this material apparently relieved the instability, and the star now appears to be in equilibrium. We will look at other examples of unstable stars in later parts of the book.

ENERGY FLOW

It appears that consideration of the force balance between pressure and gravity is not sufficient for solving the problem of stellar interiors. The outward flow of energy must also be taken into account; this energy flow is governed primarily by temperature, as our everyday experience would suggest.

If a hot object and a cool one are placed close together, experience tells us that the hot object will cool off, and the cool one will tend to warm up. This is due to the flow of heat energy, which always goes from the hot to the cool body. The same holds true in a star. As we go deeper into the star, the temperature must rise in order to have an energy flow out of the star (see Figure 14.8). The faster the temperature rises toward the core, the greater is the outward flow of energy. It is this energy flow that gives the star its luminosity.

The Sun and most stars are observed to

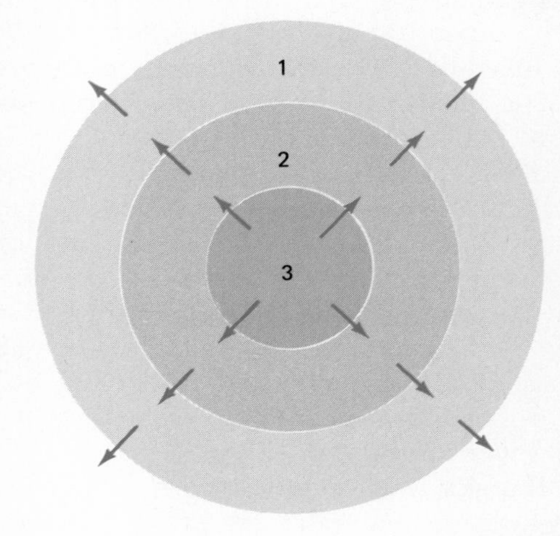

FIGURE 14.8

Energy flow out of a star.

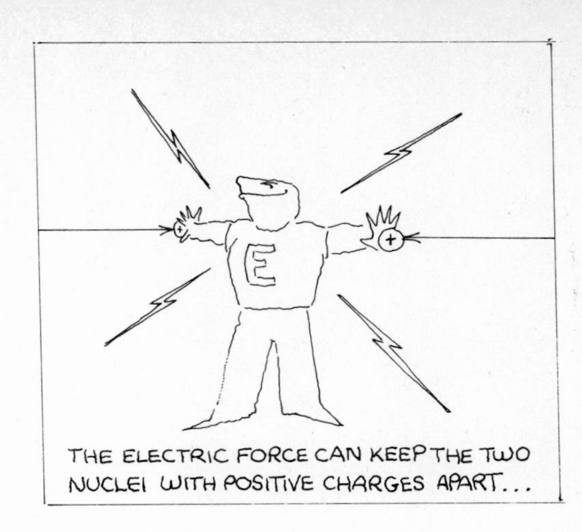
THE ELECTRIC FORCE CAN KEEP THE TWO NUCLEI WITH POSITIVE CHARGES APART...

have a steady outflow of energy since their luminosities do not change from one time to another. The different layers of such stars must have just the right temperatures, therefore, to account for that steady luminosity. To illustrate, suppose that layer 1, the outer layer in Figure 14.8, were somehow cooled to a temperature lower than it previously had. Matter emits less radiation when it is cooled, so the emission from layer 1 into space would become less, and the star would appear fainter. At the same time, layer 1 would receive more radiation from layer 2 below it. This is because the energy flow from 2 to 1 depends on the difference between their temperatures, and this difference has now been made greater. The increased energy flow into layer 1 would tend to heat it back up to the temperature it had before. This is its equilibrium temperature, the temperature for which it radiates energy into space at the same rate as it receives energy from below. In this way, a star tries to adjust its inside temperatures so that each layer has an energy balance; each layer wants to receive energy from below at the same rate that it passes the energy on to the next layer above.

If a star is to become stable, it must find a way to come into both force balance and energy balance. The pressure and gravity forces must exactly equal each other at all points in the star, and the temperature must be set up in a way that provides a steady flow of energy through the star. There is a double balance that the star must find. The situation is further complicated because the temperature and the pressure of a gas are related to each other so that any adjustments made in one will also affect the other. This means that a star cannot be in either force balance or energy balance; it must be in both, or else it will be in neither. The pulsating stars are not exactly in force balance, so they cannot be in energy balance. This causes them to have variable luminosities.

14.3 Nuclear reactions and model stars

NUCLEAR REACTIONS

In Section 2.2 it was emphasized that energy cannot be made from nothing—it must come from some previously existing form of energy. What is the origin of the tremendous supplies of energy that flow through the stars and are radiated into space? For most stars the answer is nuclear energy, already alluded to in earlier sections.

It is worthwhile to review here some of the important properties of matter. The smallest particles into which matter can be divided are

...UNLESS THEY HAVE VERY HIGH SPEEDS.

FIGURE 14.9
The electric force and nuclear reactions.

the fundamental particles. Protons and neutrons are two kinds of fundamental particles. The nucleus of an atom consists of a certain number of protons and neutrons held together by very strong nuclear forces. The number of protons in a nucleus determines which chemical element the nucleus is a part of: Hydrogen has 1 proton in the nucleus, helium has 2, oxygen 8, iron 26, and so forth. Neutrons have about the same mass as protons, but they are electrically neutral. Most nuclei have about the same number of neutrons as protons. Important exceptions are the most common form of hydrogen, which has one proton and no neutrons, and the very heavy elements, which have considerably more neutrons than protons. Nuclei of the same element having different numbers of neutrons are called "isotopes" of that element.

A nuclear reaction is any event which causes one nucleus to change into another by changing the number of protons or neutrons in it. This is usually caused by a collision between two nuclei, although radioactivity consists of nuclear reactions taking place spontaneously. The forces that hold a nucleus together are very strong, so large amounts of potential energy are gained or lost when nuclear reactions occur. (Refer to Section 2.2.) Radioactivity is generally not important in stars.

Nuclear reactions are usually very difficult to bring about. One reason for this is that nuclear forces are important only over very small distances; if two nuclei are more than about 10^{-13} or 0.0000000000001 inch apart, they are not close enough together for their nuclear forces to have much effect on each other. Remember that all nuclei have positive electrical charges due to the protons they contain. Electrical forces try to keep all positive charges away from each other (opposite charges attract, like charges repel). This makes it very difficult for nuclei to come close enough together for the nuclear forces to overcome the electrical forces and cause a nuclear reaction.

Neutrons can solve this problem: Neutrons have no electric charge, so they have no difficulty getting close to nuclei since electric forces do not affect them. The trouble is that free neutrons do not usually occur in stars. If a neutron is alone and not with other fundamental particles in a nucleus, it is unstable. Within a few minutes it breaks up into a proton and an electron. With no neutrons around, nuclear reactions can be produced only by nuclei having enough energy to overcome the electrical forces (see Figure 14.9). For most stars only the regions near the center have temperatures that are high enough for collisions among the nuclei to produce significant numbers of nuclear reactions.

There are an almost limitless number of

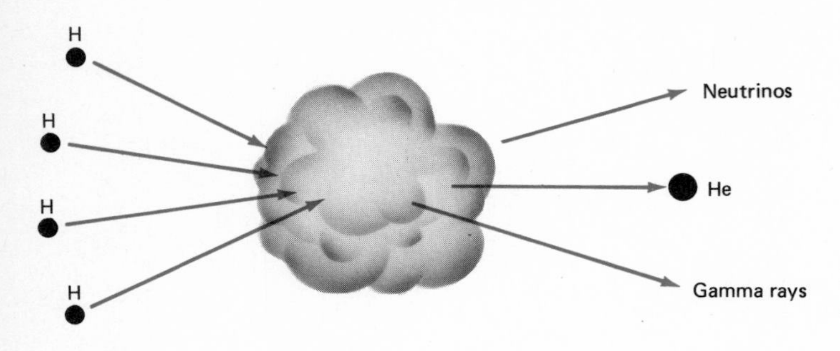

FIGURE 14.10
Converting hydrogen into helium.

possible nuclear reactions; however, since hydrogen and helium are the most abundant elements, it is reasonable to look to them as the source of nuclear energy in stars. Also, hydrogen has fewer protons than the other elements, so reactions with hydrogen have smaller electrical forces to overcome. In fact, the nuclear reactions that supply the energy for most stars, including the Sun, are those that have the effect of changing hydrogen into helium. There are many complications in how this is done, but we shall not be concerned with the details. The important fact is that there are nuclear reactions that convert hydrogen into helium and release large amounts of energy. The energy released is partly in the form of very high energy gamma rays and partly in the form of other elementary particles called neutrinos. This process is generally referred to as "hydrogen burning" (see Figure 14.10), although it is not the chemical process of forming molecules with oxygen to which the term "burning" usually refers.

ENERGY BALANCE

How much nuclear energy is released at the center of a star? The answer depends on how fast the nuclear reactions take place. The higher the temperature and pressure at the center of a star, the more violent are collisions that take place among nuclei, and this leads to more reactions and more energy release. These reactions are an important part of the energy balance in a star.

Remember that a star tries to adjust the temperatures of its different layers so that a steady outward flow of energy is set up. For complete energy balance to occur, the temperature and pressure at the center must provide nuclear reactions at just the correct rate to account for this steady outflow of energy. For a star in equilibrium, the luminosity is equal to the nuclear energy released each second at the center of the star.

If a star is in equilibrium, there are very strong restrictions on what the temperature and pressure can be at different points in the interior. It can be shown that if the mass and composition of a star are known, then all other properties of the star can be calculated from the assumption of perfect force and energy balance at all points in the star. This statement is known as the Russell-Vogt theorem. A star with a certain mass and chemical make-up has forces and processes acting in it that require the star to have a definite size, luminosity, temperature, pressure, and so forth. The laws of physics do not allow it any choice in the matter.

MODEL OF THE SUN

Figure 14.11 shows a scale model of the Sun which was calculated from the assumption of force and energy balance. (Further details of this calculation will be described in Section 15.2.) The temperature is 28,000,000°F in the core, and it cools down to 10,000°F at the surface. The pressure has a maximum value of about 250 billion times the Earth's atmospheric pressure at sea level, and it drops to a very low value in the solar atmosphere. The matter near the center is strongly compressed, as is shown by the fact that one-half of the mass is contained within the small dotted circle. The gases in the core are squeezed to about 160 times the normal density of water, a density far greater than any substance known on the Earth.

It should be pointed out that there is one important respect in which theoretical model calculations for the Sun do not agree with observations. As mentioned earlier, nuclear reactions release part of their energy in the form of gamma-ray photons and part in the form of elementary particles called *neutrinos*. The neutrinos escape directly from the Sun, and calculations indicate that enough of them should be given off by the Sun for them to be detected at the Earth. Unfortunately, attempts to detect neutrinos from the Sun have so far been unsuccessful, showing a significant discrepancy between theory and observation. If the calculations give the wrong amount of neutrinos coming from the Sun, how far wrong are they in other respects? At present no one can say, but you can be sure there are persons who are seriously concerned with this problem.

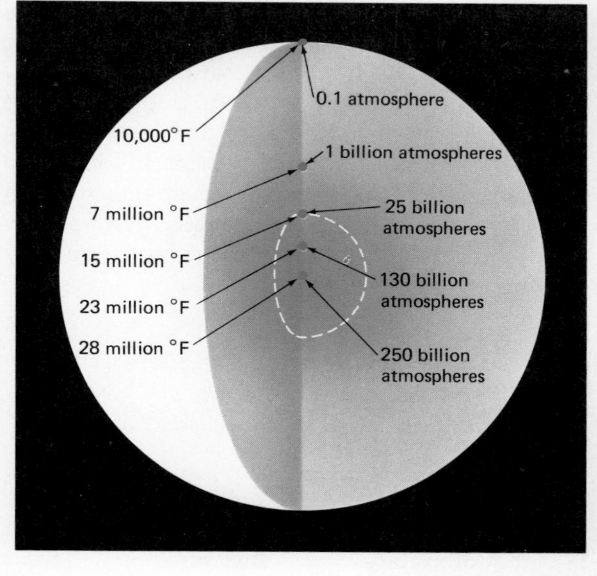

FIGURE 14.11
Model of the inside of the Sun. The figures indicate the calculated values of the temperature and the pressure at the designated points. [Based on the calculations of R. L. Sears, *The Astrophysical Journal* 140 (1964), 477, published by the University of Chicago. Copyright © 1964 by the University of Chicago.]

As a star radiates energy into space, it must eventually run low on its nuclear fuel, the hydrogen that is being converted into helium. The rate of release of nuclear energy will be affected by this, so the properties of a star must change with time if it is to remain in force and energy balance. These aging effects are the topic of the next two chapters. Meanwhile, perhaps we could update the familiar nursery rhyme along the following lines:

Twinkle, twinkle little star,
I think I know now what you are:
A huge self-gravitating mass
Of luminescent hydrogen gas.

IMPORTANT WORDS

Stellar atmosphere	**Variable star**
Equilibrium	**Russell-Vogt theorem**
Pressure	**Neutrino**
Pulsating star	

REVIEW QUESTIONS

1 What types of photons are emitted by the nuclear reactions in a star?

2 Why don't photons come straight out of a star as soon as they are created?

3 What is meant by the atmosphere of a star?

4 About how long does it take radiation energy to "eat its way" out from the center of the Sun?

5 If two stars are made out of the same kind of material, how can they show different spectral lines?

6 What is the basic problem of stellar atmospheres?

7 Hydrogen and helium make up about what percent of all atoms in the Sun and most other stars?

8 What are the two main forces that are exerted on the material in a star?

9 How do we know that gravity and pressure balance each other inside a star?

10 What happens to a star for which the force of gravity is stronger than the pressure force?

11 What is a pulsating star?

12 What does energy balance mean?

13 What conditions must exist in a star if it is to be stable?

14 What provides the energy that stars radiate into space?

15 What is a nuclear reaction?

16 Which nuclear reactions are important in most stars?

17 What is the Russell-Vogt theorem?

18 What is the problem that theoretical models of the Sun have with neutrinos?

QUESTIONS FOR DISCUSSION

1 Suppose that a strong spectral line occurs precisely at frequency f_1 in a star. Frequency f_2 is a frequency of radiation very close to f_1 but without

any line. Will photons of frequency f_1 or those of frequency f_2 have greater difficulty in passing through the atmosphere of the star without being absorbed? Of the photons that escape into space, which will tend to come from deeper layers in the star, f_1 or f_2 photons? Do deeper layers tend to be hotter than shallow layers? Do hotter layers tend to emit more radiation than cooler layers? Put this chain of reasoning together to understand why the line at frequency f_1 should be an absorption line, not an emission line.

2 The Earth formed from material that originally had the abundances listed in Table 14.1. Human protoplasm has roughly the following abundances by numbers of atoms: hydrogen, 63 percent; oxygen, 25 percent; carbon, 10 percent; nitrogen, 1.3 percent; calcium, 0.25 percent; phosphorus, 0.2 percent; and potassium, 0.06 percent.

What are some of the more obvious similarities and differences between these two sets of chemical abundances? Can you think of any worthwhile comments on this?

3 Some persons argue that astronomers can't possibly know anything about the inside of a star since they have never been there. How would you answer this argument?

4 Compare the bouncing ball of problem 5 in the Questions for Discussion in Chapter 2 with a pulsating star. The energies in the two cases are closely analogous.

5 A pan of very hot water gives off radiation and heat to its surroundings. This drains heat energy from the water and cools it off. A burning log in the fireplace also gives off radiation and heat to the surroundings, but it does not cool off as long as it keeps burning. Why not? Is a star more analogous to the pan of hot water or to the burning log? Why?

6 If helium were just as plentiful as hydrogen in a star, do you think that helium-burning reactions would be as important as hydrogen-burning ones? Why?

7 We hear that hydrogen is the lightest element. Yet the gas at the center of the Sun, which is nearly pure hydrogen, has a density that is more than 10 times the density of solid lead. How can this be explained?

REFERENCES

Two interesting, popular books that have much material that is related to this chapter are *Astronomy of the Twentieth Century* by O. Struve and V. Zebergs (Macmillan, New York, 1962) and *Atoms, Stars, and Nebulae*, 2nd ed., by L. H. Aller (Harvard University Press, Cambridge, 1971). The following articles are also relevant: "Why Does the Sun Shine?" by L. Mestel in *The Emerging Universe*, edited by W. C. Saslaw and K. C. Jacobs (University of Virginia Press, Charlottesville, 1972); and "Pulsating Stars" by J. R. Percy in *Scientific American* [232 (June 1975), 66].

FIFTEEN
EVOLUTION I: YOUNG STARS

In this chapter and the next we look into the question of how stars age. The lifetimes of the stars are too long for us to be able to watch them change as they grow old, so we must again employ largely theoretical methods. This chapter is concerned with the early stages of stars: how they form and how they spend their youth. Chapter 16 will describe stars in their old age.

15.1 Lifetimes

THE ENERGY SUPPLY

The stars seem to go on forever. Certainly they appear to change very little, if at all, in the lifetime of a person or even over recorded history. The "fixed and immutable" stars are a popular subject for those who wish to find a contrast to anything unpredictable or otherwise changeable. Yet our knowledge of the laws of physics should readily convince us that the stars, too, are changeable and that they have finite lifetimes. The supply of nuclear energy in a star is very large, but it is limited. As the stars keep shining, sooner or later the nuclear energy will run out. For the same reason, the stars we see could not have been shining for an indefinite time into the past, as the energy supply would not allow it. We are thus led on quite general principles to the conclusion that stars, like people, are born, live, and die.

It does seem impossible for the stars to shine forever, but couldn't they still exist forever, shining only part of the time? Let us first look at this question from the viewpoint of the stars' past. Could they have existed forever but only started shining in recent ages? Stars do not have switches by which they can be turned off and on. Stars do not shine because it is somehow convenient; they shine because the laws of physics require it. If a star existed in the extremely remote past in a condition anything like the stars of today, the laws of physics require that it must have been shining at that remote time, and it would have burned itself out by now. If a star that we see today did exist in the remote past, it must have been in a form that is unlike a star; thus it was born as a star only in more recent times. The laws of physics might have been different in the past, but any special assumptions that would allow a star of today to be infinitely old are too artificial to be seriously considered without further evidence.

Granted that stars will eventually burn out, won't they continue to exist after they stop shining? The material will continue to exist, of course. The definition of a star is so strongly connected with the idea of shining, however, that we will say that the object ceases being a star when it stops shining. This, of course, is simply a matter of definition rather than a question involving the laws of physics. We again conclude that stars have finite lifetimes.

It must be emphasized that there is no suggestion here that all stars were formed at the same time or that they will burn out together. Furthermore, although no individual star can last forever, it is a completely different question to ask whether stars in general can last forever. Might not new stars continue to form and take the place of old ones, so that there always have been and always will be stars? This is a question in cosmology, and we will consider it from several points of view in Chapter 20.

CALCULATING LIFETIMES

How long will it take for stars to burn themselves out? This length of time is very long compared to human history, but it is surprisingly easy to calculate from energy considerations. If we know the mass of a star, we know how much nuclear fuel it has available. The luminosity of a star tells us how fast its nuclear energy is being used up. Putting these two quantities together will indicate about how long the star's energy supply can last. It is no different from trying to estimate when a car will run out of gas. If you know how much fuel is in the gas tank and how fast you are using it up, it is easy to determine how long the supply will last.

Consider the Sun, for example. Each time four hydrogen nuclei in the center of the Sun are converted into a helium nucleus, a certain small amount of energy is released. This amount is so tiny that a mosquito uses a million times as much energy just to get off the ground; yet this is the source of energy that keeps the Sun shining. In order to supply the luminosity of the Sun, 500 million tons of hydrogen must be changed into helium every second. This is a fantastic rate in which to use up hydrogen, but the Sun has such a huge supply available (2×10^{27} tons = 2 billion billion billion tons) that it could continue converting hydrogen at the same speed for 100 billion years before running out. But not all hydrogen in the Sun is available for nuclear reactions since the Sun's temperature is high enough only close to the center, so 100 billion years reduce to about 10 billion years. Thus the Sun can be no more than 10 billion years old since that is the length of time needed to use up its fuel supply. This is the lifetime or life expectancy of the Sun.

A similar calculation can be made for any star whose mass and luminosity are known. A star with twice as much mass as the Sun will have twice as much fuel to burn, while one with three times the solar luminosity is using up its fuel three times faster than the Sun. Now

TABLE 15.1

Lifetimes of the stars

Mass (Sun = 1.0)	*Luminosity (Sun = 1.0)*	*Lifetime*
0.1	0.0001	10,000 b.y.
0.5	0.04	125 b.y.
1	1	10 b.y.
2	20	1 b.y.
5	600	80 m.y.
10	5000	20 m.y.
50	1,000,000	0.5 m.y.

b.y. = billion years; m.y. = million years

the more massive a star is, the more luminous it is. In fact, the luminosity changes from one star to another much more than the mass does, and this causes the luminosity effects to be stronger than the mass effects. The result is that the greater the mass of a star, the shorter is its lifetime. It is as though the richer a person is, the sooner he will go broke because of the rapid rate at which he spends his money.

As mentioned previously, the most massive stars known have about 50 times the solar mass. These stars are extra luminous supergiants having nearly a million times the luminosity of the Sun. What happens to someone who has 50 times as much money as you do but spends it a million times faster? His money doesn't last very long—in this case he runs out 1,000,000/50 or 20,000 times more quickly than you do. In the same way, the most massive stars have lifetimes 20,000 times shorter than the Sun, and they burn themselves out in about one-half million years.

At the other extreme are small stars with about 1/10 the mass of the Sun. These are very feeble stars with about one ten-thousandth the solar luminosity. These faint stars are very stingy about spending their small energy supply, so their supplies will last 10,000/10 or 1000 times longer than the Sun's. The lifetimes in this case are about 1000 × 10 billion or 10 trillion years.

Stellar lifetimes range from less than a million years to about 10 trillion years. The Sun, as usual, is somewhere in the middle. These figures are not exact. In the first place, the masses and luminosities are often not well known. In addition, no star has a luminosity that is strictly constant over its lifetime, and there are other sources of energy that can be important in stars in addition to the conversion of hydrogen to helium. These points make only rather small modifications, however, so the figures here and in Table 15.1 are fairly accurate. It is indeed remarkable that such significant numbers can be obtained so easily. But note that the lifetime of a star is not the same as its age.

15.2 Stars in their prime

MODEL STARS

Most of our knowledge about the stars cannot be obtained as easily as the approximate lifetimes we just discussed. For more precise and detailed information, theoretical calculations of model stars of the kind outlined in Section 14.3 are necessary.

To see how astronomers can calculate detailed life histories of stars, we shall refer again to the Russell-Vogt theorem mentioned in Chapter 14. This theorem states that if the mass and composition of a star are known, then all other properties of the star can be calculated from the assumption of force and energy balance in the star. In order to carry out these calculations, we must know in detail precisely how matter absorbs and emits radiation, how materials under different conditions exchange energy, how nuclear energy is converted into other forms, and so forth. Since all of these things are understood only imperfectly, the calculations of a model star are subject to a corresponding uncertainty. These inaccuracies, however, do not change the meaning or the validity of the Russell-Vogt theorem; it is still true that if two stars differ from each other in any respect, they must have different masses or compositions.[1]

To illustrate the calculations, suppose that we wish to find a theoretical model of the Sun. Then we would choose for our model a mass of about 2×10^{27} tons and a composition like that listed in Table 14.1: mostly hydrogen and helium plus tiny amounts of the heavy elements. If we then carry out the computations according to the best-known equations and theories, we might be in for a disappointment; the model would look only roughly like the Sun. Its surface temperature would be a little too low, and its luminosity would be only about 70 percent of that of the Sun. Are the inaccuracies in the theory that large? If so, the models are not very useful.

[1]This statement is not true if rotation, magnetic fields, or outside influences are important. These complications are ignored here.

AGING EFFECTS

But we have forgotten something! We found earlier that the Sun has a lifetime of about 10 billion years. This means that today the Sun could have an age of anywhere from about zero up to a maximum of about 10 billion years. But the model we calculated is actually for the Sun at age zero, for we assumed that the model has the same composition throughout as is found for the atmosphere. As the Sun ages, however, hydrogen at the center is changed into helium, so eventually it will have a helium-rich composition in the core while retaining its original hydrogen-rich composition outside the core.[2] We need an older model to represent the Sun of today.

Is there a way we can take into account this aging process and compute a model of the present-day Sun? A little thought should convince us that the answer is yes. According to the Russell-Vogt theorem, we need to know the mass and composition in order to calculate a model. This composition does not need to be the same at all points in the model as long as it is known at all points. We already have com-

[2]The material in the Sun is not well mixed.

Age (b.y.)	Luminosity (Sun = 1.0)	Surface temperature (°F)	Radius (Sun = 1.0)	Hydrogen at center (percent)
0.00	0.71	9380	0.96	0.91
0.5	0.74	9420	0.96	0.89
1.5	0.79	9560	0.97	0.86
2.5	0.85	9710	0.98	0.82
3.5	0.92	9860	0.99	0.76
4.5	1.00	10000	1.00	0.69

[From R. L. Sears, *Astrophysical Journal* 140 (1964), 477.]

TABLE 15.2
The Sun at different ages

puted the model for the zero-age Sun, so we know all about that model. In particular, we know how fast hydrogen is being changed into helium in that model. Suppose we want a model of the Sun at the age of 1 billion years. Then we need to calculate how much hydrogen will be changed to helium over the billion-year interval between the two models, and we can do this from the known rate of conversion. In this way we can find the new composition at each point in the new model. Since the mass of the new model is the same as that of the first model, which is known, the Russell-Vogt theorem says that we have enough information to calculate the new model and find what the Sun was like when it was 1 billion years old.

If we calculate this second model, we find that it does not differ much from the first model, but the differences that do exist are all closer to the characteristics of the present-day Sun. The luminosity and surface temperature are slightly higher, and of course there is a small depletion of hydrogen at the center. The Sun apparently didn't show many effects of aging during its first billion years. Since this model is still rather different from the Sun we observe, we conclude that the age of the Sun is well over 1 billion years.

By the process outlined here, we could let the 1-billion-year-old model serve as the starting point for calculating a still older model. In this fashion, a whole series of models of different ages could be constructed. These models would show how all of the characteristics of the Sun have changed with age. We could find out how old the Sun is by selecting the model that most closely fits the Sun. This is how the model Sun described in Section 14.3 was constructed.

Table 15.2 lists some of the properties of the solar models at different ages. The models are seen to come to the proper values of luminosity, surface temperature, and radius at an age of about 4.5 billion years, and we conclude that this must be very close to the present age of the Sun.[3] Our Sun is a middle-aged star and still has about one-half or more of its life ahead of it. Note that when the Sun was first formed, about 91 percent of its atoms were hydrogen. Today this is still true of the Sun's outer parts, but in the core only 69 percent of the atoms are hydrogen. The nuclear reactions have raised the helium abundance at the center from 9 percent to 31 percent of the atoms.

We will consider the future of the Sun in the next chapter, but for now let us look again at the Sun at age zero. As Table 15.2 shows, it was somewhat smaller, cooler, and fainter than it now is. This was the homogeneous Sun, when it first started to burn hydrogen but was

[3]Note that this is in excellent agreement with the age of the solar system as determined by radioactive dating of the meteorites. (Refer to Section 6.1.)

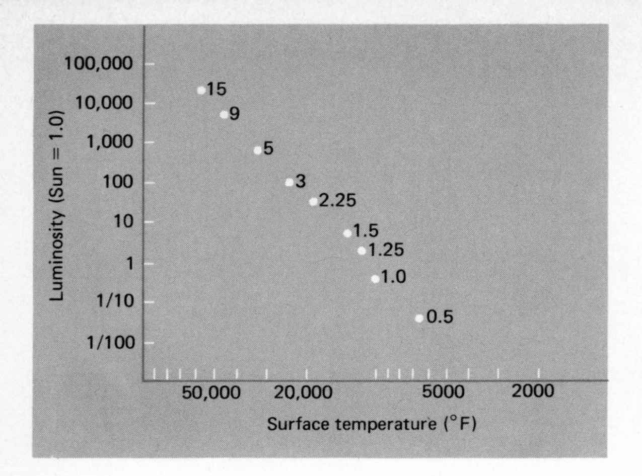

FIGURE 15.1
The H-R diagram for zero-age model stars. The numbers next to each point indicate the mass of the given model in terms of the mass of the Sun. [Data based on I. Iben, *Annual Review of Astronomy and Astrophysics* 5 (1967), 571.]

not yet old enough to have piled up very many "ashes" of helium in the core. The Russell-Vogt theorem states that the complete structure of the zero-age Sun is determined by its mass and the composition of the material it was made from. Any other type of star must be made of a different amount or a different kind of material.

ZERO-AGE MAIN SEQUENCE

Next suppose we decide to calculate some models for stars having the same composition as the Sun but different masses. We find that the greater the mass we start with, the greater is the radius, surface temperature, and luminosity of the zero-age model star. By knowing the surface temperatures and luminosities of the zero-age model stars, we can plot them in an H-R diagram (see Figure 15.1).

Each point in Figure 15.1 corresponds to a theoretical model star whose mass is indicated in the figure. The points lie along a narrow line running from the upper left to the lower right side of the diagram, that is, from hot, luminous stars to cool, faint ones. If we compare this theoretical H-R diagram to one that represents real stars, we find that the calculations fall along the main sequence. This, then, tells us that main sequence stars are those that are young enough to have not yet significantly depleted their store of hydrogen at the center. The model stars of Figure 15.1 do not extend to the regions covered by the giants, supergiants, or white dwarfs. We conclude that non-main sequence stars must either be old enough to have used up most or all of their core hydrogen or else that they are so young that they have not yet come into equilibrium as main sequence stars.

The line described by the points in Figure 15.1 is called the *zero-age main sequence*. The Sun is about 4½ billion years old and has used up part of its core hydrogen; however, it has not yet changed a great deal from what it was like at age zero, so it is still considered a main sequence star. Only when stars have changed considerably from their zero-age properties are they said to be non-main sequence stars. When we observe stars of all ages mixed together, the main sequence will appear to be a band with a certain width, as we saw in Figure 12.6, instead of as a narrow line, as in Figure 15.1.

ORIGINAL COMPOSITION

Composition also can affect zero-age stars. There appear to be only small variations in original composition among stars. All stars seem to start their careers with hydrogen and helium as the overwhelming parts of their make-up. In the Sun and in most stars around us, only about one atom in 500 is anything else. If we call this the normal composition of stars,

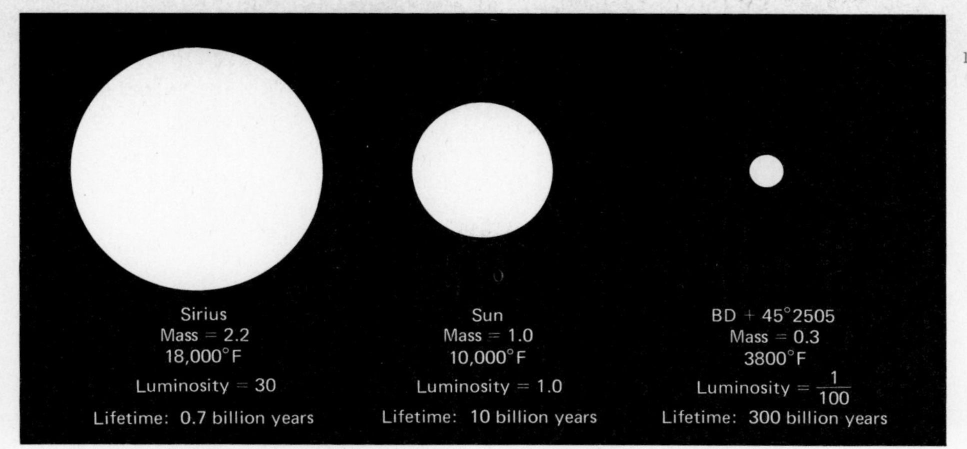

FIGURE 15.2
Three main sequence stars.

then we find that most stars with abnormal compositions have about the usual amounts of hydrogen and helium; their abnormality comes from having even fewer of the heavy elements than one atom in 500.

But these differences in the abundances of the heavy elements have only rather small effects on the properties of the stars, and we shall not consider them here.

THREE MAIN SEQUENCE STARS

Sirius, the Sun, and BD+45°2505 are three main sequence stars, all made out of approximately the same kind of material.[4] Sirius is an impressive star with about 30 times the luminosity of the Sun. BD+45°2505 is a cool, red star only one-fourth the size of the Sun, and its energy output is a feeble 100 times less than the Sun's. How did these differences come about? They all result from Sirius's being made from over twice as much material as the Sun, while BD+45°2505 (let's call him BD for short) has to get by with only one-third of the Sun's mass. Figure 15.2 gives some data on these stars, including their approximate lifetimes. These lifetimes are figured from the masses and luminosities, as explained earlier in this section. Because of BD's low luminosity, its lifetime is about 400 times longer than that of Sirius.

Consider all the stars like Sirius that exist in the Milky Way today. If a star is like Sirius, then it can be no older than about 0.7 billion years since those stars which formed earlier have burned out by now. BD type stars, on the other hand, do not burn out until they are about 300 billion years old. Later we will see that our Milky Way galaxy did not have any stars until about 15 billion years ago, so any BD-type star ever made is still around. Since only the "recently" formed Sirius-type stars are still with us, we can understand one reason why faint, low-mass stars are much more common in space than the luminous stars of high mass.

These arguments also lead us to the conclusion that stars are being formed even today. Sirius can be no older than about 0.7 billion years, and there are other stars that astronomers know must be less than a million years old. To the Milky Way with an age of about 15 billion years, a million years is a very brief period of time, like a few days in the life of a person. If conditions were suitable for star formation only a million years ago, they must still be suitable today. This brings us to the question of what stages a star must pass through prior to its burning hydrogen and joining the main sequence.

[4]The strange name BD+45°2505 comes from the fact that the star is number 2505 in the 45-degree zone of a catalog called the "Bonner Durchmusterung." Faint stars are usually known by their ID number in some catalog. Only conspicuous stars like Sirius have special names.

15.3 To make a star

WHERE IT COMES FROM

Can it be that stars just grow out of nothing? This would greatly violate what we believe are the laws of physics, particularly belief that matter and energy cannot be created out of nothing. Astronomers base this belief on a great deal of experience, but none of them has had very much firsthand experience in the formation of stars. Maybe stars do appear from nowhere; however, this is a very special assumption that we should not take seriously unless there is evidence for it. Scientists do not further their knowledge by making wild guesses but by seeing how present ideas might be expanded or generalized to include new situations. This point can be a bit subtle, and in Chapter 20 we will see that one of the serious cosmological theories requires matter to be formed continuously out of nothing. The difference is that the creation of matter is not proposed on its own merits but is a necessary consequence of what its proponents believe are quite plausible scientific assumptions.

Except under very unusual circumstances, matter seems to be conserved. If we wish to retain this idea in the creation of stars, we must look around to see if there is some place for the matter to come from. It shouldn't take long to think of the interstellar matter—the gas and dust that were the subjects of Chapter 13. When we note that very young stars are always observed in or near regions where there are large amounts of interstellar matter, the evidence becomes quite strong. Somehow the tenuous gas and dust between the stars must occasionally contract to a very small size and condense into stars.

CONTRACTION UNDER GRAVITY

Gravity is the main agent in getting an interstellar cloud to condense into a star, but there is a complication. Most nebulae have too much energy to contract; the atoms, molecules, and dust particles have enough kinetic energy to resist the pull of gravity, and so the clouds do not contract. Only if a cloud can lose its excess energy, by radiating it away, for example, will gravity be successful in bringing on a collapse.

Once a nebula has started to contract under its own gravity, as will happen if it becomes cool and dense enough, then it is past the main hurdle, and further contraction will automatically follow. The object will continue to get smaller, and the internal pressures will build up. Eventually the pressures become large enough to balance gravity, and the contraction will tend to stop. Meanwhile, energy exchanges are taking place that very much affect our star-to-be or *protostar*.

It takes energy to lift an object against the force of gravity. The energy used to do the lifting goes into the potential energy of the object (refer to Section 2.2.) This potential energy is due to the force of gravity, and it is customary to call it *gravitational energy* to distinguish it from other types of potential energy. If the body is released and falls under gravity, the gravitational energy is released and changed to kinetic energy.

It is exactly the same in a nebula. When a nebula contracts, the different parts of it are falling under the force of gravity, and the nebula is changing gravitational energy into kinetic energy. This kinetic energy of the contracting cloud is changed into heat energy and radiation as the atoms and molecules collide with each other. As a result, a contracting cloud will automatically become hot as it grows smaller. This is why stars are hot: Their contraction from very large interstellar clouds to much smaller stars releases great amounts of gravitational energy, and part of the released energy goes into heating the material.

ON TO THE MAIN SEQUENCE

In the birth of a star we can imagine a large, cool, tenuous nebula slowly contracting. As the nebula gets smaller, its internal pressures and temperatures rise. The hotter it becomes, the faster it radiates energy into space. By the time the pressures have increased to the point where they can support the material against the force of gravity, the protostar is already hot enough to have a large luminosity. Does the pressure–gravity balance then halt further contraction? It does slow it down, but it cannot quite stop it completely.

Under most conditions the pressure of a gas depends on its heat energy or temperature. If you drain heat energy from a gas, the pressure will drop unless the lost energy is replaced. The luminosity of a protostar does drain away heat energy, and unless the energy is replaced, the pressure will drop, gravity will slightly overbalance it, and the contraction will slowly continue. If the temperature at the center of the protostar becomes high enough to support nuclear reactions, these reactions will replace the energy lost by radiation. When the nuclear-energy release exactly balances the luminosity, there is no longer an overall loss of heat energy, the pressure no longer drops, and the contraction comes to a halt. The object is now in perfect force and energy balance as a full-fledged, hydrogen-burning, main sequence star.

Astronomers have discovered a number of small, dense dust clouds that emit large amounts of energy in the form of infrared rays. These infrared sources are thought to be nebulae that surround and are in turn heated by protostars. There are also some so-called *T Tauri stars* which are always associated with large amounts of interstellar matter. T Tauri stars lie somewhat above the main sequence in the H-R diagram, and they are probably in the last stage before settling down as main sequence stars. Thus examples of protostars are observed. A star spends far less time in the pre-main sequence stages than it does on the main sequence, however, so protostars are rather rare in space.

Let us define a star as any object that arrives in force and energy balance on the main sequence. Objects having less than about 1/12 the mass of the Sun can never get hot enough to release much nuclear energy. (We will discuss the reason for this in Chapter 16.) These bodies cannot come into energy balance so they cannot become stars, even though they may shine for a long time. According to our definition, therefore, no star can have a mass less than about 1/12 the mass of the Sun.

A protostar is very much like a leaky balloon. The stretched rubber takes the part of gravity, trying to collapse the balloon. The air pressure keeps it extended, just as the gas pressure does for the star. A small leak in the balloon takes the role of the luminosity of the protostar: It undermines the supporting pressure, causing a slow contraction. If you pump air into the balloon at the same rate that it is leaking out, then obviously the internal pressure will be maintained, and the balloon will not collapse. This pump is analogous to the nuclear reactions in a star that supply the energy lost by radiation. As long as the nuclear reactions and the air pump continue to function, the star and the balloon will remain in equilibrium.

Eventually the nuclear fuel will become exhausted, as will the person operating the air pump. What happens then is the topic of Chapter 16.

IMPORTANT WORDS

Evolution

Zero-age main sequence

Protostar

Gravitational energy

T Tauri star

REVIEW QUESTIONS

1 Why do we believe that a star cannot be infinitely old?

2 What could make a star stop shining?

3 What is the total lifetime of the Sun?

4 Why do highly massive stars have shorter lifetimes than stars of low mass?

5 What are some of the longest and shortest lifetimes among the stars?

6 How do we calculate models for the Sun at different ages?

7 What is the age of the Sun according to the theoretical models?

8 Is the Sun more or less luminous now than when it first became a star?

9 What types of stars does the main sequence consist of?

10 What "accident of birth" caused Sirius to be a different type of star from the Sun?

11 Where do stars come from?

12 Why don't all nebulae collapse and form stars?

13 Why does a contracting cloud heat up?

14 How do nuclear reactions allow a star to stop contracting?

15 Are any stars observed in a pre-main sequence stage?

16 Why can't objects with less than a certain mass become main sequence stars?

17 In what way is a contracting star like a leaky balloon?

QUESTIONS FOR DISCUSSION

1 Suppose astronomers measure the mass and luminosity of a star and conclude that the star has a lifetime of 1 billion years. Later they discover that the star not only emits radiation, but it also uses up energy by blowing off matter in a "stellar wind." If the energy loss per second through the stellar wind is the same as that lost through the luminosity, what would be the new estimate of the lifetime of the star?

2 If the material in a star were always kept well mixed so there could not be any differences in composition between the core and the outer layers, in what general ways would the evolution of the star be changed? Some degree of mixing does, in fact, take place in most stars.

3 In effect the Russell-Vogt theorem states that two essentially identical protostars must evolve into two essentially identical stars.[5] What arguments can you think of that would support this statement as being "reasonable" or would oppose it as being "unreasonable"?

4 If a "Russell-Vogt theorem" were made up to be applied to people instead of stars, it might read as follows: "The entire life history of a person is, in principle, fixed by the complete structure of that person." Even if we pretend that people have no more free will than stars do, the "People Russell-Vogt theorem" is very obviously not valid. What other differences between people and stars have been overlooked here?

5 If the force of gravity were magically made stronger, show that this would cause the luminosities of stars to increase.

6 It is common to think of stars being hot because nuclear reactions release energy, making them hot. Actually the reverse is true: Stars have nuclear reactions because they first were made hot. Does this make sense to you?

REFERENCES

References for all phases in the evolution of the stars will be given at the end of Chapter 16.

[5]The Russell-Vogt theorem is not totally valid, but the exceptions to it are not important for real stars in equilibrium.

SIXTEEN
EVOLUTION II: OLD STARS

Here we conclude our two-part story of stellar evolution with a description of the post-main sequence phases and the ultimate fate of stars. These are areas under heavy attack from current research, and presently held ideas are subject to sudden changes. We will finish by bringing in some of the important observations that have a bearing on the topic.

16.1 Old age

WHEN THE CORE HYDROGEN IS GONE

When a main sequence star runs out of hydrogen in its core, the energy lost from the star through radiation is no longer being replenished by nuclear energy. The star, of course, cannot stop radiating energy since any hot body will radiate as long as it remains hot. It follows that the heat energy leaking out of the star lowers the pressures inside. Pressure can no longer maintain the balance against gravity, and the star must begin contracting once again. It is much like the leaky balloon whose air pump has stopped; the balloon must start collapsing.

We have already seen that a contracting star is necessarily releasing gravitational energy and that this energy release must heat the stellar material. We thus have the surprising fact that a star actually becomes hotter after the nuclear reactions have ceased. A hotter star generally means a higher luminosity, so we also find that the luminosity increases after the nuclear energy sources have been exhausted. These facts appear to be paradoxes, but we should remember that all energy sources have not been turned off, only the nuclear ones. Gravitational energy simply takes over in place of nuclear energy; if you dump energy into some material, the temperature of the material will rise, and it doesn't make any difference where that energy came from. The nuclear-energy sources allow a star to radiate for a long period of time before it has to go back to the gravitational sources.

The detailed structure of a star changes in a very complicated way when the main sequence phase is ended. While the central and middle parts of the star contract as described, the increasing luminosity causes the outer parts to actually expand and cool off. This behavior is by no means obvious, and it results from a very complicated interaction between the matter and radiation in the outer layers of the star. We have already seen that the luminosity, size, and surface temperature of a star are related to each other in the sense that both increasing size and increasing temperature tend to increase the luminosity. In the case under discussion, the size of the star

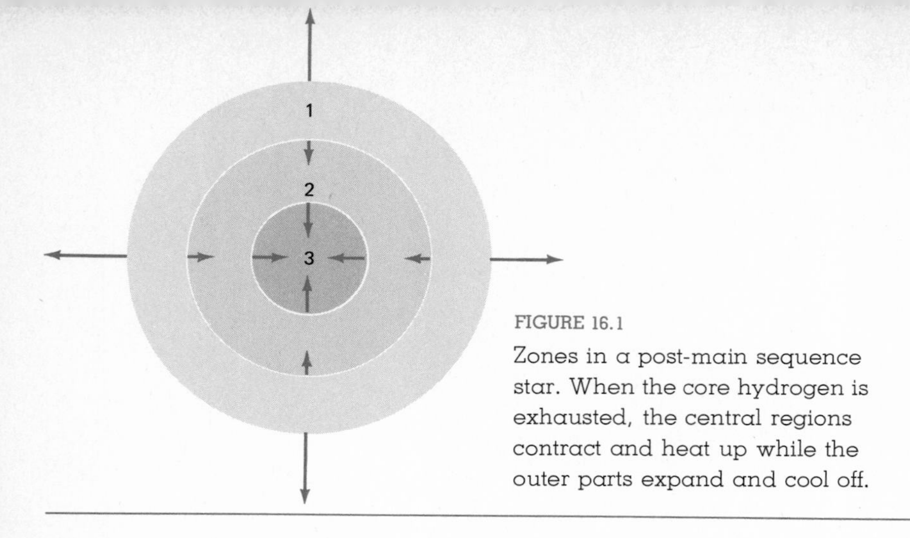

FIGURE 16.1
Zones in a post-main sequence star. When the core hydrogen is exhausted, the central regions contract and heat up while the outer parts expand and cool off.

increases enough to more than provide for the greater luminosity; thus the surface temperature becomes lower.

The inner and outer parts of our star are going in opposite directions. Figure 16.1 illustrates the different zones that are in a star shortly after it has left the main sequence. Region 1 is expanding and cooling, giving the star a larger radius and lower surface temperature. At the same time the inner zones 2 and 3 (which contain most of the mass) are contracting and getting smaller. Zone 3 is the core in which all of the hydrogen has been used up while the star was on the main sequence; thus it is almost pure helium. Zone 2 is farther out and still has unburned hydrogen in it.

FURTHER NUCLEAR REACTIONS

As time goes on, region 2 in Figure 16.1 will continue to become hotter as the inner parts of the star contract. Finally it will become hot enough for hydrogen-burning nuclear reactions to be ignited, and the star will then have a hydrogen-burning shell surrounding its helium core. This will bring about a pause in the contraction of the star, but the contraction will continue when the hydrogen is exhausted in that shell. In this way more and more hydrogen in subsequent shells will be brought to the high temperature at which it will undergo nuclear reactions. Each time a shell loses all of its hydrogen, it joins the growing core of helium-rich material at the center.

While this shell burning of hydrogen continues farther and farther out in the star, the helium core is continuing to grow hotter and more dense due to the gravitational contraction. When the core reaches a temperature of about 400 million degrees Fahrenheit, the helium will ignite in nuclear reactions. By now the star is in the red giant part of the H-R diagram in which it has a large luminosity, a large size, and a cool surface temperature.

GIANTS AND SUPERGIANTS

All nuclei will take part in nuclear reactions if conditions are suitable, and helium is no exception. Carbon, oxygen, and neon are the main products of helium burning, and these reactions also release large amounts of energy. The star has once again attained an equilibrium of perfect force and energy balance, but this time the energy is provided by core helium burning plus shell hydrogen burning, and the star is a cool giant instead of a main sequence star. (Cool here refers only to the surface.)

The star cannot remain in equilibrium long, as the core helium will soon be depleted. Following this will be more contraction and release of more gravitational energy. Internal temperatures will become hotter and hotter, shell burning of hydrogen and now helium will occur, and eventually the carbon-rich core will ignite. We can imagine the processes of nuclear reactions in the core and in shells interspersed between intervals of further contraction and heating of the interior. Heavier and heavier nuclei are built up in the core and in shells around the core. (See Figure 16.4 in Section 16.2.) Can anything stop this cycle from continuing indefinitely? Yes—the details are the topic of Section 16.2.

There are some rather large uncertainties about the luminosities and surface temperatures of stars while they are in these advanced stages of nuclear burning and contraction. There is no doubt that giants and supergiants are in such late stages; they are stars that were once main sequence stars and have by now used up all of their core hydrogen. This does not mean that giants and supergiants are necessarily old in years, for we have seen that massive stars age more rapidly than stars of small mass.

Most stars become giants after their core hydrogen has been burned up, while the most massive ones become the supergiants. The details of the giant and supergiant phases of evolution are much more uncertain than those for stars on or near the main sequence. This is because the model calculations become more complicated and because the important physical processes are less accurately known. An important point is that a star passes through the giant or supergiant phases much more quickly than it does the main sequence stage. More energy is released in the hydrogen-burning nuclear reactions than in those reactions involving the heavier elements, which means the main sequence energy supply is larger. Also, stars are usually much more luminous in the giant or supergiant stage, so their energy is being used at a more rapid rate. For these reasons, stars spend perhaps 90 percent of their lifetime on the main sequence and only about 10 percent as giants or supergiants.

16.2 Death

To understand the possible end stages in the life of a star, we must study the properties of matter at very high temperatures and densities. This requires that we digress to a short discussion about the sizes of the fundamental particles.

DEGENERACY

For everyday objects that we are familiar with, size has an obvious meaning. For particles on the scale of an atom or less, the term "size" becomes somewhat ambiguous. If you say that the size of a particle is the distance from one side of it to the other, then you must be able to tell precisely where the sides are. This doesn't mean that we can't apply the idea of size to tiny particles, but we will have to be careful about applying ideas from everyday experience to places where they may be invalid.

Certain elementary particles, including electrons, protons, and neutrons, are observed to have the following properties: If you add more and more of them to a fixed volume of space, the volume will eventually appear to fill up. After a certain limit, you do not seem to be able to squeeze in any more particles. When that limit has been reached, we can determine the average distance between the particles and call that the size of the particles. This seems reasonable, so let us accept this definition of size.[1]

Suppose we have packed as many electrons into a volume as the volume will hold, and it is now full. But if we give the electrons more energy, by heating them to a higher temperature, for example, then we find the surprising result that the volume is no longer full of electrons; more electrons can be put in before the volume is again filled up. Since there are now more electrons in the volume than before, and since the volume was filled up both times, the electrons must have somehow gotten smaller when we heated them. The strange fact is that the size of the electrons depends on their energy, in other words, on how fast they are moving. The same is true for protons and neutrons: The faster they move, the smaller is their size. If we keep making the particles hot enough, there is no limit to the number that can be forced into a given volume. *Degeneracy* is the name given to the state of the fundamental particles when they are packed together as tightly as possible. There are various degrees of partial degeneracy when the matter is compressed almost, but not quite, as much as possible.

It would seem reasonable that if a volume were completely filled with degenerate electrons, there wouldn't be any room left for anything else. This is not the case. The fact that all electrons possible have been forced into a volume does not limit the number of protons or neutrons that can be put into the same volume.[2] Degeneracy only affects particles of a given type. The idea of fundamental particles being solid spheres that keep other particles

[1]There are other ways to define the size of fundamental particles, but this definition has the advantage of being relevant to an understanding of certain late stages in the lives of the stars.

[2]Of course it doesn't limit the number of electrons either, if the new electrons are given enough energy, that is, made "small enough."

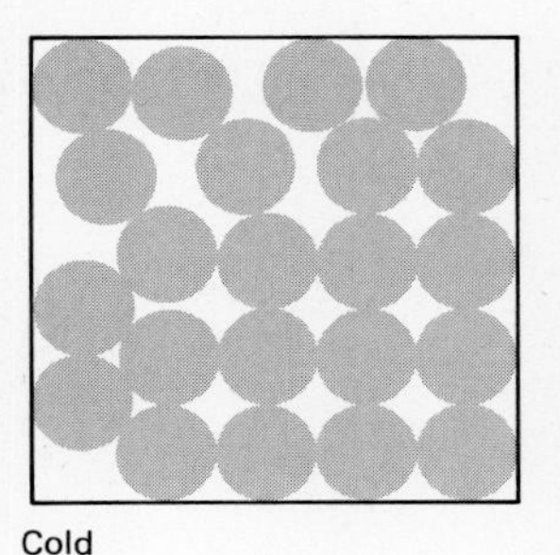

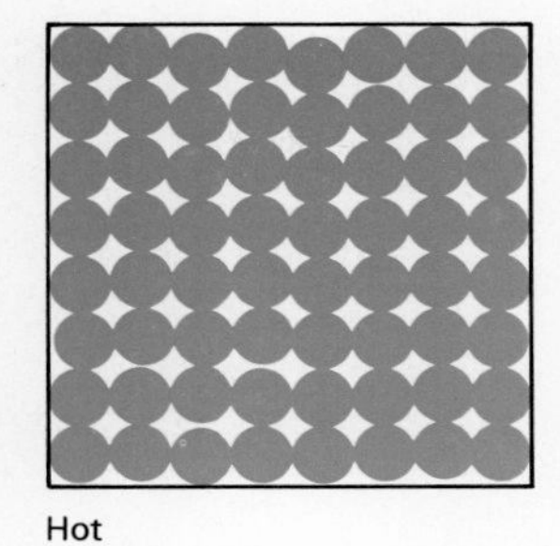

FIGURE 16.2
Degeneracy. When particles are degenerate, more of them can fit into a given volume when they are hot than when they are cold.

from penetrating their surfaces is obviously of limited accuracy (see Figure 16.2).

DEGENERACY IN STARS

What does degeneracy have to do with stars? As we have seen, stars without nuclear-energy sources will normally contract. This is because the luminosity drains heat energy from the star, lowering the pressure of the gas. A gas that is completely degenerate, however, does not need heat energy to provide its pressure. It simply cannot contract because of its degeneracy, and so it can support any amount of weight without budging an inch, providing its temperature is not raised. If the temperature is raised, the particles will become smaller, and the degeneracy will be at least partially removed, permitting more contraction. Barring this, there is no way for a gas to contract if it has become completely degenerate. A possible final state for a star, therefore, is a degenerate gas.

Consider a star that is in one of its contraction phases. The shrinking squeezes its particles closer together, but at the same time contraction heats the material, making the particles smaller. Will the particles ever get completely packed together, or will they shrink in size fast enough to keep avoiding degeneracy? The answer depends on the mass of the star. For large-mass stars, the particles shrink in size fast enough that they never become tightly packed, no matter how high the density becomes; large-mass stars cannot become completely degenerate. Whatever small volume the star contracts to, the particles become small enough to easily fit into that volume. For small-mass stars, the reverse is true. As the star contracts, the particles come closer together more rapidly than they shrink in size, so the star can become degenerate. The largest mass that can become degenerate is about 1½ times the solar mass, the precise value depending on the composition of the material. The Sun can become a degenerate star because it is under the degenerate-mass limit; Sirius cannot do so without losing material because it is over the limiting mass.

SMALL-MASS OBJECTS

Consider a low-mass protostar in the process of contracting. It has not had any nuclear reactions, and it has the usual hydrogen-rich chemical composition. If this object is to become a main sequence star, its core must become hot enough to burn hydrogen. This requires a temperature of close to 10 million degrees Fahrenheit. At this temperature the "size" of the electrons is such that they are degenerate if the matter has a density above about 30 pounds per cubic inch, or 1000 times the density of water. Protons and neutrons are much more difficult to make degenerate, which means that they are much smaller than electrons.

As the protostar contracts, both the temperature and density rise. But which comes first, the temperature needed to ignite the hydrogen or the density needed to make the electrons degenerate? If the mass of the body is less than about 1/12 the mass of the Sun, the density gets there first, and the electrons become completely degenerate before the temperature is high enough for nuclear reactions. The electrons can contract no further. The protostar is a mixture of protons and electrons in addition to a few other things, and the protons are not degenerate, only the electrons. The protons could contract more, but the electric force between the electrons and them won't let them. The electric force is strong enough to prevent the electrons and the protons from completely separating, so the degeneracy of the electrons halts the contraction of the whole protostar. Since it is the contraction that causes the temperature to rise, the interior stops getting hotter. The body is caught short, unable to get its core hot enough to burn hydrogen. The would-be star still has its heat energy, and it will continue to feebly emit radiation and to slowly cool off until it has gotten rid of essentially all of its excess heat. This will take a long time, and eventually the object will become a cold, dark, degenerate mass. There are probably many such objects in space that are now in the process of cooling, but they have luminosities too low to be easily detected.

We might ask whether the planets resist gravitational contraction because of degeneracy, and the answer is no. When matter becomes solid or liquid, it is able to exert a high pressure without the help of its heat energy. In this respect solids and liquids are similar to degenerate gases. The planets condensed to the liquid or solid state before their densities were high enough for degeneracy.

HELIUM BURNING

Objects with more than 1/12 solar mass will be able to ignite the nuclear reactions that convert hydrogen to helium before their densities are large enough for degeneracy. They settle down in force and energy balance and join the main sequence as regular stars. After the central hydrogen is exhausted, there will be further contractions that are interrupted by hydrogen burning in shells around the helium core. Once again the question arises: Will the core temperature become hot enough for helium burning before complete electron degeneracy sets in?

Helium burning needs temperatures of around 400 million degrees Fahrenheit. At these temperatures the size of the electrons is such that they are degenerate at a density of several tons per cubic inch. (*Now* we are getting some of those impressive densities mentioned in Section 12.2.) Calculations indicate that in this case the limiting mass is about ⅓ the mass of the Sun. A star with a mass in the range 1/12 to ⅓ the Sun's will join the main sequence, but degeneracy will overtake it before it can achieve helium burning. It will develop a helium core in which the electrons are degenerate as the hydrogen-burning shell eats its way out through the star. When the hydrogen burning has ended, the electrons in nearly the whole star will be degenerate, and the star will slowly fade away to a small, dark mass of helium nuclei and degenerate electrons.

There shouldn't be any objects like this in the Milky Way, however, because our galaxy isn't old enough. Stars of the type we are discussing stay on the main sequence for 200 billion years or longer, while the Milky Way began star formation only some 15 billion years ago. All stars in the mass range we are considering are still on the main sequence.

THE WHITE DWARFS

Precisely the same arguments can be applied to the next step after helium burning, which is carbon burning. The temperature needed for carbon burning is about 1½ billion degrees Fahrenheit, and the electrons will be degenerate if the matter has a density of about 100 tons per cubic inch. The limiting mass here is calculated to be about 1.5 solar masses, meaning that stars with masses below this limit will become degenerate before they get hot enough to ignite carbon in nuclear reactions.

Our interest is aroused in this case because degenerate stars of this kind probably do exist. Their lifetimes are less than the age of our galaxy, so we might look to see if any stars are known that fit their description. The extremely high densities of the degenerate stars remind us of the white dwarfs mentioned in Section 12.2. Today there is no serious doubt that white dwarfs are stars in which the electrons have become degenerate, halting further contraction. They have collapsed to a small size and are slowly radiating away their available heat energy. They are dying stars in the sense that they are approaching the end of their ability to shine.

MASSIVE STARS

The same process of making white dwarfs cannot be extended to higher masses because we have reached the limiting mass of a degenerate star. Stars with more than the degenerate limit of about 1.4 solar masses do exist. They cannot become degenerate unless they shed

FIGURE 16.3
The Helix Nebula, an example of what astronomers call a planetary nebula. Figure 14.7 and Plate 22 are also planetary nebulae. The name comes from the appearance, and they have nothing to do with planets. The central star ejected the shell of material that is visible, and it will soon become a white dwarf star. (The Cerro Tololo Inter-American Observatory)

part of their material. So what happens to them?

If a star is too massive to have its career ended by degeneracy, nuclear burning (followed by more contraction and heating of the deep interior) will build up heavier and heavier elements in the core and in the surrounding shells. Such a layered star can have difficulty remaining stable. Also, some of these heavy elements can act explosively when they are ignited in nuclear reactions. The highly evolved stars, that is, stars in which many heavy elements have been built up from the original hydrogen, may become unstable enough to eject matter back into space. If enough matter is lost through instabilities, the mass of the star may go below the limit for electron degeneracy, and the star can end as a white dwarf after all (see Figure 16.3).

UNSTABLE IRON CORE

If a massive star doesn't blow itself apart or otherwise fall below the degenerate-mass limit, it will eventually run out of nuclear fuel. Large though the nuclear supplies may be, they will sooner or later be used up. The point is that nuclear reactions that build heavier elements out of lighter ones will release energy only as long as the nuclei involved are lighter than iron. The reactions in which iron is involved absorb energy instead of emitting it. When a star has built up a core of iron, the nuclear-energy supply has been exhausted, except for the shells of lighter elements surrounding the core. The star is like the leaky balloon whose air pump has quit operating; further contraction cannot be prevented.

What happens as the iron core contracts and heats even more? Nuclear reactions with the iron will take place, and these remove heat energy from the surrounding material. When the core iron is ignited, suddenly huge amounts of heat energy are removed from the center of the star. The core is not completely degenerate, so this reduces the central pressure drastically, the same pressure that has been supporting the whole star against the force of gravity. This action is similar to removing the bottom three or four stories from a tall building: The rest of the building soon comes crashing down, which is exactly what happens to the rest of the star. This catastrophic collapse of the star suddenly heats up the outer layers of the star, bringing on all sorts of nuclear reactions that occur at a furious pace. These outer layers contain elements lighter than iron, so the reactions release energy very quickly. The end of all of this is a major explosion in which much of the material is thrown off into space, leaving a small remnant behind. This course of events is illustrated in Figure 16.4.

FIGURE 16.4
Possible evolutionary phases of a star with large mass. The sizes of the star and the various shells are not drawn to scale, and the different shells will not all burn at the same time. (a) On the main sequence. (b) Helium-burning core. (c) Carbon-burning core. (d) Silicon-burning core. (e) Iron core. (f) Supernova.

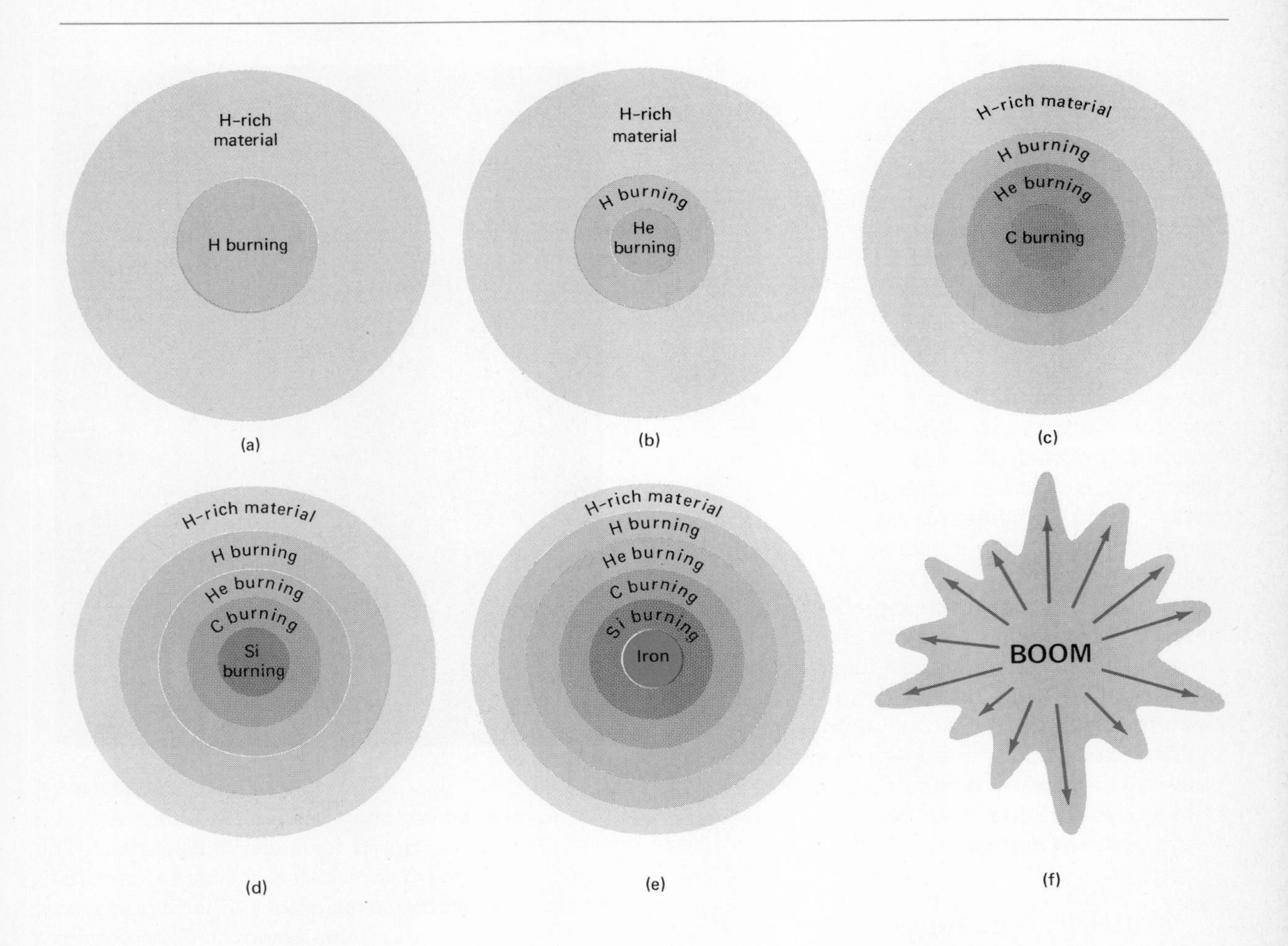

FIGURE 16.5
A supernova explosion in a distant galaxy. The supernova is billions of times brighter than the Sun for a short period of time, but it loses most of its brilliance in a few months. (Lick Observatory photograph)

SUPERNOVAS

Can stellar explosions of this magnitude really occur? In fact, tremendous explosions like this are observed. They are called *supernovas*, and Figures 16.5 and 16.6 are photographs of two of them.

During a supernova explosion, a star very quickly flares up in brightness until it is perhaps a billion times more luminous than the Sun. Over the next few weeks and months its brightness slowly fades. Three supernovas have been observed in our galaxy over the past thousand years. Whether supernovas are actually caused by the events previously described is open to some question, but they certainly do indicate an extremely unstable condition into which some stars can get themselves.

When a supernova explodes and blows much of its mass off, there will be a part of it left behind. These remains can become a white dwarf if their mass is less than the electron-degeneracy limit. For many years astronomers believed that all stars ended as white dwarfs one way or another, and this picture seemed to hold together rather well, although there were many obscure details.

FIGURE 16.6
A supernova remnant. The circular nebulosity in this photograph is the material blown off in a supernova explosion that occurred several thousand years ago in our galaxy. (The Cerro Tololo Inter-American Observatory)

NEUTRON STARS AND PULSARS

Then an alternative to the white-dwarf stage as the final condition of stars suggested itself:

nuclei becoming degenerate instead of electrons. The theoretical existence of such *neutron stars*, as they were called, has been known for many years, but it was not known whether a normal star could evolve into a neutron star. It was also uncertain whether neutron stars had anything to do with reality since they represented a rather long extension of the laws of physics beyond the limits of what had been experimentally confirmed. Then, unexpectedly, they were actually observed: *Pulsars* were discovered.

In the late 1960s, astronomers with radio telescopes discovered a number of objects that give off radiation in pulses instead of the continuous way stars usually shine. About one pulse per second is typical, although one, the Crab Nebula pulsar, gives off 30 pulses each second. In most cases the pulsed radiation is limited to radio waves, but again the Crab pulsar is exceptional: It has been observed to blink off and on 30 times per second in visible light and in x rays, as well as in radio waves. Figure 16.7 shows photographs of this pulsar blinking off and on. This star has been known for many years, but the knowledge that it is pulsing has only come in recent years.

What could cause a star to emit pulses of energy? It is inconceivable that a star could become alternately hot and cold in a fraction of a second, and a star could not pulsate that rapidly. (A pulsar should not be confused with

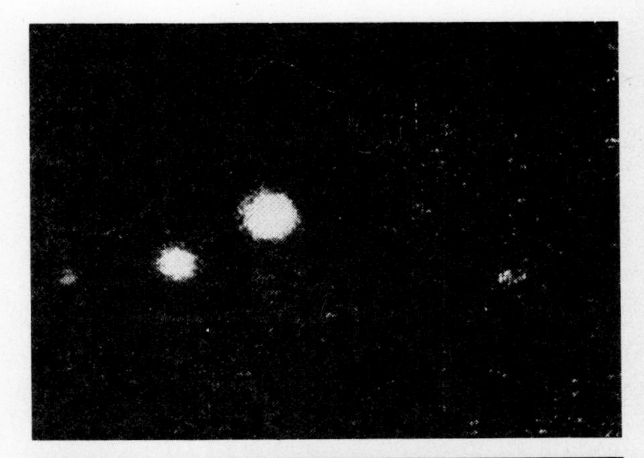

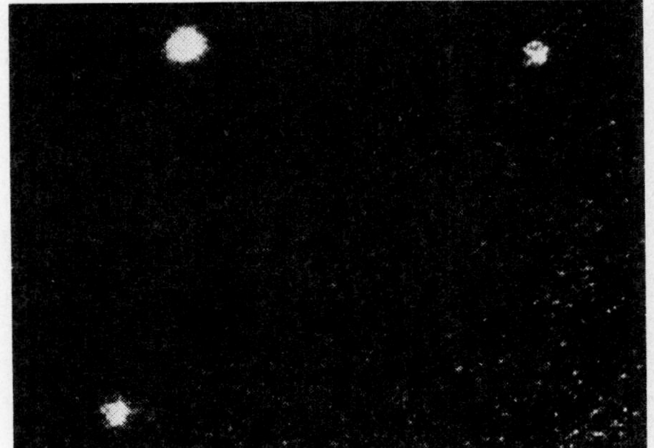

FIGURE 16.7
The Crab Pulsar. The two photographs show the pulsar in the "on" and in the "off" conditions. (Lick Observatory photograph)

FIGURE 16.8
Model of a pulsar.

a pulsating star.) The pulses come at very regular intervals, suggesting a rotating star. If a rotating body gives off a narrow beam of radiation and if this beam comes only from one (or two) points on the surface, a distant observer would see one (or two) pulses of radiation flash by each time the body rotates, like the rotating beacon of a lighthouse (see Figure 16.8). Why pulsars should radiate from only one or two points on their surfaces is uncertain. The radiation is not heat radiation that comes from the heat energy of the objects, as in the case of ordinary stars. Pulsars emit synchroton radiation. (Refer to Section 3.3.) The radiation comes from electrically charged particles being expelled from the bodies along the directions of very strong magnetic fields.

The details of the radiation of pulsars are not understood, but our main interest at present is in what kind of objects the pulsars can be. Specifically, what kind of object can spin around 30 times each second without breaking apart? The answer is a very small one. A grain of sand could be made to spin 30 times a second with no difficulty, but a grain of sand could not be the source of the large amount of energy the pulsars are observed to be emitting. An object must have a large mass, comparable to the mass of the Sun, if it can in any way account for the luminosity of the pulsars. But that mass must be compacted to a very small size to be able to survive a spinning rate of 30 times per second. The only candidate that scientists have been able to come up with is a star that is far more dense even than the white dwarfs: a neutron star. These fantastic stars have densities of several billion tons per cubic inch.

The evidence for the neutron-star interpretation of pulsars is so strong that today there are no serious doubts expressed concerning it, although new evidence could change that, of course. The direct source of energy of pulsars is the rapid rotation. The rotational energy comes from gravitational energy that was released in the past, since the star was forced to rotate more rapidly as it decreased in size.

The Crab Nebula pulsar gets its name from its location in a cloud of interstellar material called the Crab Nebula (see Figure 16.9). It has been known for a number of years that the Crab Nebula is the debris left over from a supernova explosion that was observed in the year A.D. 1054. The Chinese recorded a bright star that appeared for a short time in that year. The Crab Nebula is in the same region of the sky as that star, and the nebula is observed to be expanding at such a rate that we know the expansion began about the year 1054.

We thus have direct evidence for a star passing through various stages of evolution and culminating in a violent explosion that blew off most of the mass, leaving behind a superdense neutron star.

PRODUCING NEUTRON STARS

So how can a neutron star be produced? As we have seen, nuclear reactions tend to build heavy elements out of lighter ones in the cores of stars. When temperatures and densities become too extreme, however, the heavy elements cannot stand up under the violent collisions and energetic radiation to which they are exposed. The heavy nuclei are broken into smaller pieces, these are broken into smaller ones yet, and eventually we are back to the electrons and protons that we started with. At sufficiently high densities, a proton and an electron will combine with each other to produce a neutron, and the core of the star will finally have been broken down to nearly pure neutrons. (At very high densities, neutrons are not unstable.) If the density is high enough for the temperature, the neutrons will become degenerate. If the outer parts of the star are stripped away, a neutron star will be left. The supernova explosion apparently provides the proper conditions, both for producing a degenerate neutron core and for driving off the layers surrounding this core. It should be pointed out that just as there is an upper limit to the mass of a white dwarf, so there is also an upper limit to the mass of a neutron star, although the value of that limit is not well known. It seems that supernova explosions generally bring the neutron core down to a mass below this limit. Although the chain of events leading to supernova explosions and neutron stars is somewhat speculative, it does account for most of the observed facts. This is one of the active fields of current astronomical research.

FIGURE 16.9

The Crab Nebula. This material was ejected by the exploding star over 900 years ago, and it is still observed to be expanding away from the star. Photograph by W. F. van Altena. (The Kitt Peak National Observatory)

A star of large mass begins as nearly pure hydrogen, builds up all sorts of complicated nuclei in its interior, and ends up by breaking them back down again into neutrons. The building up of the heavy elements releases large amounts of nuclear energy, and the same amount of energy must be supplied from somewhere in order to break them down again. And guess what—once again gravity must come to the rescue The net effect of the nuclear reactions is essentially zero, except for those that took place outside the neutron core. We thus have the situation where practically the entire energy output of the star was supplied by gravitational contraction. Nuclear reactions get the credit while gravity does *all* the work.

BLACK HOLES

We still have left a basic question concerning stars of large mass. *If* they get rid of the excess mass, they can end as white dwarfs or as neutron stars. But do we know for sure that they *always* shed the proper amount of material? What happens if they don't?

According to presently held ideas, such a star would keep contracting until its surface gravity became so large that *nothing* could escape from it, matter or radiation. The very light waves themselves would be pulled back by this gravitational monstrosity. This type of object could not be seen because it could emit no radiation, so the term *black hole* has been coined for it. A black hole can make its presence felt through gravity. It attracts exterior objects in the ordinary way—as long as they maintain their distance. If they come too close, however, they will be swallowed up, never to appear again to the rest of the Universe.

Black holes are a long way from anything that has been experimentally checked, and their existence may or may not be real. No one can deny that there may be as yet unknown effects that will considerably change our ideas. Still, we must continue by plausible extrapolations of what is known now; by this criterion, black holes are quite possible.

X-RAY SOURCES

Scientific satellites in recent years have detected a number of strong sources of x rays in space. These sources cannot be detected from the ground because of the Earth's atmosphere. In several cases the x rays have been found to be coming from binary stars (two stars moving in orbits around each other). The x-ray sources have a variable brightness, and some of them send out x-ray pulses every second or so in a way that is reminiscent of the pulsars. Maybe one of the stars in such a binary system has passed through its life cycle and become a neutron star. It has been suggested that the x rays come from matter that is streaming onto the neutron star from its companion.

As mentioned in Section 12.2, we can find the mass of a star if it is a binary star. One of the binary x-ray sources, called Cygnus X-1 (see Figure 16.10), is composed of a hot supergiant plus a smaller star, and both stars probably have too much mass to become degenerate, although there is some question about this. If the smaller star really is compact as is suggested but not demanded by the observations, it is a good candidate for being a black hole.

THE FUTURE OF THE SUN

Let us now take a brief look at the probable future of the Sun. Over the next 4 to 5 billion years, it should increase slightly in surface temperature and become about twice as luminous as it is now as it slowly uses up its core hydrogen. Then contraction of the interior and hydrogen burning in shells around the core will cause the surface to expand and become cooler. Some 6 billion years from now the Sun will approach a rather violent onset of helium burning at the center. This will make some sudden changes in the Sun's internal structure, and the difficulty of calculating this phase makes it impossible to know exactly what will occur. It is estimated that the Sun will become nearly 1000 times more luminous than it is now, and its surface layers will cool and expand to a huge size, perhaps engulfing some of the inner planets. It is possible that this phase will be violent enough for the Sun to lose a small amount of mass, but in any event, it will certainly destroy any life that might be left on the surface of the Earth.

The Sun will then be a red giant, convert-

FIGURE 16.10

Cygnus X-1. The star indicated is a binary star. One of the stars is a supergiant, and the other, too faint to be seen, might be a black hole. Photo by W. R. Forman at the Agassiz Station. (Harvard College Observatory)

FIGURE 16.11
A schematic view of stellar evolution.

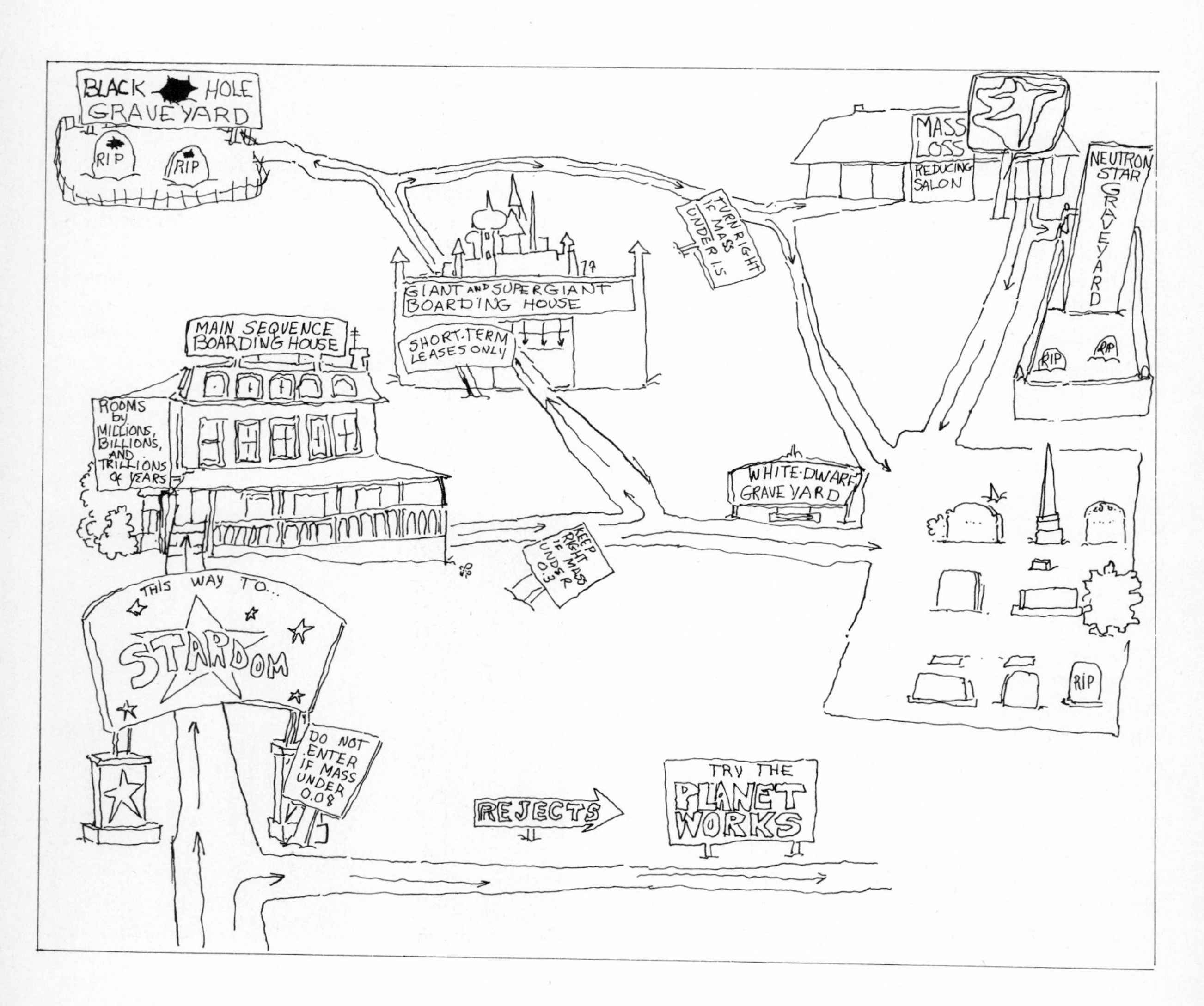

FIGURE 16.12
The double cluster in Perseus. These two clusters contain very young stars. The H-R diagram of these stars is shown in Figure 16.14. (Lick Observatory photograph)

ing helium to carbon in its core. The central helium will be used up in only a few tens of millions of years, and more core contraction will bring it to the verge of carbon burning. Before the core can ignite its carbon, however, electron degeneracy will set in and keep it from getting any hotter. Shell burning of hydrogen will continue until the degenerate helium core has grown to include most of the star. It will then be a white dwarf, very small and very faint but very hot. This will occur about 7 billion years from now.

The Sun will spend the following several billion years cooling and growing fainter, approaching the "white dwarf graveyard", and carrying whatever planets managed to survive earlier ordeals along with it to oblivion.

16.3 The observations

CHECKING THE THEORY

How trustworthy are the theoretical results just outlined? Experience has often shown that if theory strays very far from observational checks, it deserves some suspicion. It is nice to know that such things as main sequence stars, white dwarfs, supernovae, neutron stars, and so forth, actually do exist, but can we be sure that they are correctly accounted for by our theories of stellar evolution? Fortunately, there are more detailed observational checks that can be made.

The Russell-Vogt theorem plays an important part in understanding stellar evolution. It states that all properties of a star are fixed by its mass and composition. The internal composition is changed by nuclear reactions in a way that depends on how old the star is. Therefore, the theorem can be restated in the more useful form: All properties of a star are fixed by its mass, original composition, and age. This means that if two stars differ from each other in some way, the difference is due to the stars' having different masses, original compositions, or ages, or a combination of these characteristics.

STAR CLUSTERS

The stars around us are a mixture of all masses, all compositions, and all ages. Trying to untangle the observed data can be very difficult.

In this respect, a great amount of assistance is rendered astronomers by clusters of stars. A *star cluster* (see Figures 16.12 and 16.13) is a group of stars held together by the gravity of the stars. Star clusters are similar to binary stars except that a cluster can have any number of stars in it, up to 100,000 or more for the largest ones. There are good reasons for

FIGURE 16.14

Schematic H-R diagrams of four different star clusters. [Based on data of H. L. Johnson and W. A. Hiltner, *Astrophysical Journal* 123 (1956), 267; H. L. Johnson, A. R. Sandage, and H. D. Wahlquist, *Astrophysical Journal* 124 (1956), 81; and A. R. Sandage, *Astrophysical Journal* 135 (1962), 333; published by the University of Chicago. Copyright © 1956 and 1962 by the University of Chicago.]

FIGURE 16.13

The globular cluster M 3. The stars in this large cluster are all very old. (Lick Observatory)

believing that all members of a star cluster are formed at about the same time and out of the same material. The different stars in one cluster, therefore, are of the same age and original composition. They differ from each other only in mass. By studying the members of a single cluster, we can obtain data concerning the effects of mass differences on the properties of stars. By comparing two clusters with each other, we can directly observe the effects of age and original composition.

As was mentioned before, however, the effects of variations in original composition are rather small, so we shall confine our attention to mass and age for the remainder of this section.

AGES OF CLUSTERS

Figure 16.14 shows schematic H-R diagrams of four star clusters. Individual stars are not shown, only lines that indicate where the points representing stars occur. Note that this figure is based on actual observations of real stars, not on theoretical models.

First is the H-R diagram of the cluster named NGC 2362. The stars in this cluster appear only along a narrow line running from the lower right to the upper left, in other words, from cool, faint stars to hot, luminous ones. There are no other points in the H-R diagram of this cluster. The diagonal line is, of course, the

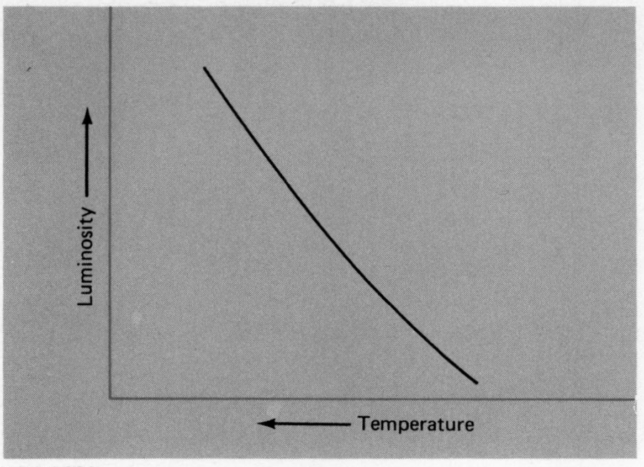

NGC 2362

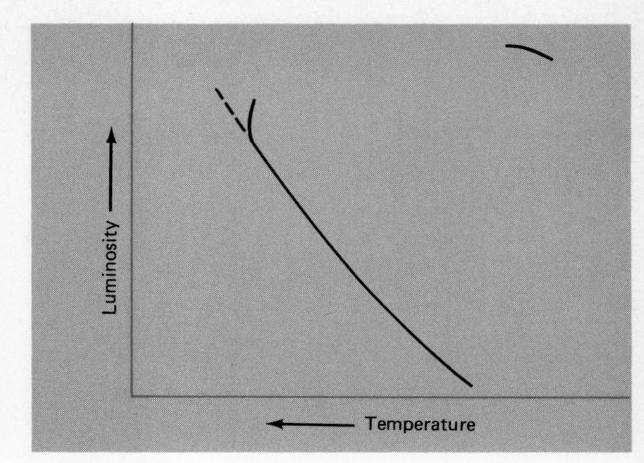

h and χ Persei

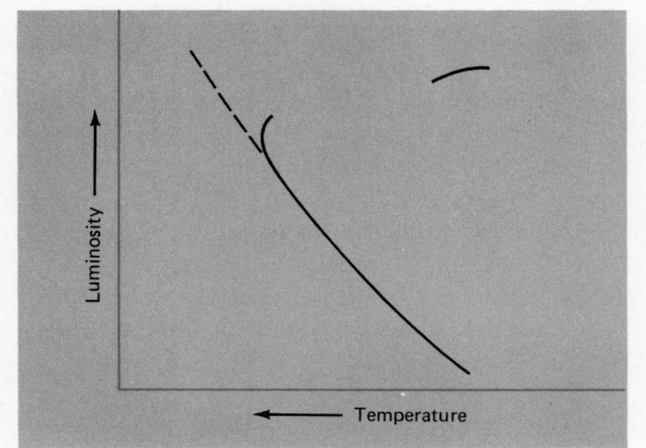

M 11

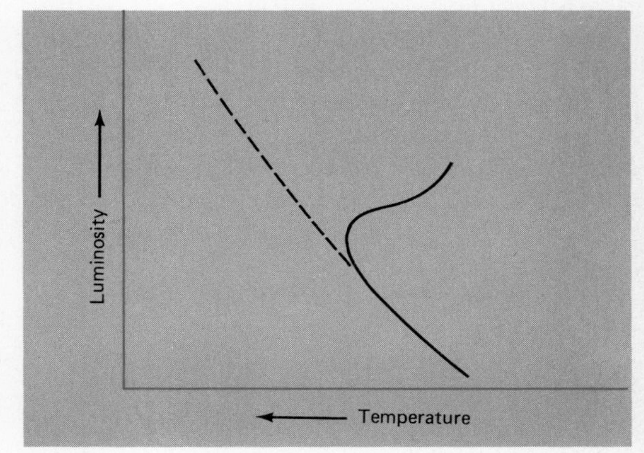

NGC 188

main sequence. Since not even the luminous stars of high mass have evolved off of the main sequence as yet, NGC 2362 must be extremely young, perhaps only a million years old. Since the cluster is so young, the line in the figure is not just the main sequence but the zero-age main sequence (ZAMS), representing stars that still have almost all of their original core hydrogen. Because an H-R diagram like this is actually observed, any doubts that stars begin their careers on the main sequence should be dispelled.

Next in Figure 16.14 is the H-R diagram of the well-known double cluster in the constellation of Perseus. The clusters are named h and χ Persei, where χ is the Greek letter *chi*. The ZAMS found from NGC 2362 has been added as the dotted line. The clusters h and χ are seen to have a main sequence plus a few cool supergiants at the upper right side of the diagram. The main sequence lies right along the ZAMS most of the way, but at the upper left end, h and χ extend a little above and stop a little before the end of the ZAMS. How do we interpret this? The answer is that h and χ are slightly older than NGC 2362 but still quite young. Remember that the lifetime of stars of small mass is very long, while large-mass stars burn up their nuclear fuel very rapidly. Since the large-mass stars occur in the upper left part of the H-R diagram, it is just this end of the main sequence that will first show the effects of age. The very large mass stars in h and χ Persei have already changed from hot, luminous main sequence stars into the red supergiants we now see. Stars a little bit less massive have used up part of their core hydrogen but not all of it; thus, they are a little above the ZAMS but still in the general region of the main sequence. In another few million years they will also evolve off to the right of the diagram and become cool supergiants. Stars of still lower mass change so slowly that they have not yet had time to move appreciably away from the ZAMS.

The cluster M 11 is the next to be considered. The main sequence of M 11 is much shorter than the main sequences of the previous clusters. Also, M 11 has a rather large number of cool giants but no supergiants. This cluster is much older than both NGC 2362 and h and χ Persei. The stars of very large mass that used to be on the top left part of the main sequence and later evolved into cool supergiants have all burned out by now in M 11; today they are either white dwarfs, neutron stars, or black holes and are much too faint to be seen. The M 11 red giants have changed from moderately hot main sequence stars only in the past few tens of millions of years. Note that as stars of fairly large mass use up their core hydrogen, they evolve in a way that carries them across the H-R diagram from left to right into the cool, red-giant region. This part of the evolution takes place very rapidly, so very few stars are actually observed halfway between the main sequence and the red-giant region. The stars in M 11 that are slightly less massive than its giants are still on the main sequence, a little above the ZAMS and ready to "jump across" to the red-giant section. Stars of still lower mass have not evolved off of the ZAMS.

NGC 188 is the oldest of the four clusters. All stars of moderate to high temperature have evolved away from the main sequence and have burned out by now. The giants in NGC 188 have lower mass, lower luminosity, and come from farther down on the main sequence than the giants in younger clusters. This cluster is probably about 10 billion years old. We see that the giant branch for such an old cluster is directly connected to the main sequence. This is an indication that stars of moderate mass take much longer to change from the main sequence to red giants than do stars of larger mass.

It again should be emphasized that Figure 16.14 is based on the data from observations of real stars. One can determine a great deal about the evolution of stars purely from observations. When theory and observation are combined, our understanding increases correspondingly. Weak points in the theory can be strengthened by forcing the models to agree with the observations. Ambiguities in the interpretation of observations may be cleared up by appealing to the theoretical models. The H-R diagram is the field (sometimes the *battle*field) where theory and observation are brought together. Small wonder that the life of a stellar astronomer, to a large extent, revolves around the H-R diagram.

This completes Part Three, but we are not yet finished with the life stories of the stars. The galaxies, which will be discussed in Part Four, are huge collections of stars, and understanding them will require many applications of our knowledge of stellar evolution.

IMPORTANT WORDS

Degeneracy
White dwarf
Supernova
Neutron star
Pulsar
Crab Nebula
Black hole
X-ray source
Star cluster

REVIEW QUESTIONS

1 When a star exhausts its core hydrogen, what happens to its luminosity? Why?

2 Does the whole star or only part of it contract?

3 What is meant by shell burning?

4 Can helium burn?

5 Do stars spend a large fraction of their time as giants or supergiants?

6 How might we measure the size of fundamental particles?

7 What does degeneracy mean?

8 How is degeneracy important in stellar evolution?

9 What is special about an object having $1/12$ the mass of the Sun?

10 What types of stars will become degenerate before they can ignite helium in their cores?

11 What is a supernova, and how does it fit into the evolutionary picture?

12 Where does the name "pulsar" come from?

13 How are supernovas related to neutron stars?

14 What is a black hole?

15 About how long will it be before the Sun becomes a red giant?

16 What is a star cluster?

17 Why are star clusters important in studying stellar evolution?

18 How can the H-R diagram of a star cluster tell us the age of the cluster?

19 What does ZAMS stand for?

20 About how old is NGC 2362, a very young cluster?

21 About how old is NGC 188, a very old cluster?

QUESTIONS FOR DISCUSSION

1 The interstellar gas and dust lose material to stars when star formation takes place. They gain material when stars blow off mass at various stages of their evolution. Can you say anything about the question of whether stars and interstellar matter can keep on existing together and exchanging material indefinitely?

2 Suppose a star were formed out of pure carbon instead of almost pure hydrogen and helium. In what general ways would its structure and evolution be affected by this?

3 The star Procyon B is a white dwarf and is in a binary system with the bright star Procyon. (Refer to the star maps in Appendix 7.) If the mass of Procyon B is 0.6 that of the Sun, what can be said about the possible past history of Procyon B? Remember that no stars in the Milky Way can be older than about 15 billion years. (Note also the lifetimes listed in Table 15.1.)

4 The white dwarf Sirius B has a mass about the same as that of the Sun. It is a binary companion of the star Sirius. If Sirius has a lifetime of 0.7 billion years, what can you say about the past of Sirius B?

5 Suppose there was once a binary system with one star having 3 solar masses and the second star 1.0 solar mass. The more massive star rather quickly burnt itself out, possibly throwing off much of its mass and becoming a white dwarf. Suppose that some of the mass lost by the first star fell onto the second star, making its mass 2.2 solar masses instead of the original 1.0.

How is the evolution of the second star affected by this mass exchange? Could this possibly be what happened to Sirius? Do you think this possibility of mass exchange could change your answer to question 4 here?

6 There are a number of energy and angular-momentum calculations that can be applied to the observations of pulsars, and they all indicate that pulsars are very high density neutron stars; however, these arguments are much less direct than those from which we conclude, for example, that Sirius B is a white dwarf. One might say, "There is no reasonable doubt that pulsars are neutron stars." At the other extreme, one could say, "Neutron stars are the best available interpretation right now, but we should keep an open mind since we do not have nearly enough evidence to claim that it is proven."

Which statement do you think gives the better assessment of the situation? Is it relevant that neutron stars were predicted long before the pulsars were discovered? (Most astronomers would probably side with the first statement rather than the second.)

7 Draw an H-R diagram showing the ZAMS. After studying Figure 16.14, draw arrows from different parts of the main sequence to the appropriate areas in the giant or supergiant regions which show the approximate paths of evolution stars follow after their main sequence lives. Each arrow should connect the main sequence position of a star with the later giant or supergiant position of the same star.

REFERENCES

A number of popular level books cover the subject of stellar evolution, including: *Atoms, Stars, and Nebulae*, 2nd ed., by L. H. Aller (Harvard University Press, Cambridge, 1971); *Violent Universe* by N. Calder (Viking, New York, 1969); *The Structure of the Universe* by E. L. Schatzman (McGraw-Hill, New York, 1968); and *Astronomy of the Twentieth Century* by O. Struve and V. Zebergs (Macmillan, New York, 1962). Numerous *Scientific American* articles on the subject are reprinted in *New Frontiers in Astronomy*, edited by O. Gingerich (Freeman, San Francisco, 1975).

There are numerous articles on the more exotic parts of stellar evolution, such as: "The Crab Nebula and Pulsar" by V. Trimble in *The Emerging Universe*, edited by W. C. Saslaw and K. C. Jacobs (University of Virginia Press, Charlottesville, 1972); "X Rays From Supernova Remnants" by P. A. Charles and J. L. Culhane [*Scientific American* 233 (December 1975), 38]; "After the Supernova, What?" by J. C. Wheeler [*American Scientist* 61 (January-February 1973), 42]; "Supernovas in Other Galaxies" by R. P. Kirshner [*Scientific American* 235 (December 1976), 88]; "Black Holes" by P. C. Peters [*American Scientist* 62 (September-October, 1974), 575]; "Historical Supernovas" by F. R. Stephenson and D. H. Clark [*Scientific American* 234 (June 1975), 100]; and "The Gum Nebula" by S. P. Maran [*Scientific American* 225 (December 1971), 20].

PART FOUR
GALAXIES AND COSMOLOGY

Here we are concerned with those huge aggregates of stars known as galaxies and, finally, with modern methods of investigating the structure of the Universe as a whole. Chapters 17 and 18 cover the Milky Way system, our own galaxy. Chapter 19 describes other galaxies, and Chapter 20 is a presentation of cosmology.

SEVENTEEN

THE MILKY WAY I: SIZE AND SHAPE

Our position in the Milky Way makes it difficult for us to discover its over-all properties, but in this chapter we find out how this difficulty can be overcome.

17.1 Star counts

All of the stars we can see at night belong to one huge system. All of the stars that can be seen or photographed with the largest telescopes, except for some of the brightest stars in the nearest galaxies, also belong to that one system. The system is called the *Milky Way*, and it is the galaxy that the Sun belongs to. Figure 17.1 is a photograph of an incredible number of stars. Each little point of light, remember, is a large, hot mass of the general size and brightness of the Sun. Even all of these stars are only a very tiny part of the Milky Way. In addition to the stars, of course, our galaxy also contains interstellar matter, the gas and dust that exist between the stars.

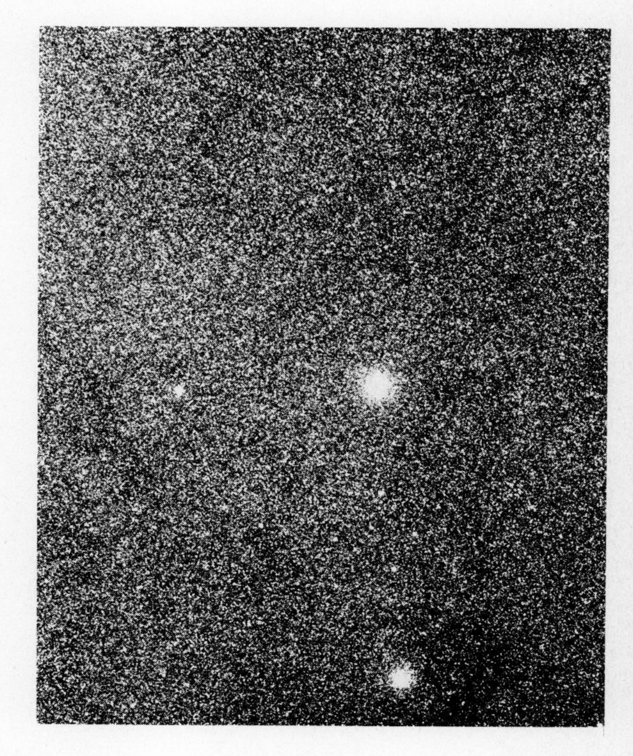

FIGURE 17.1
Even all of the stars shown in this photograph are but a very tiny part of our Milky Way. Shown is a region in the constellation Sagittarius. (The Kitt Peak National Observatory)

ONE SYSTEM

It should be understood what is meant by the statement that the Milky Way is one system. Just as all of the planets, comets, and meteors are held together by the gravity of the Sun, so all the material in the Milky Way is held together by gravity. Stars are seen to come and go

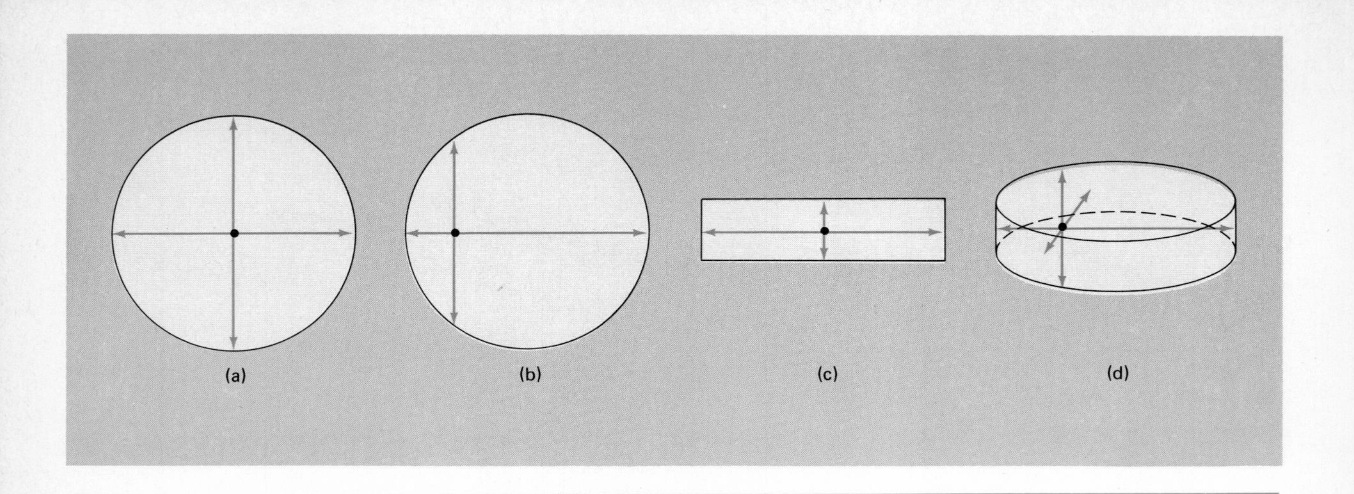

along their separate paths, but all of these paths are controlled by the gravity of the Milky Way as a whole. In the solar system, the force of gravity is strongly concentrated in the Sun, since the Sun has most of the mass; in the Milky Way, gravity is spread over a very large volume since no one star has enough mass to dominate all of the others. Our galaxy is a physical system held together by its own gravity. The stars and interstellar matter move around within this system, but they don't have enough energy to escape from it.

How do we know that the Milky Way is one isolated system? Can we be sure that the stars don't just go on forever in all directions, filling the Universe with matter? The answers to these questions have been very difficult to come by, and it has been only in the past half century that astronomers have understood the general features of our galaxy.

POSSIBLE SHAPES

Finding the size and shape of the Milky Way is a matter of finding the positions of large numbers of stars. The best way to find the position of a star is to accurately measure the direction in which we see it and its distance. We have already seen that it is very difficult or impossible to measure accurate distances for all except the very closest stars; however, we can get at least some information without knowing the distances. If we sight along a direction in which our galaxy is very thick, we should see a large number of stars; if we look along a direction in which the Milky Way is quite thin, there will be a much smaller number of stars to be seen. We ought to be able to find the approximate shape of the system simply by comparing the numbers of stars visible in different directions.

Figure 17.2 illustrates some possible shapes of the Milky Way. If it were a ball with us near the center, as in (a), we should see about equal numbers of stars in all directions. If we were located near the edge of a sphere, as in (b), the number of visible stars would be small as we look toward the edge, and it would increase smoothly as we look further and further from the edge. A bar-shaped galaxy, as in (c) would exhibit a maximum number of stars along the length of the bar, and fewer stars would be seen in other directions. In a galaxy shaped like a disk, as in (d), few stars would be seen above and below the disk, and many would be visible in directions along the plane of the disk. For any other shape you wish to imagine, you can figure out how the number of stars seen will depend on your location and on the direction in which you look.

APPEARANCE OF THE MILKY WAY

So many people today are living among the city lights and smog that they rarely, if ever, have a clear night sky in which only the stars

FIGURE 17.2
Possible shapes of the Milky Way. (a) Center of a sphere. (b) Edge of a sphere. (c) A long rod. (d) A thin disk.

themselves provide the main illumination. Even the Moon must be excluded if we wish to see the fainter stars without interference. Under these conditions, a person has a good chance of noticing a white, cloudlike band stretching across the sky. It is most conspicuous during July and August, although parts of it are visible any time of the year. This band is known as the Milky Way because of its appearance, and it gives its name to our galaxy. The Milky Way band is a circle extending all around the sky. Galileo, in 1609, was the first person to look at this band with a telescope. He saw that it is composed of very large numbers of stars, most of them too faint to be seen separately.

Figure 17.3 shows four photographs contrasting the numbers of stars that occur in a direction within the band of the Milky Way and the numbers that occur in a direction away from this band. The band lies in a flat plane

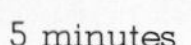

FIGURE 17.3
The two photographs at top were taken near the galactic pole, and the two at bottom were taken near the galactic equator. The greater numbers of stars near the equator are apparent. We can also see the effects of increasing the exposure time from 5 minutes to 50 minutes. Photographs by B. J. Bok and C. C. McCarthy. (Steward Observatory, University of Arizona)

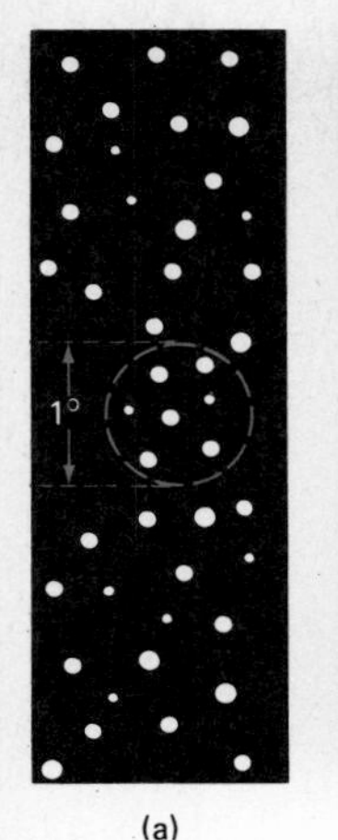

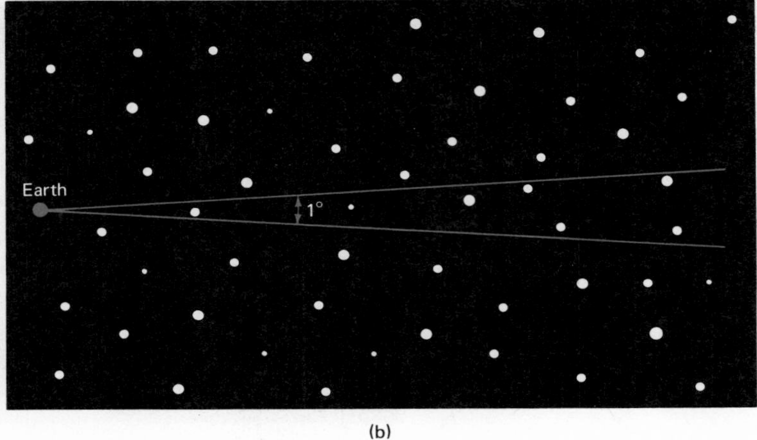

FIGURE 17.4
Counting the stars in space. (a) Hypothetical photo of stars. (b) Actual positions of stars in space.

called the *galactic plane*. Very simple observations thus show that the stars around us are concentrated into a flat disk or plane and that we are located in that plane. Figure 17.2(d) seems to best represent the shape of the Milky Way.

FINDING THE EDGE

We would next like to know how big our galaxy is. This is a more difficult problem than simply finding the shape, but we will find a way to discover the approximate answer. We have found that the nearest star to the solar system is about 4 light years distant from us. (Remember that a light year is a distance of about 6 trillion miles.) The stars near us average about 5 or 6 light years apart, so let us assume that throughout the Milky Way the stars average this distance apart. This means that there would be approximately one star for each volume of 200 cubic light years.

Figure 17.4(a) shows a hypothetical photograph of some stars, and in 17.4(b) we see the actual positions in space of the same stars. The stars that appear inside the circle of diameter 1° on the photograph are actually somewhere within the long cone of opening angle 1° and centered on the Earth. The number of stars that occur within this cone obviously depends on how closely packed the stars are and on how far the system of stars extends. But we know how closely packed the stars are, or at least we are pretending to know: one star for each 200 cubic light years of space. If we measure the number of stars inside the cone, therefore, we should be able to determine how far the stars extend in that direction.

We will first consider the direction up and away from the galactic plane. This direction (and the one opposite) is known as the *galactic pole*, and toward the galactic poles there are relatively few stars. Suppose we take a photograph of the region using a large telescope. If a short exposure time is used, only the brighter stars will appear on the photograph. By taking photographs with longer exposure times, fainter stars will appear in addition to the brighter ones. The number of stars that can be seen in a 1° circle on the photograph will increase as the exposure time is made longer. After a certain limit is reached, however, longer exposure times are found to add only a very few additional stars. This limit suggests that we have reached the edge of the system of stars, so that practically all of the stars in the 1° field cone now appear on the photograph.[1]

If we count the limiting number of stars in our 1° circle on the photograph of the galactic pole, we find it to be about 700. Thus, the 1° cone extending in this direction from the Earth out to the edge of the Milky Way contains about 700 stars. As we are assuming that each star

[1]There are other effects that can cause this type of limit, but we shall pretend that they are not important here.

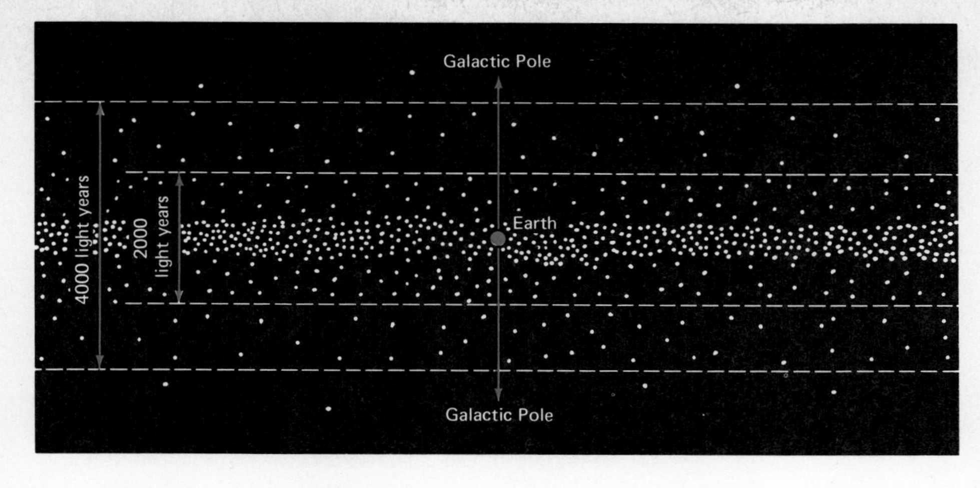

FIGURE 17.5
A cross section of the Milky Way near the Earth.

has an average of 200 cubic light years to itself, this cone must have a volume of about 700 × 200 or 140,000 cubic light years. Now, a cone having an angle of 1° and a volume of 140,000 cubic light years must have a length of nearly 1000 light years. The conclusion is that the Milky Way extends about 1000 light years in the direction of the galactic pole. About the same numbers would be obtained if we had considered the opposite direction (toward the other galactic pole), so our numbers indicate that our galaxy has a total thickness of about 2000 light years at the position of the Sun.

This analysis is, as usual, oversimplified. The Milky Way does not have a sharp edge beyond which there are no stars; instead, the stars thin out gradually at great distances. More elaborate investigations count the numbers of stars at different apparent brightnesses. These investigations indicate that about ⅔ of the stars are contained in a thickness of 2000 light years, while a few stars extend more than twice this far toward the galactic poles.

17.2 Overall properties

WITHIN THE GALACTIC PLANE

If we try to find the size of our galaxy along the galactic plane, in other words, in the directions toward the band of the Milky Way, we find that counting stars does not work. There are two reasons for this:

1 The extent of the Milky Way in these directions is too great for even the longest exposure photographs to record most of the stars out to the edge.

2 Interstellar dust occurs mostly in the plane of the galaxy, and it blocks off starlight so well that we cannot see very far in this plane.

One way to study the Milky Way along the galactic plane is to observe other galaxies and try to make analogies with our own. Of the several types of galaxies that exist, only those showing a conspicuous spiral pattern are in the shape of thin disks, as the Milky Way seems to be. It is thus reasonable to expect our galaxy to look much like the many known spiral galaxies, and detailed observations, including those with radio telescopes, show that this is the case.

Figures 17.6 and 17.7 are two photographs of spiral galaxies that we have reason to believe are quite similar in shape to the Milky Way. In the edge-on view, note the dark lanes in the plane of the galaxy that obscure the light of the stars in the disk. This photograph indicates that the galaxy is some 10 to 20 times as wide as it is thick, and other spirals have similar proportions. Since we found the thickness of the Milky Way at the position of the Sun to be somewhere around 4000 to 5000 light years

FIGURE 17.6
The spiral galaxy NGC 4565. We are viewing it edge on. Note the dark lane of dust that occurs in the disk. (Lick Observatory photograph)

FIGURE 17.7
The spiral galaxy M 101. This galaxy would look very much like the one in Figure 17.6 if we could see it from the edge. (Lick Observatory photograph)

(see Figure 17.5), we can expect its diameter to be somewhere in the neighborhood of 50,000 to 100,000 light years. More accurate studies confirm 100,000 light years as a good estimate. The Milky Way is indeed a huge system of stars.

HOW MANY STARS?

Let's try to find out how many stars are in the Milky Way. It is shaped like a thin circular disk. Such a disk that is 100,000 light years in diameter and about 5000 light years thick has a total volume of 40,000,000,000,000 or 4×10^{13} cubic light years. We noted before that in the space around us, there is an average of about one star for each 200 cubic light years. Once again let's assume that this average holds for the entire Milky Way. If each star takes 200 cubic light years, and if our galaxy has 5×10^{13} cubic light years available, then it is apparent that we have room for $(5 \times 10^{13})/200$ or 2×10^{11} stars. This is 200 billion stars.

Can this number really be correct? Think what this means: Throughout history the phrase "as many as there are stars in the sky" has been used to denote a number beyond compare, too large to be counted. Yet we claim to have counted all of the stars in the sky plus a lot more, and we aren't even breathing hard from the exertion. Is this some kind of swindle?

"Oho!" you might exclaim. "You haven't actually counted the stars. You have only estimated their number, using assumptions that are not completely correct. You don't even expect your answer to be *exact*. What's the matter, isn't astronomy an exact science?"

Of course we haven't counted the stars one-by-one, and of course the assumptions used are not rigorously correct. We have already seen that the stars are not spread evenly throughout space, so our figure of 200 cubic light years per star cannot be exact. Also, the dimensions of the Milky Way are also approximate. Thus our final answer might be wrong by a factor of 2, or even much more.[2] There are many cases in which astronomers hope that the errors are no greater than a factor of 10.

No, Virginia, astronomy is *not* an exact science.

OUR LOCATION

We now have a fairly good idea of the size and shape of the Milky Way and about how many stars it contains. Let's next see if we can find our position in this system. Are we near the center or far out in the disk? The occurrence of a special type of astronomical object helps us to find the answer. This object is the *globular cluster* (see Figure 17.8; see also Figure 16.13).

Star clusters are groups of stars that are

[2]This means that the correct figure could be as large as 400 billion or as small as 100 billion stars.

FIGURE 17.8
The globular cluster Omega Centauri. Compare this photograph with Figure 19.4, in which the globular clusters are just barely visible as tiny dots. (The Cerro Tololo Inter–American Observatory)

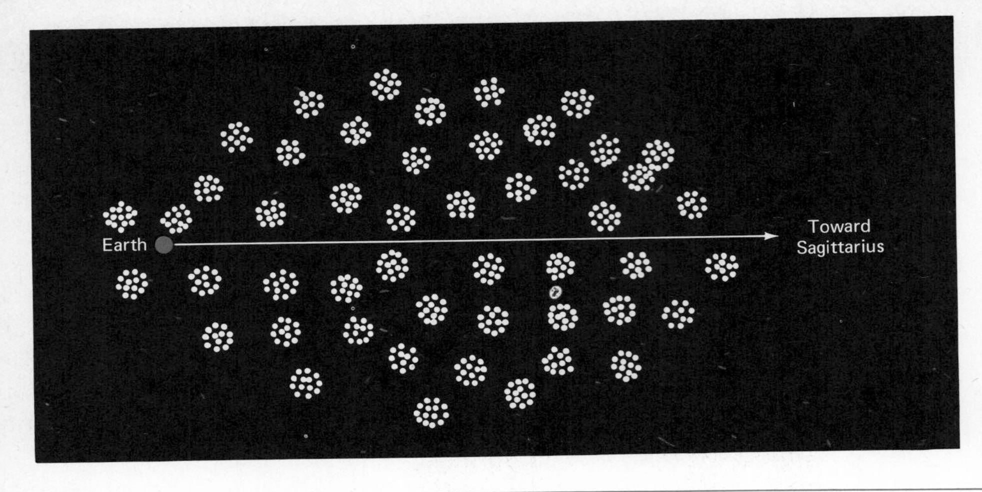

FIGURE 17.9
Distribution of the globular clusters.

bound together by their own gravity. Galaxies contain many star clusters, as well as individual stars. Globular clusters are the largest kind, each containing up to 100,000 or more stars. Such huge clusters are rather like small galaxies in their own right, although they are not known to occur separately in space.

There are about 100 globular clusters known to be part of the Milky Way. These clusters are all very far away from us, but we can see them because, unlike most individual stars, globular clusters can occur far above and below the galactic plane in which the obscuring dust lies. We can observe globular clusters in other galaxies, and they appear symmetrically distributed around the centers of those galaxies. It thus seems reasonable to assume that our own globular clusters are randomly spread around the center of the Milky Way; if we can find the center of the system of globular clusters, that is, the average position of all the clusters, then this should locate the center of our galaxy.

If we are at the center of the Milky Way, we expect to see the globular clusters spread around us evenly in all directions. What is observed is actually quite different: In a list of 119 known globular clusters, 109 of them appear in the half of the sky that is centered on the constellation Sagittarius, and only 10 are seen in the opposite half of the sky.[3] It is thus apparent that we are a long distance from the center of the Milky Way (see Figure 17.9). The direction toward Sagittarius, which seems to be the center of symmetry for the globular clusters, is in the galactic plane; it is undoubtedly the direction toward the center of our galaxy.

THE DISTANCE TO THE CENTER

If we are not at the center of the Milky Way, how far away from it are we? We could answer this if we could find the distances to the globular clusters, since the galactic center should be at the average position of the clusters. Unfortunately, it is not easy to find accurate distances to the globular clusters.

The best way to find the distances to the globular clusters is by means of a certain type of pulsating or "variable" star as it is more commonly called. This type is known as the *RR Lyrae variable*. Like all variable stars, RR Lyrae stars fluctuate in luminosity over a period of time, the period in this case being anywhere from several hours to a day or so. RR Lyrae variables can be distinguished from other types of variable stars by the special way they change with time (see Figure 17.10). The

[3]Data from H. C. Arp, *Galactic Structure*, edited by A. Blaauw and M. Schmidt, University of Chicago Press, Chicago, 1965, p. 401.

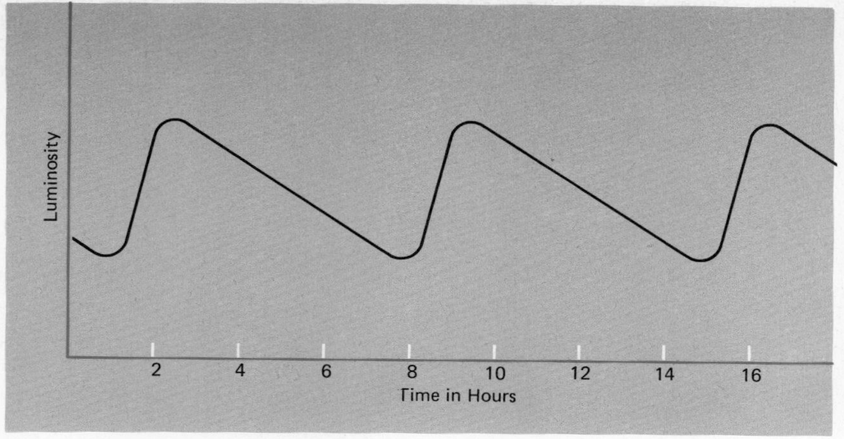

FIGURE 17.10
The light curve of a typical RR Lyrae type of variable star. The star illustrated changes from bright to faint and back to bright again in 7 hours.

important point is that all RR Lyrae variables appear to have very nearly the same average luminosity. If this average luminosity is known, it is a simple matter to find the distance to any observed RR Lyrae variable. We first measure the average apparent brightness of the star, and then we calculate how far away a star of the known luminosity must be in order to have the measured value of the apparent brightness. This is the reverse of the process used in Section 12.1 to find the luminosity of the star Sirius. There we used the measured distance and apparent brightness to find the luminosity; now we are using the known luminosity and apparent brightness to find the distance.[4] This illustrates the simple manner in which luminosity, distance, and apparent brightness are related to each other. If we know the values of any two of them, then we can calculate the third.

Unfortunately, the average luminosity of RR Lyrae variables is not easy to find with high accuracy, and this places an uncertainty on the calculated distances. We ought to be used to having uncertainties in our numbers by now, however, so we shall be satisfied with approximate values of the distances to the RR Lyrae variables and to the clusters that contain them.

Most globular clusters contain large numbers of RR Lyrae variables. When we find the distances to these stars, we also know the distances to the clusters that contain them. In this way the average distance to all of the globular clusters is found to be about 30,000 light years. We conclude that we are about this distance from the center of our galaxy. As long ago as 1917, Harlow Shapley used this procedure to conclude that we are far removed from the center of the Milky Way.

Other spiral galaxies show a bulge at the center called the nucleus. We cannot see the nucleus of the Milky Way in visible light because there is too much interstellar dust between us and the center. Large amounts of infrared and radio radiation can be detected coming from the direction of the galactic center, however, since these types of waves are not strongly absorbed or blocked by the dust. There is no doubt that the Milky Way has a large nucleus at the center, similar to the other spiral galaxies we see.

We have considered counts of stars, globular clusters and the variable stars they contain, and analogies with other galaxies to conclude that we live in a spiral galaxy shaped like a flat disk, about 100,000 light years across

[4]See problem 6 in the Questions for Discussion at the end of Chapter 12.

and about 4000 or 5000 light years thick at the position of the solar system. It contains somewhere around 200 billion stars, and we are located in the disk about 30,000 light years from the central nucleus. Figure 17.11 shows what the Milky Way might be like if you looked down on it from above.

FIGURE 17.11

An imaginary view of the Milky Way from above. The position of the Earth is indicated on the left. Practically all of the stars we can see with the naked eye at night lie within the small dotted circle around the Earth.

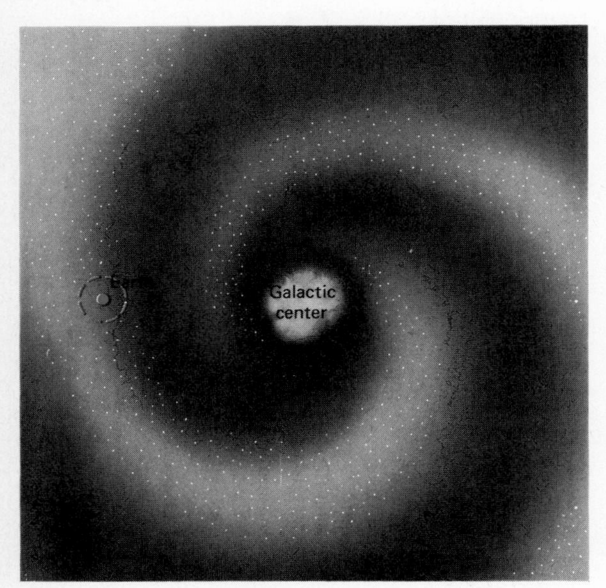

IMPORTANT WORDS

Milky Way | **Globular cluster**
Galactic plane | **RR Lyrae variable**

REVIEW QUESTIONS

1 In what way is the Milky Way like the solar system?

2 How can counting the numbers of stars we see in different directions tell us about the shape of our galaxy?

3 What causes the white, cloudlike band of the Milky Way?

4 How do we know that our galaxy is shaped like a disk?

5 How do we measure the thickness of the disk?

6 Why don't counts of stars help us find the size of the Milky Way along the disk?

7 Why do we expect the Milky Way to look more or less like the distant spiral galaxies we can see?

8 About how far is it across our galaxy?

9 How do we determine the number of stars in our galaxy?

10 How do globular clusters help us find our position in our galaxy?

11 How do we measure the distances to the globular clusters?

QUESTIONS FOR DISCUSSION

1 Why is it so much easier to determine the size and shape of a distant galaxy than that of our own galaxy?

2 By counting stars we concluded that the Milky Way has a thickness of several thousand light years. Do you think it is possible that our galaxy is actually much thicker than this, only we cannot see the more distant stars because of large amounts of interstellar dust blocking our view?

3 It is obvious that no science is exact, in the sense that measurements always have some errors in them. What do people usually mean when they call something an exact science?

4 Suppose astronomers discover they have made an error in studying RR Lyrae stars, and they realize that these stars are actually four times more luminous than had been thought. How would our estimate of the distance to the center of the Milky Way be affected by this discovery?

REFERENCES

All of the references to the Milky Way will be listed at the end of Chapter 18.

EIGHTEEN
THE MILKY WAY II: EVOLUTION

Just as individual stars form, live, and burn out, so we should expect the same of the Milky Way and of other galaxies. In this chapter we examine the evidence that relates to the origin of the Milky Way, how our galaxy probably arrived at its present state, and what is in store for it in the future.

18.1 Origin

Our galaxy is made up mostly of stars, with only a small percentage of its mass contributed by the interstellar matter. Since we know something about the origin and evolution of stars, perhaps we can use this information to discover a little about the past and future of the Milky Way.

AGE OF THE MILKY WAY

Consider first the age of the Milky Way. It must be at least as old as the stars it contains, so how old are the oldest stars around us?

We have seen that model calculations indicate that the Sun is nearly 5 billion years old. The globular star clusters are found to have ages between 10 and 15 billion years. Some individual stars show signs of much greater ages than this, but the evidence is probably misleading. The RR Lyrae variable stars, mentioned in Chapter 17, are good examples. Theoretical studies of their structure suggest that they have small masses, perhaps as small as one-half the mass of the Sun; however, they are in the giant region of the H-R diagram, having aged well past the main sequence. Since a star of one-half solar mass needs close to 100 billion years to use up its core hydrogen and evolve away from the main sequence, one might estimate that the RR Lyrae stars are that old. However, many of these types of stars are actually members of globular clusters, and globular clusters have the much smaller ages of 10 to 15 billion years. We cannot believe that the RR Lyrae stars are older than the clusters they are members of, so we must find another explanation for this apparent discrepancy. Astronomers believe that the RR Lyrae stars had larger masses in the past when they were main sequence stars and that they have lost part of their material by shedding it back to the interstellar matter. (Remember that stars with large masses age more quickly than those with small masses.) The true ages of the RR Lyrae stars are thus only 10 to 15 billion years, instead of the 100 billion years originally suggested.

Further searching reveals that there are no reliable age determinations for stars that are significantly greater than about 15 billion years. We will, therefore, take this figure as the age of the oldest stars in the Milky Way.

FIFTEEN BILLION YEARS AGO

Before the first stars formed, our galaxy must have been pure gas and dust at very low density or, as we will see later, probably pure gas. Something happened about 15 billion years ago to cause star formation to begin, and it has been taking place ever since. What could have caused the stars to begin forming out of the gas 15 billion years ago?[1] Apparently conditions have been just right for star formation during the past 15 billion years, but not earlier. In other words, the Milky Way must have been in quite a different condition 15 billion years ago than it was, for instance, 20 billion years ago.

In what ways could our galaxy have been changing just before star formation began? We can take some clues from the stars themselves. A cloud of gas can contract to become a star only if its density is great enough. The density of the gas in the Milky Way apparently was never great enough for star formation before 15 billion years ago, but then it suddenly did become great enough. This strongly suggests that the Milky Way was in the process of contracting; if a gas cloud can contract under its own gravity, why not an entire galaxy?

The contraction of a gas cloud will cease when it reaches force and energy balance and becomes a main sequence star. What can stop a galaxy from contracting? There are three known possibilities:

1 A force balance similar to that which occurs in ordinary stars

2 *Angular momentum,* or spin of a galaxy

3 The formation of stars out of the gas

Each of these possibilities warrants further examination.

SUPERMASSIVE STARS

The first case is that the gas in a galaxy simply continues to contract until its pressure becomes large enough to balance its gravity. In this way the galaxy forms one huge *supermassive star*. When we note that the Milky Way and most other galaxies around us seem to be made of ordinary stars, plus interstellar gas and dust, the evidence appears to be against the formation of supermassive stars. Certain galaxies, however, such as radio galaxies and quasars, are known to emit huge amounts of radiation that are quite different from the heat radiation given off by ordinary stars. Many of these galaxies show signs of explosions of some kind

[1] Of course the stars are also gaseous, but the distinction should be made between the very high density gas that is concentrated in the stars and the very low density gas that pervades the entire Milky Way.

FIGURE 18.1
The galaxy M 87. In this short time exposure, the jet of material pointing out from the center of the galaxy is visible. It is difficult to avoid the conclusion that some sort of explosion at the center of this galaxy gave rise to the jet. (Lick Observatory photograph)

having taken place within them (see Figure 18.1). The explanation for these phenomena is unknown, but it is apparent that these strange galaxies have much of their energy stored in forms different from those of ordinary stars and interstellar matter.

It has been suggested that a supermassive star embedded in the center of an otherwise normal galaxy could explain the observations. The quasars could be supermassive stars, either by themselves or as parts of much larger but fainter, galaxies. Even our own Milky Way has its suspicious aspects: There are significant amounts of nonthermal radiation coming from the center of our galaxy that could possibly signify the presence of a supermassive star that is in a nearly quiescent state. These points are discussed more fully in Chapter 19. As far as is now known, supermassive stars may or may not exist. If they do exist, they could be an important part of many, or even most, galaxies.

ANGULAR MOMENTUM

The second way to stop the contraction of a galaxy is through angular momentum, or spin of the galaxy.

All objects in a galaxy, be they stars or atoms of gas, feel a gravitational force pulling them toward the center. If an object is not moving at all, it will simply fall toward the center. If the object has some sideways motion, the combination of this motion and the gravitational force causes the object to swing around the center in an orbit similar to the orbit of a satellite around the Earth (see Figure 5.14). It is this sideways motion that gives angular momentum to the object.[2] The more angular momentum an object has, the farther from the center it will be, on the average. It is in this way that angular momentum can halt contraction: If the atoms of gas have a certain amount of angular momentum, they will move in an orbit of a certain size that keeps them a certain distance from the center. The gas then cannot contract to a still smaller size.

Suppose a gas cloud has a rather large amount of angular momentum, which means that the cloud has a tendency to rotate around an axis passing through the cloud. Figure 18.2 illustrates what happens as the cloud tries to contract under its own gravity. The angular momentum does not affect motions that are parallel to the axis of rotation, so the cloud can contract in that direction without difficulty. Motions inward toward the rotation axis are hampered by the angular momentum, how-

[2]The angular momentum of the object around the center is defined as the product of the mass of the object, its distance from the center, and its sideways speed. The motion toward or away from the center does not contribute to the angular momentum.

FIGURE 18.3
Collisions between clouds and between star groups. (a) Two clouds approaching. (b) After collision they cancel each other's motion. (c) Two star groups approaching. (d) They pass through each other without effect.

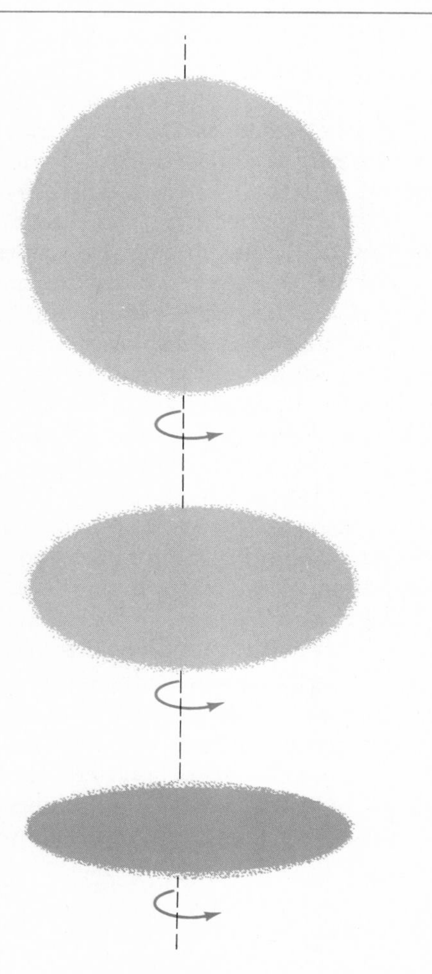

FIGURE 18.2
The collapse of a rotating gas cloud.

ever, so the cloud cannot contract very much in these directions. As a result, the cloud becomes a flat disk as the gas continues to rotate on its axis. The greater the amount of angular momentum the cloud started with, the flatter the disk it collapses to. If the cloud had no angular momentum to begin with, then it would contract equally in all directions and have a spherical shape.

STAR FORMATION

The third way in which the gravitational contraction of a gaseous galaxy can be halted is for the gas to condense into stars. Collisions between stars are almost nonexistent because of their extremely small size compared to gas clouds. When a star is formed, it will have whatever motions the parent gas had. If the gas collapses to a flat disk before any stars are formed, then the stars that are subsequently formed will have orbital motions that keep them moving in circles around the center. They will remain in the disk with essentially the same motions as the gas.

Suppose that the gas condenses into stars before it has collapsed onto the flat disk. Then the stars will not be in the disk when they are formed, and they will have motions that carry them alternately above and below the disk of the galaxy as time goes on. The gravity of the galaxy will not let these stars escape, but it cannot force the stars to remain in the flat disk.

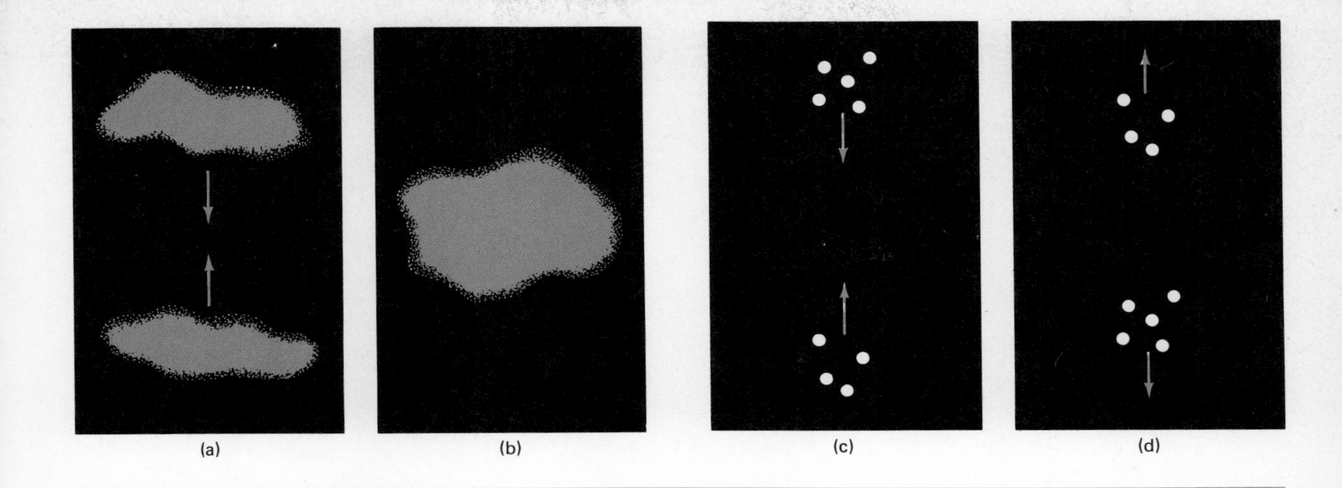

Consider Figure 18.3. If two gas clouds are falling onto the disk of a galaxy from opposite sides, they will collide with each other. The energy of collapse that they had will be changed into heat, and the gas will soon radiate away this heat energy, leaving the total mass of gas collected together in the disk. The only motion the gas can have left is that which gives rise to its angular momentum and carries it around the center of the galaxy, keeping it in the flat disk. The situation will be quite different if the gas forms into stars before the clouds collide with each other. In this case, the two groups of stars will pass through each other with almost no effects. The groups of stars will go to a large distance above and below the disk of the galaxy before gravity can pull them back again, and they will continue oscillating back and forth for very long periods of time.

18.2 Past, present, and future

THE DISK

How can the theoretical ideas outlined in Section 18.1 be applied to the Milky Way and to other galaxies?

The first thing that we might notice is that the Milky Way has a very flat shape. In view of the previous discussion, therefore, we should expect that our galaxy has a large amount of angular momentum and that this is an important reason why gravitational contraction has stopped. If this is correct, then the gas, dust, and stars observed in the plane of the galaxy must be moving more or less together in large circular orbits around the center. Also, most stars are in the thin disk of the Milky Way which means that they were formed after the gas had collapsed onto the plane.

Observations show that most stars around us, including the Sun, are moving approximately together in a large circular orbit around the center of the Milky Way. This motion is analogous to the motion of the Earth around the Sun, except that the radius of this circular orbit is about 30,000 light years instead of 93 million miles. While it takes the Earth 1 year to go around the Sun, it takes the Sun, traveling at 160 miles per second, about 200 million years to travel once around the center of the Milky Way. Since most stars in a small volume of space move nearly together in their orbits, astronomers refer to this motion as a rotation of the Milky Way. This rotation is due to the angular momentum which our galaxy had when it was formed and which it still has today. Just as the inner planets move faster around the Sun than the outer planets, so stars near the center of the Milky Way generally move faster than those further out. Thus the inner regions of the

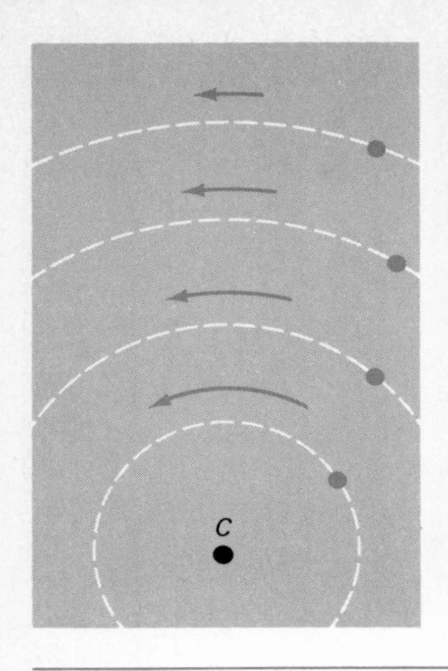

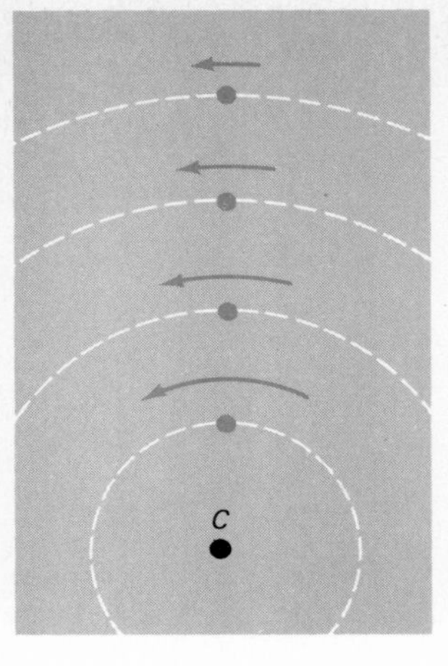

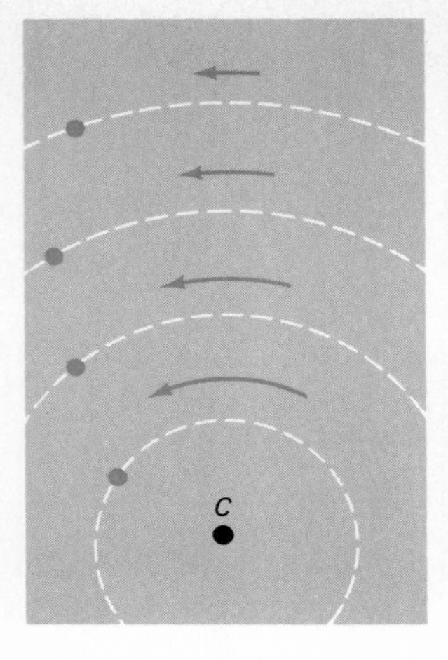

FIGURE 18.4
Rotation of the Milky Way. The stars near the center *C* move faster than those further out.

Galaxy tend to rotate faster than the outer regions (see Figure 18.4).[3]

THE HALO

Another interesting observation is that while most stars are in the disk of the Milky Way, not all of them are. As noted earlier, the globular clusters occur far above and below the galactic plane, and there are some individual stars that do the same. These objects make up what is called the *halo* of the Milky Way. According to our theoretical ideas, the halo stars were formed very early in the life of the Milky Way, before the gas had collapsed onto the galactic plane. These ideas gain more credence when it is noted that the globular cluster stars and other stars in the halo are the oldest stars known. Detailed studies of the H-R diagrams of globular clusters indicate that these clusters have ages of about 15 billion years. Figure 18.5 shows the main parts of the Milky Way.

Stars in the disk of the Milky Way are observed to be of all ages: Some are as young as a million years or less, while the oldest have ages that approach 15 billion years. The halo, on the other hand, does not contain any young stars. All of the hot main sequence stars and all of the supergiants, stars known to be very young, occur only in the disk. It is notable that all of the halo stars seem to be about 15 billion years old and that none of them are significantly older or younger than this. We conclude that star formation in the halo took place only for a short time about 15 billion years ago, and no halo stars have been formed since. In the disk, however, stars have been forming continuously over the past 15 billion years, and they are forming today.

This shows that it did not take long for the gas to collapse onto the thin plane of our galaxy. The halo stars formed during this collapse, and they have retained the motions of the collapsing gas ever since. This explains why all halo stars are of about the same age.

The gas that survived the formation of the halo stars continued to fall toward the disk of the Milky Way. When it arrived at the disk, it collided with gas falling in from the other side. In this way, nearly all motions in the gas cancelled each other out except those carrying the gas around in circular orbits within the disk. The stars that formed after the gas collapsed onto the disk naturally remained in the disk where we find them today.

HALO AND DISK COMPOSITIONS

There is another interesting observation: Halo stars appear to have a much lower abundance of the heavy elements (that is, elements heav-

[3]There are, however, certain parts of the Milky Way in which this statement is not correct.

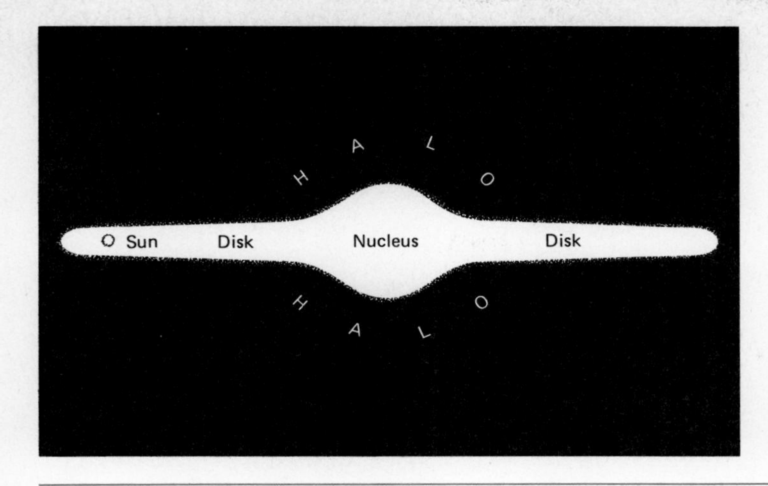

FIGURE 18.5
Schematic edge view of the Milky Way.

ier than hydrogen and helium) than disk stars. As we found in Section 14.1, about 499 atoms out of every 500 in the Sun are either hydrogen or helium. All of the heavy elements together amount to only that one atom in 500. The sun is a disk star, which means that it is one of the stars that are relatively rich in the heavy elements. The halo stars are almost completely devoid of anything except hydrogen and helium. Here we are discussing the original compositions of the stars; the interiors, of course, could have much different compositions brought on by the nuclear reactions in the deep interiors of the stars.

The composition differences between disk and halo stars do not seem to be very great: One atom in 500 doesn't appear to be much more than no atom in 500. That one heavy atom in 500 does have a significant effect on the structure of disk stars, however. Also, the heavy elements are the main constituents of the Earth and all things on it, including ourselves. We have reason to rejoice for that one atom in 500 of the disk stars.

Why should the heavy elements be more abundant in disk stars than in halo stars? Apparently the gas which condensed into the halo stars was nearly pure hydrogen and helium, while that which condensed into disk stars had and does have the small amount of heavy elements just indicated. The disk stars were formed after the halo stars, but why should this make a difference? We have already seen that nuclear reactions in stars have the general effect of making heavy elements out ot lighter ones. We have also seen that stars often blow off part of their material. If some of this blown-off material has been exposed to nuclear reactions while it was inside the star, it will carry back to the interstellar medium a higher abundance of heavy elements than it originally had. When future stars are made of this material, they will be correspondingly richer in heavy elements.

The fact that halo stars are observed to be nearly devoid of heavy elements suggests that the Milky Way was originally made out of pure hydrogen and helium. This is why the original material would have had no dust—because the dust grains are made primarily of heavy elements. All or nearly all of the heavy elements have been produced by nuclear reactions inside of stars. This also suggests that the Sun, the planets, and our own bodies are composed of material processed by nuclear reactions inside some long-dead star.

We might imagine each generation of stars made out of the debris that is processed and expelled by earlier stars, causing the abundance of heavy elements to gradually increase with time. Such is apparently not the case, since there is little or no difference in composition between the youngest and the oldest disk stars. The oldest disk stars are nearly as old as the halo stars, yet they have about as much heavy elements as the stars forming today. We conclude that a very large number of high-mass stars were formed very early in the halo of the Milky Way. Because of their large masses, these stars evolved very rapidly and shed large amounts of heavy element-enriched material back into the interstellar medium. By the time that the gas had collapsed onto the galactic plane, it already had received much of this processed material, and even the earliest formed disk stars had moderate amounts of the heavy elements. Star formation has apparently slowed down considerably since then, for the abundance of heavy elements has not increased very much since the first disk stars appeared.

SPIRAL ARMS

All galaxies that are sufficiently flat, including our own, show a spiral structure in their disks (see Figure 18.6). It was once thought that stars occur only in the spiral arms rather than throughout the disk, but it seems that only the very young stars are confined to the spiral arms. Older stars do occur throughout the disk. The spiral arms are prominent because the very high luminosity stars are all very young in years. One supergiant gives off as much radia-

FIGURE 18.6
The spiral galaxy M 51. The spiral arms wind outward from the nucleus, and in this case, a companion galaxy occurs at the end of one of the arms. (The Kitt Peak National Observatory)

tion as nearly a million stars like the Sun. It is understandable how a small number of supergiants and very hot main sequence stars strung out in a spiral pattern will make that pattern very conspicuous. But how do the young stars get strung out in a spiral pattern?

Since the very young stars have not had time to have moved far from where they were formed, the question really concerns why star formation apparently occurs only in regions that lie along a spiral pattern. C. C. Lin and others have shown that the gravity from the material in the disk of a thin galaxy should produce a spiral-shape wave that moves around the galaxy. The wave is like a sound wave moving through the air, in that it compresses the gas as it passes through it. The wave is simply a region in which the interstellar gas has a higher than average density, and the region of high density moves through the gas in the disk of the galaxy. The compression of the gas may be just what is needed to get a cloud of gas to condense into a star or a group of stars. In this way, a lot of young stars will appear wherever the wave occurs. Since the wave has the shape of a long spiral extending out of the nucleus of a galaxy, the young stars will be lined up in the same spiral pattern.

Both the *density wave* and the matter are moving around the center of the galaxy, but with different speeds. Thus the new stars and the wave will slowly part company. By the time that they are far apart, the stars will already have aged enough that the most luminous of them will be burned out. Meanwhile, still younger stars, some of which are supergiants, are being formed at the new position of the wave. In this way, stars in general occur throughout the disk of the galaxy, but the youngest and most luminous of them occur mainly along the spiral pattern outlined by the density wave.

THE FUTURE OF THE MILKY WAY

There is a continual exchange of matter between the stars of the Milky Way and the interstellar dust and gas. New stars are formed while old stars lose part of their material back into space. A star will not usually lose all of its mass, however, so the interstellar matter is slowly being used up as more and more material is being locked up in old stars. Star formation must slow down as the gas and dust are gradually depleted.[4]

The Milky Way started as pure gas, and today it is about 98 percent stars, with only a small percentage left as gas. In the early days, with lots of gas and lots of star formation, massive stars of high luminosity were far more common than they are today. The Milky Way

[4]This assumes that the interstellar matter is not replenished by additional matter from outside the Milky Way.

then must have presented a brilliant spectacle! Now it is much more subdued, although it is still a very impressive sight. Eventually, perhaps in another 15 or 20 billion years, the amount of interstellar gas will become too small to support further star formation, and there will be no more new stars to take the place of those that burn out and fade away. As always, the more luminous stars will go first.

One hundred billion years from now, the only stars still shining will be the faint, cool stars on the lower part of the main sequence. The Milky Way will appear to a distant observer as a faint, red glow. Over the following trillion years, the Milky Way will become fainter and redder as even the very low mass stars approach the ends of their lives.

In the final state, all stars will be burnt-out white dwarfs, neutron stars, or black holes. They will still continue in their orbits around the galactic center, but the Milky Way will be merely a ghost whose existence will show itself only through the force of gravity.

IMPORTANT WORDS

Angular momentum
Supermassive star
Disk star
Halo star
Density wave

REVIEW QUESTIONS

1 How old is our galaxy?

2 Why do we think the Milky Way was in the process of contraction about 15 billion years ago?

3 What processes can stop the Milky Way from contracting?

4 Does our galaxy have a supermassive star at its center?

5 What is angular momentum, and how can it stop the Milky Way from contracting?

6 Why does it make a difference whether the gas in a galaxy forms into stars very early or whether the stars are formed only after the gas has time to collapse?

7 Does the Milky Way have much angular momentum?

8 What is the halo of our galaxy, and how is it explained?

9 Where can we find the oldest stars in our galaxy?

10 Where can we find the youngest stars in the Milky Way?

11 Why are there differences in composition between halo stars and disk stars?

12 What was the gas probably made of when the Galaxy was first formed?

13 What will happen to the Milky Way when almost all of the gas and dust are used up?

QUESTIONS FOR DISCUSSION

1 What would our view of the Milky Way be like if the Sun were a halo star instead of a disk star?

2 Why does it seem likely that the Milky Way was in the process of contracting to a smaller size about 15 billion years ago?

3 Do you think it more likely that a supermassive star would form at the center of a flat galaxy or a spherical galaxy? Why?

4 Compare the origin of the Milky Way with the origin of the solar system. Angular momentum plays an important role in both.

5 Halo stars can occur anywhere in the Milky Way, including in the disk. When we observe a given star in the disk, how can we tell whether it is a disk star or a halo star?

6 At a meeting of experts held in 1964, a vote was taken on what the best determined values of some of the properties of the Milky Way should be. Is this a reasonable way to settle scientific differences, or is this carrying democracy too far?

REFERENCES

An outstanding general reference to the Milky Way is the book entitled *The Milky Way*, 4th ed., by B. J. Bok and P. F. Bok (Harvard, Cambridge, 1974).

Numerous articles from the popular journal *Sky and Telescope* are reprinted in *Stars and Clouds of the Milky Way*, edited by T. Page and L. W. Page (Macmillan, New York, 1968).

A very interesting account of how our knowledge of the Milky Way developed over history is contained in *The Discovery of Our Galaxy* by C. A. Whitney (Knopf, New York, 1971). Further background on the Milky Way is contained in: *Astronomy of the Twentieth Century* by O. Struve and V. Zebergs (Macmillan, New York, 1962); *The Structure of the Universe* by E. L. Schatzman (McGraw-Hill, New York, 1968); and *Galaxies*, 3rd ed., by H. Shapley and P. W. Hodge (Harvard University Press, Cambridge, 1972).

Radiation from the center of the Milky Way is described by R. H. Sanders and G. T. Wrixon in "The Center of the Galaxy" [*Scientific American* 230 (April 1974), 67].

The main properties of stars of the disk and halo of the Milky Way are described by G. R. Burbidge and E. M. Burbidge in "Stellar Populations" [*Scientific American* 199 (November 1958), 44].

Theories on the formation of the spiral arms are discussed by: B. J. Bok in "Updating Galactic Spiral Structure" [*American Scientist* 60 (November-December 1972), 709]; and F. H. Shu in "Spiral Structure, Dust Clouds, and Star Formation" [*American Scientist* 61 (September-October 1973), 524].

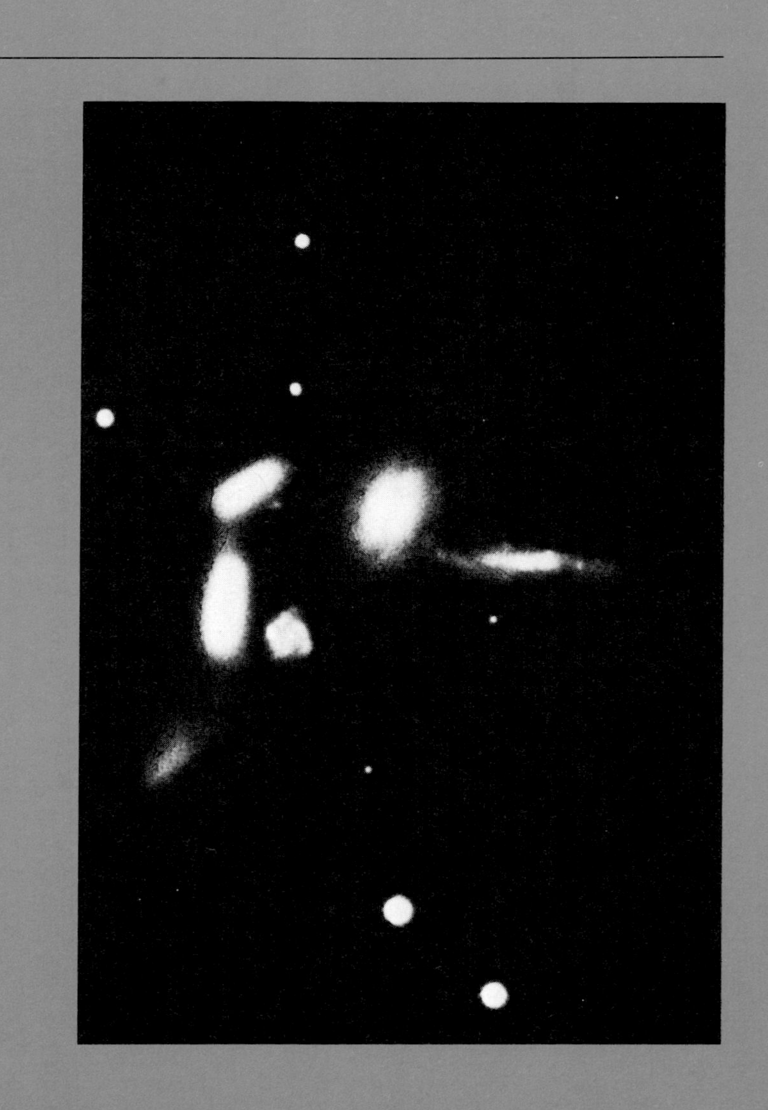

NINETEEN
OTHER GALAXIES

The Milky Way is only one of a seemingly countless number of galaxies floating in the void of space. Most of them appear to be huge collections of stars that are more or less like the Milky Way. Others contain much more than meets the eye, and they present a puzzle that keeps many astronomers very busy.

19.1 Spirals, ellipticals, and irregulars

NEARBY GALAXIES

In Chapter 18 we saw that there are many star systems or galaxies that are to some extent like our Milky Way system. They come in various sizes and shapes. The flattest galaxies show a spiral structure like our own (see Figure 19.1), while those that are less flat are called *elliptical galaxies* because they appear elliptical in shape (see Figure 19.2). A third type is the *irregular galaxies*, which do not have the regular shapes or structures of the other two kinds.

About 150,000 light years away from us are two irregular galaxies called the Large

FIGURE 19.1

The spiral galaxy M 81, viewed at a small angle. (Lick Observatory photograph)

FIGURE 19.2
The elliptical galaxy NGC 205. This small galaxy is a satellite of the Andromeda Galaxy and can be seen in Figure 1.1. (The Hale Observatories)

and the Small Magellanic clouds (see Figure 19.3). These are the galaxies nearest to us. The Magellanic clouds are easily seen with the naked eye from the Southern Hemisphere, although they cannot be seen from most of the Earth's northern latitudes. These are very small galaxies, and they are actually satellites of the Milky Way in the same way that the Moon is a satellite of the Earth. The Magellanic clouds take about a billion years to travel in orbit around our galaxy.

About 2 million light years away from us is the Andromeda Galaxy (see also Figure 1.1). The Andromeda Galaxy is the spiral closest to us, and it is very much like the Milky Way in size, shape, and mass. Two small elliptical galaxies appeared in Figure 1.1 as satellites to the main galaxy. Apparently it is common for a number of small galaxies to be formed in the vicinity of larger ones.

ORIGINS

The origin and evolution of ellipticals and spirals can probably be understood, at least in an

FIGURE 19.3
The Large Magellanic Cloud. This small galaxy is visible to the naked eye from southern latitudes on the Earth. (Lick Observatory photograph)

approximate way, in terms of discussion in Chapter 18. The spirals have a large amount of angular momentum, so much of their matter has collapsed to a flat, rotating disk. The ellipticals have very little angular momentum, so they have not become flat. Either star formation or collapse to a huge, supermassive star at the center should be the fate of the gas from which the elliptical galaxies form. Thus, most stars in elliptical galaxies ought to be old halo-type stars, and this appears to be confirmed by observations. The amount of gas and dust in ellipticals is very much less than that in spirals, and the ellipticals contain no supergiants or other very young stars. There is no indication that any other galaxy is either much older or much younger than the Milky Way, although the direct evidence bearing on this is sparce.

Irregular galaxies do not seem to fit well into this picture of evolution. They are not very flat, yet they do have large amounts of gas and dust and very young stars. There are also very old stars in irregular galaxies. We still have much to learn about the formation and aging of galaxies.

DISTANCES OF GALAXIES

Even the Magellanic clouds, the nearest of galaxies, are so far away from us that only the brightest stars in them can be individually photographed with large telescopes. In the Andromeda galaxy, only the supergiants are visible in long-exposure photographs, while most galaxies are much too far away for any of the stars to be detected individually. It is only the combined light of large numbers of stars that makes it possible for galaxies to be detected at their huge distances.

Finding distances to galaxies is difficult, and the results are often quite uncertain. If a galaxy contains supergiant stars, and if it is near enough that these supergiants can be observed, then the approximate distance can be found. Assuming that the supergiants are similar to the ones that occur in the Milky Way, we can estimate their luminosities. Then it is a simple matter of finding how far away stars of these luminosities must be placed in order for them to have the observed apparent brightnesses. Most supergiants in the Milky Way do not have well-determined luminosities, so the results are not highly accurate. Also, not all supergiants have the same luminosity, so the astronomer must try to find out what kinds of supergiants are in the distant galaxy.

Elliptical galaxies do not contain supergiants, but their distances can be estimated from their globular clusters (see Figure 19.4) if they are near enough that these are visible. This principle is the same as the preceding one: The assumption is made that globular clusters belonging to other galaxies are essentially the same as those associated with the Milky Way. Since we know the approximate

FIGURE 19.4
The elliptical galaxy M 87. This galaxy appears similar to a globular cluster (see also Figure 17.8), but it actually has its own globular clusters. They are barely perceptible in this photograph as the tiny dots clustered around the galaxy. This is the same galaxy that has the jet that was illustrated in Figure 18.1. (The Hale Observatories)

sizes and luminosities of our own globular clusters, we can figure out how far away globular clusters must be placed in order that they appear the same size and brightness as the ones in the distant galaxy. This method, of course, can be used for any galaxy containing observable globular clusters.

These methods consist of identifying some parts of a distant galaxy as objects familiar to us and seeing how big or how bright they appear. Since we understand the manner in which distance affects the apparent size and brightness of an object, we can figure the distance of the object and of the galaxy that contains it. The important thing is to be able to observe something in the galaxy which we can recognize and are familiar with.

What can we do about galaxies that are too far away for us to observe supergiants, globular clusters, or any other parts? If a galaxy appears only as a faint blur on a long-exposure photograph taken with a large telescope, is there any way to estimate the distance to that blur? Yes, there is, by using the size and brightness of the galaxy as a whole. We know roughly the size and luminosity of the average nearby galaxy. If that blur on the photograph is an average galaxy, then measuring how big and bright it appears will give us the distance in the usual way.

What if the blur is not an average galaxy? Actually, it probably won't be. The galaxies near us consist of big ones and small ones, bright ones and faint ones, and the average is somewhere in the middle. When we look off at great distances, however, the small, faint galaxies cannot be seen. Any very distant galaxy that we can see is probably larger and more luminous than the average, and this must be taken into account when estimating the distances of the very remote galaxies. In any individual case, however, we cannot tell for certain from looking at a faint image on a photograph whether the galaxy is large, luminous, and very distant or small, faint, and relatively near.

Galaxies are scattered in space in all directions. The nearest is about 150,000 light years away, and who knows how far away the most remote ones are? Can we learn anything from measuring the distances to a large number of galaxies? Do we find any important patterns related to the distances of galaxies? Surprisingly, around 1930 E. P. Hobble found that the distances to galaxies seem to be related to their speeds, that is, to how fast they appear to be moving.

THE DOPPLER EFFECT

Galaxies are so far away that it seems impossible to be able to find out how fast they are moving. Yet the properties of radiation are such that astronomers can do just that. When radiation comes to us, the frequency of the radiation is determined by the number of waves that go by us each second. If we move toward the source of the radiation, more waves pass us each second, and so the radiation appears at a higher frequency. In the same way, travel away from the source of the radiation, which is travel in the same direction that the wave is moving, means that a smaller number of waves pass us each second and a lower frequency is observed. This is known as the *Doppler effect* (see Figure 19.5). Note that sideways motion does not affect the measured frequency, only motion toward or away from the source. It makes no difference whether the source of radiation or the observer is moving—it is the relative motion between the two that matters. Note that two persons can see the same radiation and report a different frequency if these two persons have very different motions.

The late physicist George Gamow often told the story of the physics professor who was arrested for running a red light in his car. On a whim, this professor decided to argue his case on the basis of the Doppler effect. He explained the frequency shift of the radiation as a result of his speed in the car. Since green light has a higher frequency than red light, the professor stated that as he approached the signal in his car, the red light actually appeared green to him. Gamow would have us believe that the argument was so convincing that the judge was on the verge of acquitting the professor when a former student, whom the professor had failed, made his untimely appearance. Still harboring a grudge against his old teacher, the former student suggested that the judge find out how fast the professor would have to be traveling in order for the red light to appear green. The professor admitted that a speed of 200 million miles per hour would be required to produce the effect he claimed, whereupon he was fined for speeding.

If an astronomer can identify some lines in the spectrum of a star or a galaxy from the pattern they exhibit, he immediately knows what frequencies those lines normally have. When he measures the frequencies the lines actually have, he might find them to be different from what he expected. If all of the lines are of higher frequency than normal, the explanation is probably that the star or galaxy is moving toward the Earth, and the Doppler effect has increased the frequencies. If the frequencies are all lower than expected, the star or galaxy is probably traveling away from us. A measurement of the amount of the shift in frequency will immediately tell us how fast the source is moving toward or away from us. Astronomers find the Doppler effect very useful in many ways.

FIGURE 19.5
The Doppler effect.

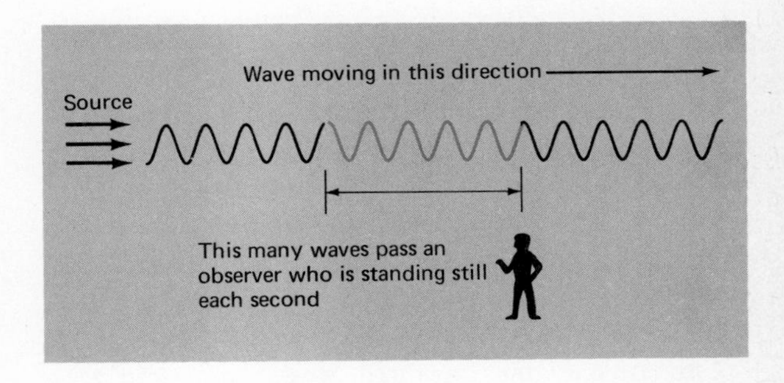

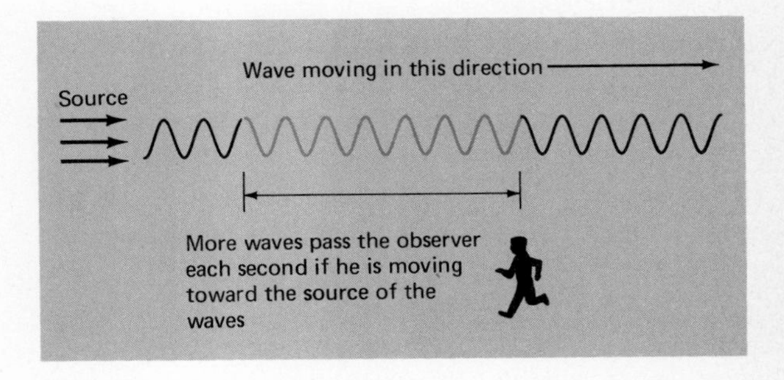

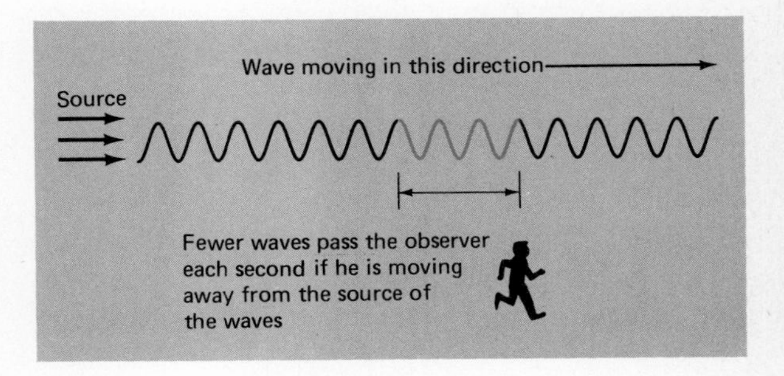

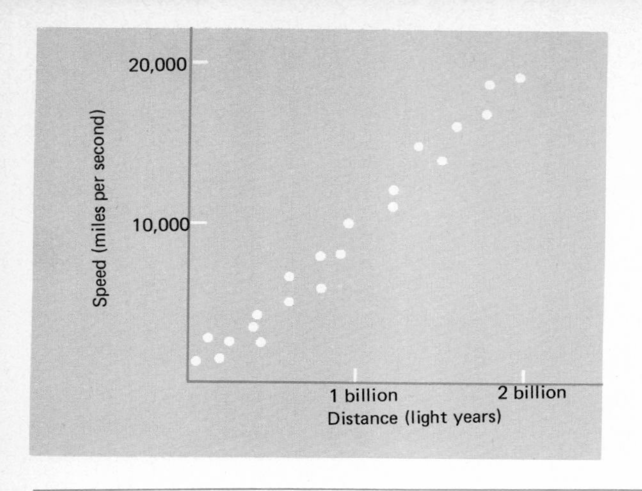

FIGURE 19.6
The velocity-distance relation.

THE VELOCITY-DISTANCE RELATION

The nearby stars show motions, through the Doppler effect, that are usually in the range of 10 to 40 miles per second. These speeds are not at all excessive when you consider that the Earth travels 18 miles per second in its orbit around the Sun. When we examine the spectra of galaxies, however, we find shifts in the frequencies of the spectral lines that indicate quite impressive speeds of thousands of miles per second, and higher. The most intriguing thing about galaxy motions is that, except for a few very nearby galaxies, all of the galaxies are moving away from us, and the farther away a galaxy is, the faster it is receding from us. These observations are interpreted as evidence that the Universe is expanding.

Since practically all galaxies are moving away from us, we see their spectral lines at smaller frequencies than normal. Red light has a smaller frequency than any other color of visible light, so the visible-light photons in galaxies have their frequencies shifted closer to that of red light. For this reason, the Doppler effect of sources moving away from us is often called a *redshift*. For the same reason, a star or galaxy moving toward us would seem to have its spectral lines shifted to higher frequencies than normal; this is called a *blueshift*. Galaxies show very large redshifts, indicating large velocities of recession. Note that a redshift or a blueshift is not a change of color.

Figure 19.6 shows a graph of the velocity-distance relation for galaxies. The distance to a galaxy is indicated at the bottom of the graph, while the speed is shown by the scale on the left. Each dot represents a separate galaxy. The dots are laid out in a nearly straight line, and we see that the more distant galaxies are moving faster than the nearer ones.

Figure 19.6 indicates that when a galaxy is 1 billion light years away from us, then for some reason it can be expected to have a velocity of about 10,000 miles per second; a galaxy 1½ billion light years away is probably receding from us at a speed of about 15,000 miles per second; and so on. Suppose we take a spectrum of a galaxy, measure the frequencies of its lines, and find that it has a redshift that corresponds to a velocity of 5000 miles per second. Then the graph of Figure 19.6 shows that this galaxy can be expected to be about 0.5 billion light years away. We can see, then, that the velocity-distance relation helps us find the distance to any galaxy if we can find its speed by measuring its redshift. This is how distances can be estimated for even the most remote galaxies.

19.2 Radio galaxies and quasars

The galaxies mentioned so far seem to be simply huge collections of stars and interstellar matter. The evidence indicates that starlight, combined with the effects of starlight on gas and dust, can explain most of the observed properties of galaxies. There are some galaxies, however, for which this is decidedly not the case.

RADIO GALAXIES

Radio telescopes have detected radiation coming from many galaxies, but there are a few that have particularly powerful radio waves. Some of these galaxies actually emit much more energy in the form of radio waves than visible light. For obvious reasons, they have become known as *radio galaxies*.

Why can't stars account for the radio waves given off by radio galaxies? After all, stars radiate at all frequencies and not as visible light alone. Although stars do emit radio waves, the fraction of the energy emitted by a star in different frequencies depends on the temperature. Visible light has much more energy than radio waves, and at higher temperatures a star emits much more energy in visible light than it does in radio waves. A very cool object has a larger fraction of its energy in radio waves than visible light. A star with a temperature of about 700°F emits equal amounts of energy in the two spectral regions but few stars are this cool. The Sun has only one part in a trillion of its energy in radio frequencies. Spectra of radio galaxies show that the visible light they emit comes from ordinary types of stars, but these stars cannot possibly account for the huge amounts of radio energy given off by the radio galaxies.

Another characteristic of the radio emission is that it is polarized. As mentioned in Section 13.2, this means that the electric and magnetic fields of the radiation vibrate more in one orientation than in another, in other words, they are not completely random. The radiation emitted by ordinary stars comes from the heat energy of the stars, or the chaotic motions of the atoms that make up the stars. The atoms are continually suffering collisions with other atoms that cause them to move first in one direction, then in another. The radiation from such a collection of atoms will not be able to prefer one direction to another, so it will not be polarized. This is again evidence that the radio radiation does not come from ordinary stars.[1]

When we photograph a radio galaxy

[1] The interstellar dust, which can polarize starlight, has almost no effect on radio waves.

FIGURE 19.8

The radio galaxy NGC 4038-4039. (The Kitt Peak National Observatory)

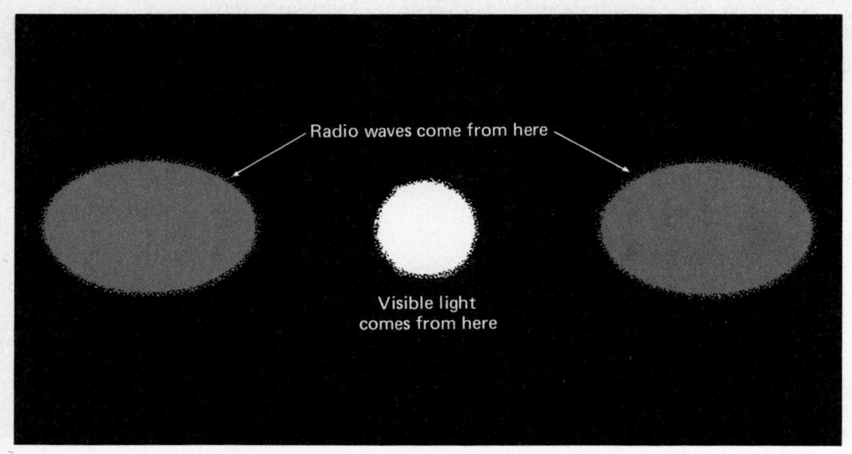

FIGURE 19.7

A radio galaxy.

using visible light, we see an image of what appears to be simply another huge collection of stars. When the radio waves from such a galaxy are carefully studied, however, we find that they often come from a separate region (see Figure 19.7). As the drawing illustrates, the radio radiation often comes from two regions on either side of the place where most of the stars are. There are few or no stars where the radio emission takes place. In some cases, there is strong radio emission from a very small region in the nucleus of the radio galaxy. It is apparent that there is much more to a radio galaxy than the stars it contains.

The discovery of radio galaxies and the determination of their properties have taken place only rather recently. The reason is that instruments for detecting and measuring visible light were developed long before we had equipment for measuring other types of radiation. In the 1950s, most astronomers thought that they recognized the main ingredients of the Universe and understood most of their properties. In the 1960s, astronomers made enough new discoveries to realize that the earlier ideas were seriously incomplete. Many of the new discoveries resulted directly from newly developed equipment that could more accurately measure radio, infrared, ultraviolet, x, and gamma rays. There are many fascinating aspects of the Universe which can be "seen" only with radiation other than visible light. Our eyes do not tell us the whole story.

SYNCHROTRON RADIATION

The radio waves from radio galaxies come from the synchrotron process (this process was described briefly in earlier sections). When an electron, or any particle with an electric charge, moves in a region where there is a magnetic field, the magnetic field will exert a force on that electron which causes it to emit radiation. The type of radiation it emits depends, among other things, on the speed of the electron. If the electron is moving extremely fast, so that its speed is very close to the speed of light, then the radiation emitted is the type known as *synchrotron radiation*.

A photon emitted by the synchrotron process is exactly the same as a photon of the same frequency emitted by a star through the thermal process. Thus there is no way to tell a "synchrotron photon" from a "thermal photon." It is only by examining many photons coming from an object that one can tell whether it is thermal or synchrotron radiation that the object is emitting. The relative numbers of photons of different frequencies, that is, the spectra, are different for the two cases. Also, synchrotron radiation is polarized, and thermal radiation is not.

It seems that radio galaxies have enor-

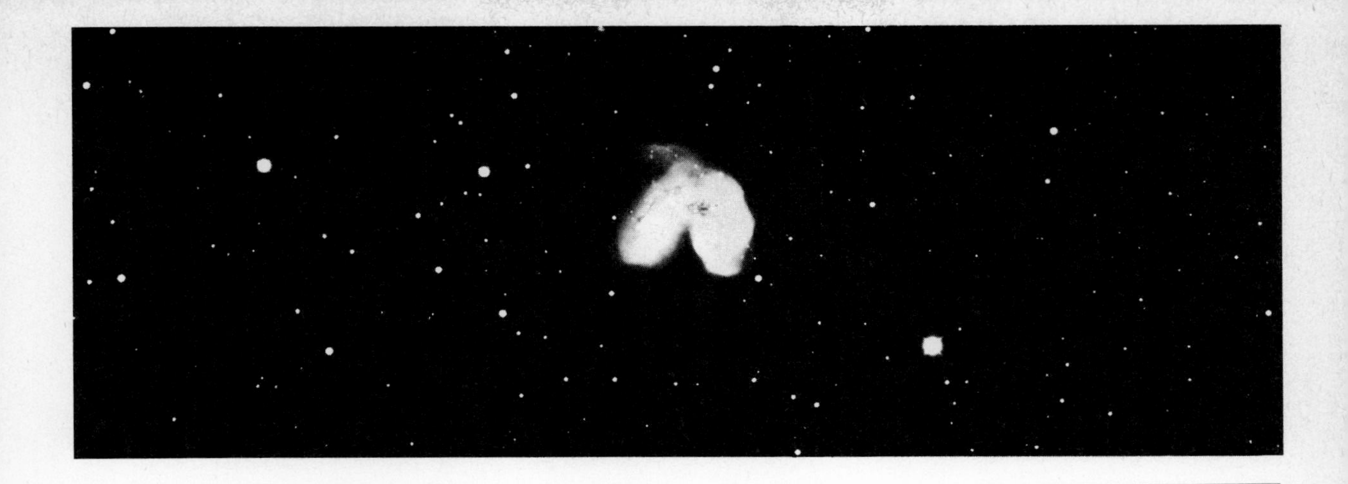

mous amounts of energy stored up in magnetic fields and in high-speed electrons and other particles. "Where do they come from?" is the natural question, and the answer is not known. Why are there so often two regions of radio emission placed symmetrically on each side of the galaxy? It certainly looks as though the galaxies have ejected in opposite directions two masses containing high-speed particles and magnetic fields, and we detect the masses through the synchrotron radiation they emit. In some cases, multiple ejections appear to take place over a period of time since several radio sources are noted around a single galaxy. The explosive energy needed to eject these masses from galaxies makes even the mighty supernovas seem weak by comparison. Another important question which still needs answering is whether radio galaxies are a special type of galaxy or whether the strong radio emission is a temporary phase of the life of many, or even all, galaxies.

A NEW DISCOVERY

By the early 1960s, many other strong radio sources had been discovered, but they differed in two important ways from the radio galaxies in that (1) they did not appear to be associated with any galaxy and (2) the radio signals came from a very tiny region of space instead of a rather large, extended area, as is usually the case for radio galaxies. In a few cases, some stars were discovered at the precise positions from which the radio waves came, and the suggestion of radio stars analogous to radio galaxies was made. These stars, however, turned out to have some very strange properties. They were very blue in color, they did not have a constant energy output, their spectral lines were almost all in emission instead of absorption, and none of the lines could be identified with a known substance.

In 1963, M. Schmidt was finally able to identify the emission lines, and the results were startling to astronomers. First, the lines were of a kind observed in emission nebulae rather than the kind observed in the spectra of stars. More importantly, the spectral lines had all been shifted to much lower frequencies than normal and it was this large frequency shift that had made the lines so hard to identify. The frequency shift had exactly the same nature as the redshift due to the Doppler effect, but it was of a far greater amount than had ever been observed previously, even in galaxies. What could all of this mean?

These strange, blue "radio stars" were obviously not ordinary stars, and neither did they look like ordinary emission nebulae. They somehow were strong emitters of synchrotron radiation in radio frequencies and in visible light, as well, but why were the spectral lines

FIGURE 19.9
The quasar 3C 273. A jet of material can be seen extending out from the main body of the object. (The Kitt Peak National Observatory)

all shifted so much in frequency? Was this a Doppler effect indicating extremely high-velocity travel away from us?

With ordinary galaxies, we noted that the farther away from us a galaxy is, the greater is its redshift. This is the effect of the "expansion of the Universe." Does this same rule also apply to these "radio stars"? If so, they are much more distant than the ordinary galaxies we have observed. We have already noted that the distance of a star or galaxy affects how bright it appears. These "radio stars" appear rather faint to us on the Earth. If they really are off at the farthest edge of the observable Universe, as the large redshifts suggest, then they would have to be tremendously luminous to be detectable at all. This would make the "radio stars" the most luminous objects known, far outshining the brightest ordinary galaxies.

QUASARS

The term "radio stars" is obviously an inappropriate one for these new beasts. This is particularly true because A. R. Sandage, in 1965, found that there are many such objects that emit only weakly in the radio region of the spectrum. Whatever these objects are, they look like stars when photographed through a visible-light telescope (see Figure 19.9) and so they soon became known as *quasi-stellar objects*, or *QSOs*, because of their appearance. The term *quasar* has also been coined for them, and both terms are used interchangeably.

The most important question concerning the QSOs is their approximate distance from us. It is not a matter of finding accurate distances for them. Astronomers would like to know whether their extreme redshifts are a true indication of their distances, as they are for normal galaxies, in which case QSOs are many billions of light years from us, or whether QSOs are relatively nearby, meaning only a few hundred million light years away. We can measure the distances of the stars that are closest to us by means of the Earth's orbital motion around the Sun.[2] We must find the distance for more distant stars and for galaxies by comparing the luminosity with the measured brightness. But we do not know what kind of objects quasars are, so we do not know their luminosities. Thus, we have no way to make a comparison with the apparent brightnesses in order to find distances. The quasars certainly lie beyond the confines of our Milky Way system; little else about their distances is known for sure.

[2]There are other methods of finding distances of stars under certain circumstances, such as using the space motion of the Sun or the rotation of the Milky Way.

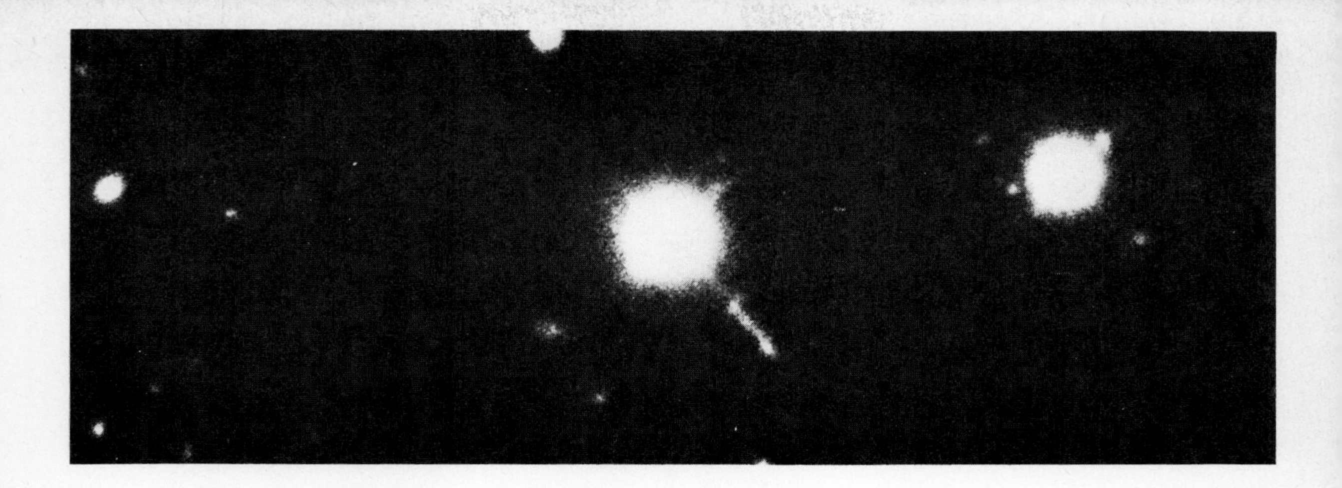

The most common opinion among astronomers today is that the distances of the quasars are indeed given by their redshifts, through the expansion of the Universe. Figure 19.6 showed the velocity-distance relation as indicated by ordinary galaxies. If the same relation holds for the quasars, and if the quasars' redshifts are due to the expansion of the Universe, then the same figure would be valid for quasars. Even the slowest moving quasar has a speed of over 25,000 miles per second, and most of them have redshifts that suggest speeds of around 150,000 miles a second away from us, so it is apparent that Figure 19.6 must be extended to much larger velocities and distances if the quasars are to be plotted on it. This is the basis for the statement often seen that the quasars are the most distant and the most luminous objects known.

The true physical nature of the QSOs is not known. One possibility that seems to hold promise is that QSOs are supermassive stars. Since very little is known about the expected properties of supermassive stars (none have been observed for certain), it is easy to convince ourselves that they should have precisely the same properties observed in quasars. Actually, the evidence is somewhat stronger than this.

As indicated in Section 18.1, we might expect supermassive stars to exist. If they do exist, then the centers of galaxies seem likely places to find them. It has been known for some time that there are certain galaxies that contain normal stars, yet that give off large amounts of nonthermal radiation from their nuclei. In 1973, A. R. Sandage showed that the radiation from such galaxies is just what we would expect if a quasar were embedded in the center of a giant elliptical galaxy. J. Kristian also found direct evidence from photographs that QSOs might occur in the nuclei of ordinary galaxies.

The visible light from a quasar probably is mostly synchrotron radiation. How can a supermassive star emit large amounts of this type of radiation? The details are not understood, but it is conceivable that a quasar is a sort of a large-scale model of a pulsar. As described in Section 16.3, a pulsar is a rapidly rotating neutron star which has a large magnetic field and which produces lots of synchrotron radiation. The much larger and more massive quasar would not rotate as fast as a pulsar, but, as in the case of a pulsar, the large gravity, rotation, and magnetic field could be the main ingredients in a complicated interaction that produces the observed radiation. These ideas are, of course, subject to considerable change as new information becomes available.

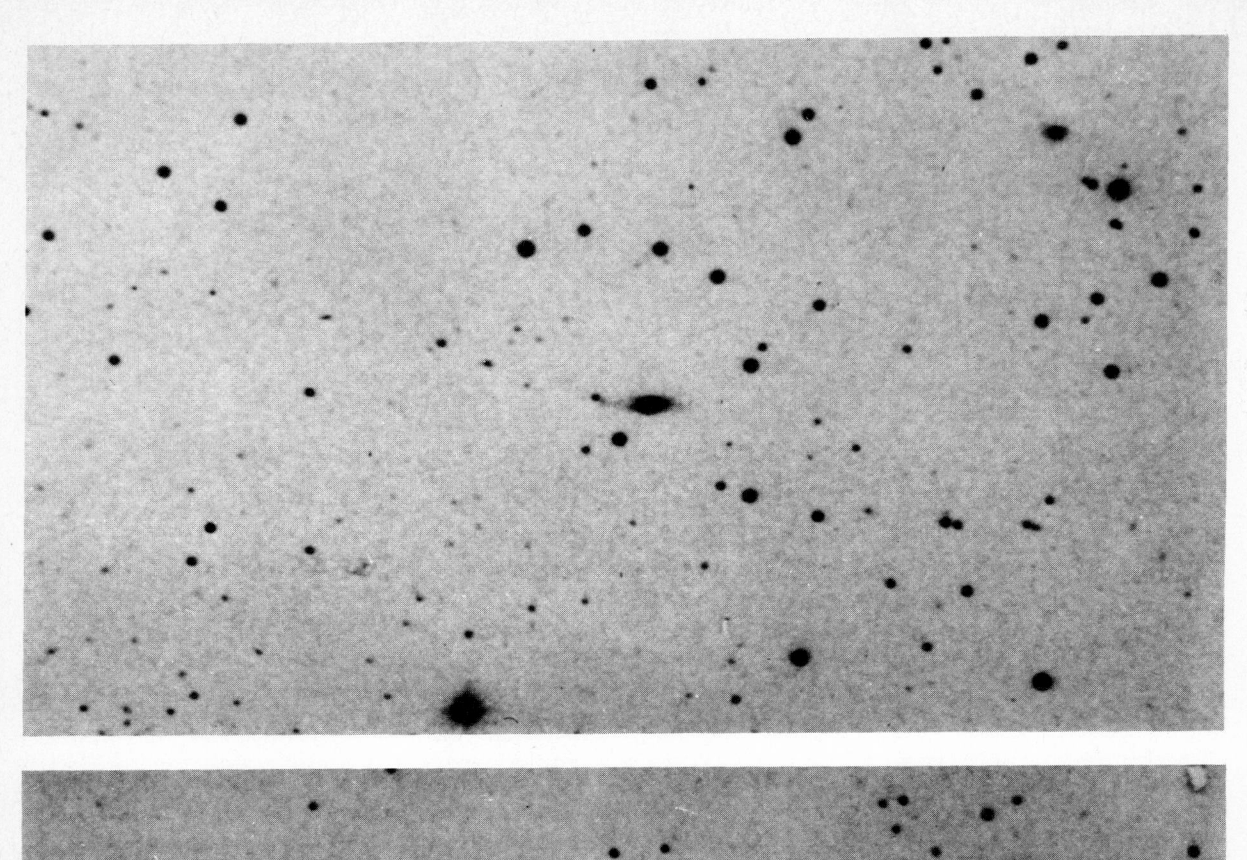

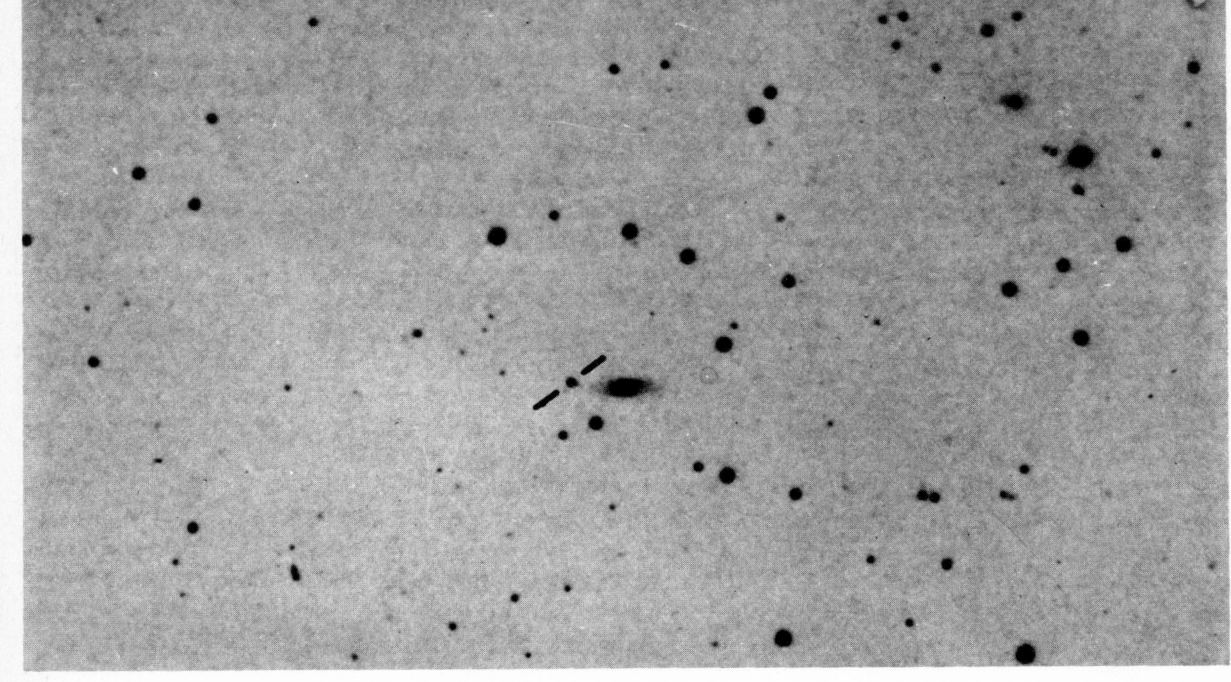

FIGURE 19.10

A galaxy-quasar pair. The quasar is between the lines in the lower photograph. Although there appears to be a bridge of material connecting the two objects, the quasar has a much larger redshift than the galaxy. The top photograph is taken in blue light, the bottom in red. Note the strong blue color of the quasar. These are photographic negatives, so the stars and galaxies appear dark on a light background. [From E. M. Burbidge, G. R. Burbidge, P. M. Solomon, and P. A. Strittmatter, *The Astrophysical Journal* 170 (1971), 233. (Copyright by the National Geographic Society–Palomar Observatory Sky Survey. Reproduced by permission from the Hale Observatories.)]

19.3 The redshift problem

DISCREPANT REDSHIFTS

Not all astronomers agree that QSOs are as far away as their large redshifts might suggest. The main reason for this is found in the redshift itself. The conventional view is that the expansion of the Universe is the cause of the redshifts, for quasars as well as for ordinary galaxies. There is evidence, however, that at least some QSOs are physically connected with galaxies that have much smaller redshifts than their associated QSOs. In other words, there may be examples of a QSO and a galaxy occurring together in space where the QSO has a much larger redshift than the galaxy (see Figure 19.10). If the two are really at the same distance from us, they ought to have the same redshift, if expansion of the Universe is the only factor involved. This evidence, therefore, suggests that something else besides the expansion of the Universe is producing at least part of the observed redshifts of QSOs and that the QSOs are not nearly as far away as the conventional view holds.

The evidence mentioned here is not strong enough to be decisive, or else the conventional view would be abandoned. This evidence is mostly statistical and difficult to evaluate, so it is not surprising that it is not universally accepted. It should also be pointed out that J. E. Gunn, in 1971, found a QSO that was associated with galaxies in which the redshifts all agreed with each other, and other examples of this nature are suspected. In this case it is difficult to escape the conclusion that the QSO redshift is entirely due to the expansion of the Universe. Perhaps there are *two* kinds of QSOs, one more conventional than the other!

POSSIBLE INTERPRETATIONS

What else besides the expansion of the Universe could cause the large redshifts observed in QSOs? One possible cause is a very large gravity, which can produce an effect that looks exactly like the Doppler effect. The spectral lines emitted by the quasars seem to rule out this possibility, however, because they indicate a much lower gravity than that which would be required.

Another suggestion is that QSO redshifts do indeed indicate high speeds but that the speeds are not due to the expansion of the Universe. In this view the QSOs are relatively nearby (within a few hundred million light years). In this case, we ask how the QSOs got such high speeds, and the usual answer is that they have been expelled from the nuclei of certain galaxies in tremendous explosions. We have already seen evidence that such explosions occur.

There is another interesting problem that

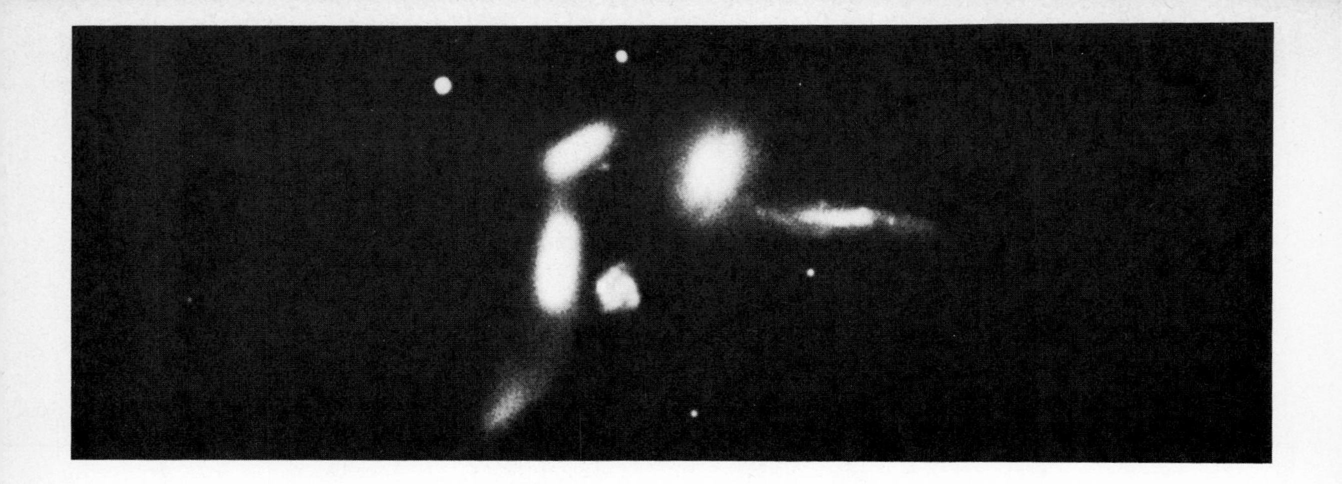

relates to this question. Galaxies are often grouped together in clusters (see Figure 19.11). Such clusters of galaxies are naturally thought to be held together by the gravity that their members exert on each other, just like a cluster of stars. In most cases, however, the motions of the galaxies within a cluster seem too large for the gravity of the cluster to be able to hold itself together. In other words, most clusters of galaxies seem to be in the process of breaking apart.

This is certainly a surprising fact. Why should a cluster break up at all? And why should the clusters all appear to be doing it just now? They are at least as old as the galaxies they contain, and there is no evidence that any galaxy is much younger than the Milky Way, that is, about 15 billion years old. It seems absurd that the clusters would get along fine for some 15 billion years and then suddenly fall apart now. If they fall apart at all, they should have done so long ago, so today there would be no, or very few, clusters left.

Not many astronomers believe that the clusters are really breaking apart, so how can the observations be interpreted? One possibility is that a few giant galaxies in a cluster have ejected smaller galaxies (and QSOs?) at high speeds through explosions in their nuclei, and these high-speed objects create the false impression that the entire cluster is breaking up.[3] Not many years ago the idea that a galaxy could explode and eject huge masses of material at high speeds was considered science fiction. Today the evidence for this occurrence is strong enough for it to be seriously considered as a possibility.

It is not difficult to imagine the existence of a rapidly rotating supermassive star within the nucleus of a galaxy. Suppose the star rotates fast enough to become unstable, which might cause pieces of it to be thrown off at high speeds. It is not as easy to imagine these pieces turning into small galaxies, but who can be certain? Perhaps satellite QSOs and regions of synchrotron emission around galaxies could be produced in this fashion. Smaller instabilities in the rotating supermassive star could account for the variable luminosities observed in QSOs.

There is a major problem with the theory that QSOs are shot out of the nuclei of relatively nearby galaxies. If this were the case, some of the QSOs would be directed at least approximately toward us, and this would cause the spectral lines to be shifted to higher frequencies than normal. Thus, some blueshift QSOs

[3]A more conventional view is that the clusters contain large amounts of "hidden" matter which is unseen but adds enough gravity to hold the clusters together. This matter could be in the form of black holes, faint stars, or matter between the galaxies.

FIGURE 19.11
A cluster of galaxies in Serpens. Note the material that appears to be connecting some of the galaxies. (The Hale Observatories)

should exist, but all of the observed ones have redshifts. Attempts have been made to explain this; one such explanation is that the QSOs all come from the nucleus of the Milky Way or from galaxies that are near enough to the Milky Way that the QSOs have already sped past us and are now receding. Since about 300 QSOs are known and all of them have redshifts, such an explanation appears very doubtful. Also, the Milky Way is far less active than many other galaxies, in terms of emitting large amounts of synchrotron radiation and having a variable luminosity, so it seems unlikely that all QSOs originate in our own galaxy.

Are there discrepant redshifts? That is, are there objects that are close together in space but have widely different redshifts? Opinions vary on this question. Suppose we try a tentative "yes" to see where it leads us. Then we must look again at the question of what else there is besides the expansion of the Universe that could cause the large redshifts observed in quasars. The possibilities that quasars are relatively nearby high-speed objects and that they have extremely large gravities seem to be ruled out. Is there anything left?

The main thing left is our incomplete understanding of the basic laws of physics. There is no question that quasars, radio galaxies, and related objects present many puzzling features. Many apparently contradictory data have been collected on them, and it may be some time before we can be confident that we understand them. The biggest question in all of this is whether the quasars can be understood on the basis of conventional physics. The majority of astronomers answer this question in the affirmative, but the number who qualify their answers or who hesitate before answering appears to be growing larger as new evidence comes in. This is indeed one of the most exciting of times in the long history of astronomy.

IMPORTANT WORDS

Elliptical galaxy
Irregular galaxy
Doppler effect
Velocity-distance relation
Redshift
Blueshift
Radio galaxy
Synchrotron radiation
Quasi-stellar object (QSO)
Quasar
Discrepant redshift

REVIEW QUESTIONS

1 What are the main differences between the three basic kinds of galaxies?

2 How far away are the nearest galaxies?

3 What is special about the Andromeda galaxy?

4 How do we find the distances to the nearest galaxies?

5 How do we find the distances to the very remote galaxies?

6 What is the Doppler effect?

7 What does the term "redshift" mean?

8 What is the velocity-distance relation?

9 What is a radio galaxy?

10 Does the radio radiation come from stars?

11 Why were radio galaxies discovered only rather recently?

12 How do we know that the "radio stars" are not really stars at all?

13 Why do many astronomers believe that the QSOs are the most luminous objects in the Universe?

14 What might supermassive stars have to do with QSOs?

15 What might QSOs and pulsars have in common?

16 What is meant by a discrepant redshift?

17 Are there any QSO redshifts that are not discrepant?

18 What else besides the expansion of the Universe could cause large redshifts?

19 What is so surprising about most clusters of galaxies?

20 Are any QSO blueshifts known?

QUESTIONS FOR DISCUSSION

1 It has been suggested that perhaps all galaxies are spirals when they are young and that they become ellipticals when they grow older. The opposite view, that ellipticals evolve into spirals, has also been suggested. The final possibility is that these two types of galaxies are formed differently and that they stay different throughout their careers.

What evidence can you think of that might relate to this question?

2 Think again of a passing freight train as analogous to a radiation wave. The distance between cars is the wavelength, and the number of cars that pass you each second is the frequency. Convince yourself that the frequency you see depends on whether you are moving in the same or opposite direction as the train, and how fast. How can you reduce the observed frequency to zero?

3 A spectrum is taken of a faint galaxy, and its redshift is measured and found to correspond to a velocity of 10,000 miles per second. From Figure 19.6 we conclude that the galaxy is about 1 billion light years distant. Now suppose that it is discovered that astronomers have made a horrible mistake all these years, and all stars within the Milky Way are really twice as far away from us as we had thought. Would the discovery of this error cause us to change our estimate of the distance of the faint galaxy? Why?

4 Spiral galaxies do not appear to be strong emitters of nonthermal (synchrotron) radiation. Can you relate this observed fact to question 3 in the Questions for Discussion in Chapter 18?

5 Suppose a person claims to have "irrefutable proof" that the QSOs are nearby and that their large redshifts cannot, therefore, be due to the expansion of the Universe. A second person then challenges the first:

SP: Then what are the QSO redshifts due to?
FP: I don't know.
SP: Until you come up with a satisfactory explanation of them, I cannot accept your proof that the QSOs are nearby.

Is the attitude of the second person justifiable?

6 Some quasars have both emission and absorption lines in their spectra. Usually the absorption lines show a slightly smaller redshift than the emission lines in the same quasar. Astronomers don't know why for sure, but there are a number of possible explanations. Can you think of any?

REFERENCES

General references are: *Galaxies*, 3rd ed., by H. Shapley and P. W. Hodge (Harvard University Press, Cambridge, 1973); *Galaxies and Cosmology* by P. W. Hodge (McGraw-Hill, New York, 1966); *The Structure of the Universe* by E. L. Schatzman (McGraw-Hill, New York, 1968); and "Galaxies" by M. S. Roberts, in *The Emerging Universe*, edited by W. C. Saslaw and K. C. Jacobs (University of Virginia Press, Charlottesville, 1972).

Some of the poorly understood properties of galaxies are emphasized by N. Calder in *Violent Universe* (Viking, New York, 1969) and H. L. Shipman in *Black Holes, Quasars, & the Universe* (Houghton Mifflin, Boston, 1976).

Many articles on galaxies and related matters from *Scientific American* are reprinted in *New Frontiers in Astronomy*, edited by O. Gingerich (Freeman, San Francisco, 1975). A similar collection of articles reprinted from *Sky and Telescope* is in *Beyond the Milky Way*, edited by T. Page and L. W. Page (Macmillan, New York, 1969).

Some of the more recent discoveries in radio astronomy are described by R. G. Strom, G. K. Miley, and J. Oort in "Giant Radio Galaxies" [*Scientific American* 233 (August 1975), 26]. An interpretation of the quasars is given by P. Morrison in "Resolving the Mystery of the Quasars?" [*Physics Today* 26 (March 1973), 23]. The redshift problem is discussed by G. B. Field, H. Arp, and J. N. Bahcall in *The Redshift Controversy* (Benjamin, Reading, Mass., 1973) and G. R. Burbidge in "The Problem of the Redshifts" [*Nature Physical Science* 246 (November 1973), 17].

An interesting account of energy in an astronomical context is "Energy in the Universe" by F. J. Dyson [*Scientific American* 225 (September 1971), 50].

TWENTY
COSMOLOGY

Cosmology is the study of the Universe as a whole. Here we see how scientists attack the problems concerned with the large-scale properties of the Universe. Unfortunately, observational data pertaining to these problems tend to be rather marginal at best, so our conclusions will be much more tentative than usual.

20.1 Basics

In previous chapters we discussed some of the more important parts of the Universe. Here we put them all together—planets, stars, galaxies, and whatnot—to see what we can tell about the Universe as a whole.

COSMOLOGY AS A SCIENCE

In studying different parts of the Universe, we look around to see which parts need or deserve studying. Stars appear to be an important part of the Universe; therefore we first observe them as well as we can, then we appeal to the laws of physics to see what type of object would best fit those observations. If all goes well, we end up with an understanding of how stars are built and how they form, age, and die. The same plan of attack applies to galaxies, mountains, insects, and so on. In most cases, of course, we are only part of the way along the road to understanding what we study.

The Universe as a whole does not lend itself directly to this method of attack. There are many stars, so we can see what range of masses, sizes, and so on, the stars exhibit. There is only one Universe, however, so it is not possible for observations to tell us what range of properties a well-behaved universe might be expected to possess. Also, we cannot observe the whole Universe but must be content with those parts that are detectable from the Earth. The problem here is that we don't know and can't know what fraction of the Universe we actually do see.

A fish in a small pond sees the water bounded above by air and below by mud and sand. What is beyond the mud and sand? How far does the air extend? The fish has no way to answer these questions, so to him the Universe has an obvious part, the water, plus a vague part about which he has almost no data. He can see things enter the water from above and, sometimes, later disappear, but he probably would not be able to collect enough information to obtain an understanding of the part of the Universe beyond his water.

Are we like the fish in that we see only a tiny part of the whole Universe? We can make guesses about the unseen part of the Universe, but pure guesses are not a part of science. Is there no way to eliminate pure speculation and apply scientific principles to study the Universe?

There is, in fact, a way to proceed scientifically, but it requires us to make an assumption that can never be proved correct. This assumption is simple enough: The part of the Universe that we can see is typical of the Universe as a whole. Since our direct knowledge is, of course, limited to what we can see, there is no way we can ever know for sure whether this basic assumption is correct.[1]

Our assumption is an extension of the Copernican idea that the Earth is only one of several planets orbiting the Sun, of Bruno's idea that the Sun is only one of many stars in the sky, and of the more recent idea that the Milky Way is only one of many galaxies in the Universe. The point is that the Earth is more or less typical of the other planets, the Sun is typical of other stars, and the Milky Way is fairly typical of other galaxies. This does not mean that all planets, stars, and galaxies are identical; it does mean that we can learn much about planets, stars, or galaxies by studying the Earth, Sun, and Milky Way. We might say in analogy that what we can see from our location is typical of what could be seen from other locations in the Universe. Thus we hope to be able to learn much about the Universe as a whole by studying the part of it accessible to us. Of course we could be wrong, as is the fish that assumes that all of the Universe is a pond.

[1]Here the term "see" means detect in some scientifically acceptable fashion.

THE COSMOLOGICAL PRINCIPLE

The assumption that there is nothing special about our position in the Universe certainly seems reasonable; it is the only way to eliminate pure guesswork about the nature of the Universe. On the other hand, it is not so easy to justify the assumption that scientists usually make when they are working in the field of cosmology. They are not satisfied with the statement that what we see is typical but insist on the much stronger statement that when we average over a sufficiently large volume of space, what we see is *identical* to what anyone else in the Universe would see. Such a strong statement about such an important item as the Universe naturally has a suitably splendid name: It is called the *cosmological principle*.

Why in the world would cosmologists believe in anything as dogmatic and as unprovable as the cosmological principle? This appears to be about as scientific as assuming that the visible part of the Universe is one big olive floating in a gigantic martini. Prove that it isn't!

The cosmological principle, however, is not really that bad. Observations are consistent with it even if they don't prove it. If the visible part of the Universe had major deviations from it, they would have been noted by now.

The cosmological principle also makes investigations into cosmology much simpler. In cosmology, as elsewhere in science, the objective is to construct a physical and mathematical model that will satisfy all observations. Use of the cosmological principle makes the calculations much easier to carry out. You might ask whether or not this is a valid reason for assuming that it is true, and the answer, surprisingly, is yes.

If there are both a simple way and a complicated way to explain something, and if both ways seem to work equally well, then the simpler way is preferred. A complicated theory that fits some observations can always be devised, but when a simple theory explains a complex set of data, then either the theorist is awfully lucky, or else the theory really is basic to the phenomena being observed. Maybe the simple theory will later be proven wrong, but don't give it up until you have to. It is like a person who loses his billfold somewhere while walking home at night. If he has no idea where he lost it, it is to his advantage to look first under the street lights where the search is easiest. Only after he is sure that the billfold is not in a well-lit place will it make sense for him to look in the darkness, where the search is much more difficult to make.

There is yet a third point for supporting the cosmological principle, and some consider it the most compelling. It is a great mistake to assume that scientists base their beliefs solely on solid scientific fact. As pointed out in Section 1.1, scientists always work in areas where the right answers are not known, where the facts are not yet strong enough to separate the correct ideas from the incorrect ones. We are often told that scientists will withhold judgment under these circumstances until further facts can clear the air. Such a statement is just not true.

All scientists are exposed to many facts and ideas over the years. This exposure builds up in them a feeling or an intuition of how things really are, or at least how they think things really are. When the facts in a case are conclusive, nobody argues. But when the facts are inconclusive, scientists always supplement these facts with their intuitive ideas of how things "ought" to be. The degree of agreement to be found among scientists on an issue depends on the relative amounts of fact and intuition involved. In cosmology, the facts are more scarce than anywhere, so intuitions run rampant, and disagreement is the order of the day.

In spite of the widely different intuitions held by scientists, there are a number of points on which almost all agree. Simplicity, as just mentioned, is one. Another is *symmetry*. Symmetry is a balance between opposites. A symmetrical shape in a figure is generally pleasing to the eye. Symmetry of opposite properties or ideas in a theory is the personification of beauty to scientists. Many times in the past, new theories have been proposed that were based mainly on symmetry arguments, and I believe that scientists themselves are surprised at how successful these theories turned out to be in predicting the behavior of nature. Symmetry appears to be an important property that is built into the very foundations of our Universe.

As an example of symmetry arguments, consider the basic structure of matter and radiation. In the simple days of the mid-1800s, scientists believed that matter is composed of solid particles and that radiation is composed of waves. Around the turn of the century, it became apparent that radiation also is composed of particles—the photons introduced in Section 3.2. The wave picture of radiation was not discarded, but it was modified to allow both particle and wave properties in the radiation. In 1924, L. de Broglie asked himself, "If radiation can be both particle and wave, why can't matter be the same?" "Ridiculous!" replied most of the scientists who bothered to give any thought to the question. Electron waves were actually discovered in 1927, and de Broglie received the Nobel Prize in physics in 1929.

With this in mind, who would dare to suggest that we live in an asymmetric Universe? Horrors, such a thought verges on heresy! We should not only insist that the Universe have symmetry, but we should try to give it as much symmetry as possible. This is precisely where the cosmological principle comes in, for it requires that after suitable averaging is carried out, all points in space have identical properties with all other points. Nothing could be more symmetrical than that.

"Not true," said H. Bondi, T. Gold, and F. Hoyle in 1948. The cosmological principle ensures perfect symmetry only for positions. Bondi, Gold, and Hoyle found a way to obtain perfect symmetry for the Universe not only for position but for time, as well. The statement they introduced says that the Universe should appear identical as seen from any position, and as seen at any time. This grand declaration is known as the *perfect cosmological principle.*[2]

Anyone who has properly entered into the spirit of the occasion must admit that the perfect cosmological principle is a thing of beauty with tremendous appeal. On the other hand, we might wonder if we haven't given a bit too much weight to intuitive methods. After all, beauty is at least to some extent in the eye of the beholder. Can we expect the Universe to conform to our ideas of beauty? Do we really live in the most perfect of possible universes (whatever that means)? Such thinking in the past has often led people astray, but we must also admit that it has occasionally been very fruitful scientifically.

Actually, the perfect cosmological principle *is* worthy of consideration, and we shall develop some of its consequences in Section 20.2. We note here, though, that observational evidence is strongly against the perfect cosmological principle, and it has been abandoned by most cosmologists as not applicable to our Universe. This reaffirms the importance of observations in keeping our ideas consistent with reality. The ordinary cosmological principle is still consistent with the observations, however, and it is the starting point for most modern theories of the structure of the Universe.

[2]Cosmological principles are apparently exceptions to the rule that nobody is perfect.

20.2 Models of the Universe

EXPANSION OF THE UNIVERSE

Probably the most basic observation relating to cosmology is the *expansion of the Universe.* This is based on observations of the shifts in frequency of the spectral lines in the radiation emitted by the galaxies. The spectral lines, remember, are all shifted to lower frequencies than normal, an effect called a redshift. (A change to higher frequencies is called a blueshift.) If the redshifts of galaxies are due to the Doppler effect, that is, if they are due to the speeds of the galaxies, then all galaxies must be traveling away from us at high speeds. The more distant galaxies are moving faster than the nearer ones. Figure 19.6 showed how the speeds of galaxies are related to their distances from us. Although there are theories that claim that the redshifts observed in galaxies have nothing to do with the Doppler effect and that the Universe therefore is not expanding, such theories are not a major part of the cosmological research of today and will not be considered here.

Doesn't the expansion of the Universe violate the cosmological principles, both ordinary

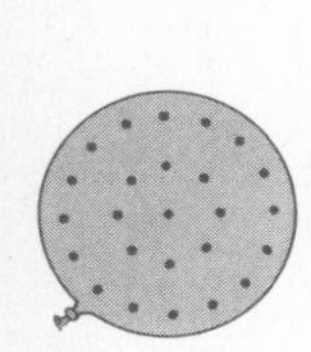
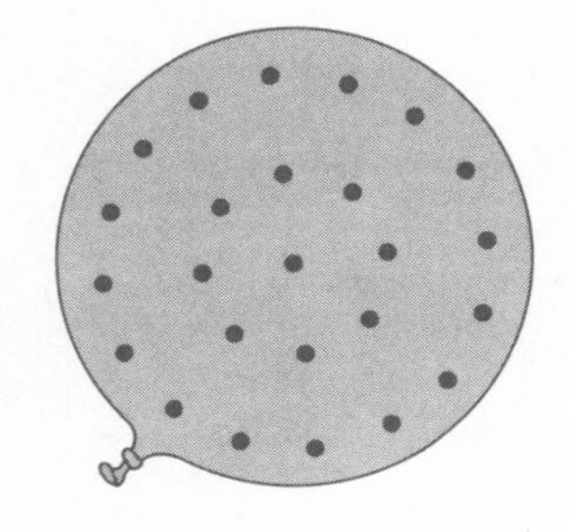

FIGURE 20.1
Dots on an expanding balloon.

and perfect? These principles say that one point in space should be just like any other point, so there shouldn't be any special position; yet we seem to be in a special position at the center of the expansion. However, a little thought shows that our position need not be special after all. Galaxies are not just moving away from us, they are moving away from each other as well. It is as though the Universe were the surface of a huge balloon in the process of being blown up, and the galaxies are dots painted on the surface (see Figure 20.1). As the balloon expands, the dots all get farther apart. A man standing on one of the dots will see the other dots going away from him, and he will probably think that he is at the center of the expansion. But a woman on any of the other dots will see exactly the same thing—all the dots seem to be receding from her position, too. All points look as though they were equally the center of the expansion, and this is the way that the expansion of the Universe can be consistent with the cosmological principles.

AN EDGELESS UNIVERSE

But what about those galaxies on the edge of the expanding Universe? From out there, we ought to see other galaxies only by looking back toward the center; in the opposite direction, there should be none to be seen. Doesn't this violate the cosmological principles? Yes, in fact, it does, for the view from the edge would be different from that as seen far from the edge. The cosmological principles, therefore, require that the Universe have no edge.

How can anything have no edge or boundary? One possibility is that the Universe extends without limit in all directions. Thus, an infinite space uniformly filled with galaxies can satisfy the cosmological principles. But there is also a finite-size Universe that can satisfy those principles, too. This can be accomplished by allowing space to bend back upon itself, much like the surface of the sphere in Figure 20.1. It is true that the sphere has a center and an edge, but if we limit ourselves to points confined to the surface, then there is no center or edge. It is not easy to visualize curved space, but mathematical equations can be written that describe it, and a respected branch of physics, general relativity, claims that curved space is a reality. In this context, space itself is not something we can measure, but rather its properties are found by certain measurements made on physical objects and radiation.

STEADY STATE THEORY

The expansion of the Universe appears to contradict that part of the perfect cosmological principle that requires the Universe to stay the same for all time. The expansion carries galaxies farther apart, so matter ought to be thinning out as time goes by. In order to allow for expansion and still keep an unchanging Universe, the perfect cosmological principle requires that new matter be created to fill the empty spaces between the receding galaxies.

And where does this new matter come from? The answer is *nowhere*. The new matter simply appears spontaneously out of nothing, and it must appear at just the right rate to exactly balance the thinning-out of the matter due to the expansion. This theory, based on the perfect cosmological principle, is known as the *steady state theory*, since the Universe is assumed to be in a state that does not change with time. Steady state theory requires that matter be created continuously throughout all space and time.

According to steady state theory, individual stars and galaxies run through the courses of evolution outlined in previous chapters. Newly created atoms eventually accumulate to form new galaxies, and these new galaxies come into being at exactly the same average rate as the old galaxies die out and move apart. The Universe is continuously provided with a supply of new parts to take the place of the old, worn-out ones, and it goes on forever without changing its main properties. Thus, the Universe is without beginning and without end. It is like a lawn sprinkler that shoots out drops of water in all directions. Individual drops disappear into the ground, but new ones are sent out at the same rate to take their place. The overall pattern of the water remains unchanged with time. Note, however, that steady state requires that new matter be created everywhere throughout space, not just at one point.

Steady state theory violates the long-established principle of the conservation of mass and energy since it proposes that new matter and energy are created out of nothing. Many people find this a rather distasteful aspect of steady state, but this is their intuition speaking, not scientific fact. But don't scientific facts back up the idea that the amount of mass and energy is fixed? The answer is yes, but with the same qualification that must always apply to experimental results: Any experiment or observation has a certain accuracy, and it cannot confirm any theory beyond the limits of that accuracy.

Now we must ask how fast new matter must be created in order to satisfy steady state, and then we can see if our answer contradicts the experimental evidence. The observed rate of expansion of the Universe requires one new hydrogen atom to be created within each cubic mile of space each year, if the perfect cosmological principle is correct. This is such a tiny amount of new matter that scientists cannot possibly detect it, no matter how hard they try. In other words, steady state is perfectly consistent with experimental evidence concerning the conservation of mass and energy; it is neither proved nor disproved by these particular data.

In the steady state model of the Universe, all space is uniformly filled with galaxies. If we consider two galaxies taken at random, we see that they are moving away from each other at a speed proportional to their distance apart (as was indicated in Figure 19.6). As time goes on and the galaxies get farther apart, they must increase their speed of separation; otherwise, there would not be a unique relation between velocity and distance as required by the perfect cosmological principle. As the distance between the galaxies increases, new matter is formed between them to "fill up the holes," and this new matter must eventually form into new galaxies. Thus space is filled with galaxies of all ages mixed together. As our two galaxies grow very old, their distance apart and their speed of separation become extremely great, and eventually they will drop out of sight of each other. The matter in a galaxy does not disappear; after the stars have all burned out, the galaxy will continue its existence as a collection of white dwarfs, neutron stars, and black holes. Such burnt-out galaxies must be very rare because the expansion of the Universe would spread them out over extremely large volumes of space. The steady state model of the Universe has no beginning or end, so according to this theory the Universe must be of infinite age. An infinite amount of matter has been created, and it has been spread over an infinite volume of space.

BIG BANG MODELS

Let us now consider models of the Universe based on the ordinary cosmological principle. They do not forbid the Universe from changing with time, so they do not require a continuous creation of matter to compensate for the expansion of the Universe. The expansion of the Universe is causing matter to thin out with time, and the average density of the Universe must have been greater in the past than it is at present. As we go back farther and farther in the past, the matter is squeezed closer and closer together until a limit is reached about 15 billion years ago. This limit represents the start of the observed expansion. It is as though a tremendous explosion occurred back then, and the Universe has been flying apart ever since. For this reason, theories of cosmology based on the ordinary cosmological principle are called *big bang theories*.[3] As we shall see, there are many models of the Universe that

[3]Actually, there are also models of the Universe based on the ordinary cosmological principle that do not start with an explosion from a very high density condition, but we will not consider them here.

are based on the ordinary cosmological principle.

The many possible big bang models of the Universe differ from each other in a number of particulars. Two of the most important questions are: (1) Is the Universe finite or infinite in size? (2) Will the expansion continue forever, or will it eventually stop and be followed by a contraction of the Universe?

The first question is one of the most intriguing of all. Does the Universe go on indefinitely, or is it somehow limited in size? And if it is limited, what lies beyond those limits? We must remember that the Universe is defined as containing all things with which we can make scientific contact. This does not preclude the possiblity of things existing outside of our Universe; it only means that we have no way of visiting them, seeing them, or otherwise detecting them by scientific methods. If the Universe is finite, the barrier that keeps us from escaping this finite region is analogous to the gravity that keeps matter and radiation from escaping from a black hole. It is as though the Universe is one huge black hole from which nothing can escape. There is no way to communicate with the outside, although it is conceivable that objects from the outside can fall in, never to get out again.

The second question is another important point over which big bang models differ with each other. The Universe is expanding, but is the expansion rate constant with time? Possible models run the complete range, from accelerating to slowing down of the expansion rate as time goes on. Of course, we expect that gravity will try to slow down the rate of expansion, but relativity introduces a new force that can, in principle, overcome gravity over large distances and cause a net repulsion between distant galaxies. This makes it possible for the expansion rate to increase with time.

The expansion of the Universe is analogous to throwing an object up in the air. If you throw an object fast enough, it will escape completely from the Earth. If its speed is less than a certain critical value, the velocity of escape, then the gravity of the Earth will pull it back again. In the case of the Universe, if the speed of expansion is greater than the escape velocity, then the expansion will go on forever, and the galaxies will get infinitely far apart. If the expansion speed is less than the escape velocity, however, gravity will eventually pull the galaxies back together again, and the Universe will come into a state of contraction.

Consider a big bang model of the Universe in which a contraction phase will eventually be reached. Since gravity is pulling everything together, how can this contraction be halted? One of the possibilities is that all matter and radiation will be squeezed together to such a great extent that the internal pressure will build up and finally stop the contraction in a way that is similar to the formation of an ordinary or supermassive star. In this case, however, the collapse is likely to occur with such violence that it will carry far past the stage in which pressure balances the force of gravity. At maximum compression, the pressure will far exceed gravity, and, like a compressed spring, it will be ready to propel the matter and radiation out into another expansion phase. In this way we can imagine an oscillating Universe alternately expanding and contracting like a pulsating star. The oscillations can go on forever, unlike those of a bouncing ball, because there is no way for the Universe as a whole to lose energy to its surroundings.

A big bang model that allows the Universe to oscillate indefinitely has great appeal for many scientists. In a model which expands forever, the moment the expansion began is a unique instant in the history of the Universe. Just as scientists are wary of the apparent occurrence of unique positions in the Universe (remember Copernicus and Bruno), so they tend to dislike unique instants of time. An oscillating Universe does not have a single moment of maximum compression but an infinite number of them; like steady state, it has no beginning and no end. It is worth emphasizing that this feeling among scientists is another example of intuition rather than scientific fact showing through. It is also worth noting how much this intuition that favors an oscillating Universe has in common with the philosophy of the steady state theory. (See question 5 in the Questions for Discussion at the end of the chapter.)

20.3 What kind of Universe do we live in?

A number of theoretical models were described in the last section, and who knows? One of them might actually be an accurate representation of the Universe. But science should not be only a theoretical recreation; one needs to collect and analyze data in order to keep in touch with the real world. Do measurements suggest that one model is more accurate than the others? Can observations tell us what kind of a Universe we live in?

There are a number of observations that are important in cosmology. By their nature, these observations tend to be measurements near the limit of what can be detected, or they require allowance for effects not completely understood. The result is sometimes tantalizingly close to pointing in a certain direction, but usually there is just enough uncertainty to keep us in doubt. Small wonder that astronomers constantly plead for more funds for bigger telescopes and more sophisticated instru-

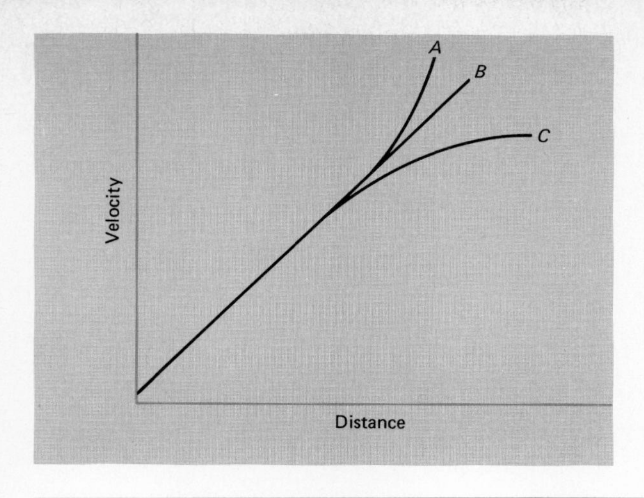

FIGURE 20.2
Possible velocity-distance relations.

ments so they might be able to peer around the next bend and see what lies a little farther down the road.

CHANGES IN THE EXPANSION RATE

One of the observations is a detail on the expansion of the Universe. Suppose that expansion is neither speeding up nor slowing down, so each galaxy moves with constant speed. In this case a galaxy that is three times farther away than another has three times greater speed, and so on. This type of relation is represented by a straight line in the velocity-distance relation—curve *B* in Figure 20.2.[4]

When we photograph or make any measurement of a galaxy, we are working with radiation emitted by that galaxy sometime in the past. How far in the past? If a galaxy is 1 billion light years away from us, it takes 1 billion years for the radiation to reach us from the galaxy; therefore, we are seeing what that galaxy was like 1 billion years ago. When we measure the velocity of the galaxy through its redshift, we find the velocity it had 1 billion years ago. There is no way to find what its velocity is today without waiting another billion years for the light it is now emitting to arrive and tell us.

Let's look again at Figure 20.2. The straight line is labeled curve *B*, and it represents the velocity-distance relation if the speeds of the galaxies do not change over an interval of time. But now suppose that the expansion rate is slowing down, so each galaxy is moving more slowly today than it did in the past. Since we see the galaxies as they were in the past, the speeds we measure are larger than the speeds of today. How much larger? That depends on how far away a galaxy is and how rapidly the expansion is slowing down. The expansion rate cannot change much in only a few million years, so the nearby galaxies will be observed with speeds that are essentially the same as their true speeds of today. The expansion could change quite a bit in 2 billion years, however, so a galaxy 2 billion light years away might have a measured velocity much larger than its true velocity of today. Thus the farther away a galaxy is, the more its measured speed would exceed the speed it should have today. This is another way of saying that the very distant galaxies would have larger speeds than are indicated by their distances according to curve *B* in Figure 20.2. We indicate larger speeds in the figure by moving the points upward; therefore, very remote galaxies should have their representative points above curve *B* if the expansion is slowing down. Curve *A* in the figure shows what the velocity-distance relation might look like if this were the case.

[4]The positions of the galaxies need to be corrected for their motions during the light travel time to the Earth for curve *B* to be valid in this case.

The argument is similar if the galaxies are speeding up with time. Looking at a very remote galaxy, we see it as it was in the distant past when it was moving more slowly than at present. The galaxy would have a lower velocity than indicated by curve *B* in Figure 20.2, and a curve like the one labeled *C* would show the proper velocity-distance relation.

What do observations really show—curve *A*, *B*, or *C*? The data are not accurately enough known for strong conclusions to be drawn, but there is evidence in favor of a curve of type *A*; in other words, the expansion rate does seem to be slowing down.[5] While a mild acceleration of the expansion rate cannot be completely ruled out by the present data, an acceleration as strong as that required by the steady state theory does appear to be strongly contradicted by the observations. The expansion rate does seem to be slowing down, but the evidence suggests that it will never completely stop. Galaxies will continue to recede from each other indefinitely, but they will move at a slower and slower pace.[6,7]

[5] A. R. Sandage and E. Hardy, *Astrophysical Journal* 183 (1973), 743.

[6] A. R. Sandage and G. A. Tammann, *Astrophysical Journal* 197 (1975), 265.

[7] J. R. Gott, J. E. Gunn, D. N. Schramm, and B. M. Tinsley, *Astrophysical Journal* 194 (1974), 543.

NUMBERS OF FAINT GALAXIES

Another observation important in cosmology is a count of the number of galaxies with different apparent brightnesses. On the average, the fainter galaxies tend to be farther away than the brighter ones, so counting the galaxies can give us information on their distribution in space. This is an important clue in cosmology.

For a number of reasons, it is easier to make this study in radio frequencies than in visible light, but the principle is the same. According to M. Ryle, counts of sources of strong radio emission suggest that radio galaxies are more common at great distances than they are near to us.[8] The easiest way to explain this is to assume that all galaxies are about the same age and that young galaxies are more likely to have strong radio emission than old ones. When we look at nearby galaxies, we see ones that are the same age as the Milky Way or about 15 billion years old. When we look at galaxies at great distances, however, we are also looking back in time, and the galaxies we see are younger than 15 billion years old. Therefore, the radio galaxies tend to be at great distances from us. This is, of course, in violation of the perfect cosmological principle that requires young and old galaxies to be

[8] M. Ryle, *Annual Review of Astronomy and Astrophysics* 6 (1968), 249.

mixed together at all distances. The apparent contradiction with the ordinary cosmological principle is resolved by the assumption that radio galaxies were more common in the past than they are now, an assumption not available to the steady state model.

APPARENT SIZES

Another check for cosmological models was reported by W. A. Baum.[9] This method relates the apparent sizes of galaxies to their distances. We know that the farther away an object is, the smaller it looks. In precisely what way the apparent size depends on distance turns out to be different for different models of the Universe.[10] Baum's observations are consistent with a model in which the expansion rate is slowing down, but not rapidly enough to eventually stop and start contracting. His observations also mildly suggest that the Universe is infinite in size. It must be emphasized that there are enough uncertainties to make these conclusions very tentative; however, they are rather strongly against steady state theory.

[9]W. A. Baum, in *External Galaxies and QSO's*, edited by D. S. Evans, Reidel, Dordrecht, Holland, 1972, p. 393.

[10]For some models, any further increase in distance after a certain limit is reached causes the object to start appearing *larger* in size.

FURTHER OBSERVATIONS

In 1965, A. A. Penzias and R. W. Wilson discovered weak radio radiation at a frequency of about 10^{10} cycles per second which seemed to come from all directions in space. Further observations have confirmed that this is just the kind of radiation one might expect to be left over if, about 15 billion years ago, the Universe was in a hot, dense state. In other words, this is fairly strong evidence that there was a big bang approximately 15 billion years ago. Needless to say, steady state theory must find another explanation for this *microwave background*, as it is called, if it wishes to stay in business.

There are yet other observations that have a direct bearing on cosmology. For example, if there was a big bang sometime in the past, it must have occurred before the formation of the Milky Way. I have been quoting 15 billion years as the ages of both the Milky Way and the Universe, but both ages are uncertain enough that there is no difficulty in insuring that the Milky Way is younger than the Universe. Determining the ages of the oldest stars in the Milky Way does place an important limit on the possible age of the Universe.

The abundances of the chemical elements provide another important cosmological clue. Theories exist that indicate how the elements were formed, both in the early stages of a big

FIGURE 20.3
With large telescopes we can photograph almost countless numbers of distant galaxies. Are these galaxies destined to continue receding from each other until each is alone in an essentially empty Universe? (The Hale Observatories)

bang model of the Universe and in the cores of stars. By forcing the elements to agree with the observed abundances, we can obtain some restrictions on what type of Universe we inhabit.

How do we summarize the evidence that is in so far? Certainly it is strongly against steady state theory, and the latter has been almost completely abandoned as a model for the real world. Also, it seems likely that a big bang did occur sometime around 15 billion years ago. There is little else one can say with much confidence, although the model Universes that best fit the observations are infinite in size, and will expand forever. If the infinite model really does represent the Universe, then we will continue to become more and more distant from other galaxies as time goes on, and the ultimate fate is complete isolation in a cold, dark Universe. The Universe as we know it (see Figure 20.3) would be gone forever.

When we note how few questions the observations seem to be able to answer and how tentative those answers are, we might tend to become pessimistic. Will we ever know the answers for sure? Can we ever find out what kind of a Universe we live in?

What do *you* think?

IMPORTANT WORDS

Cosmological principle

Symmetry

Perfect cosmological principle

Expansion of the Universe

Steady state theory

Big bang theory

Microwave background

REVIEW QUESTIONS

1 Can observations tell us what range of properties a universe should be expected to have?

2 Do we know that the part of the Universe we can see is typical of the Universe as a whole?

3 What is the cosmological principle?

4 Other things being equal, why is a simple theory considered preferable to a complicated one?

5 What does symmetry have to do with science?

6 What is the perfect cosmological principle?

7 Why doesn't the expansion of the Universe violate the cosmological principle?

8 If the cosmological principle is valid, why can't the Universe have an edge?

9 How does the perfect cosmological principle make itself compatible with the expansion of the Universe?

10 Does the steady state theory violate the principle that the amount of matter and energy in the universe must remain constant?

11 How old is the Universe if the steady state theory is correct?

12 Why are most of the models of the Universe that satisfy the ordinary cosmological principle called big bang theories?

13 If the Universe is finite in size, could there be other matter outside of our Universe?

14 What is meant by an oscillating model of the Universe?

15 Why is looking out at great distances also looking back at remote times in the past?

16 Why might different model Universes have different velocity-distance relations?

17 Does the velocity-distance relation seem to indicate that the expansion speed is increasing or decreasing with time?

18 In what way do counts of the numbers of sources of radio radiation appear to contradict the steady state theory?

19 What is the microwave background, and what does it seem to indicate?

20 What kind of Universe would you like to live in? Why?

QUESTIONS FOR DISCUSSION

1 What would be your reaction to someone who claims that the Universe is one huge custard pie? Assume that he is being serious about it.

2 Few things seem more complicated than the physical-mathematical theories scientists devise in order to explain nature; yet the scientists claim to be guided by simplicity. Just what do they mean by simplicity?

3 Many persons do not like the idea of matter being created continuously, but some of the same persons are not bothered by the thought that maybe all matter was created at one instant in the past. Does either idea appeal to you more than the other? Do you understand why?

4 The velocity-distance relation (look again at Figure 19.6) must be the same for all times in the future if the steady state theory is correct. Show that this requires a galaxy to speed up as time goes on.

5 Although an oscillating model of the Universe is based on the ordinary cosmological principle, show that it also satisfies a modified form of the perfect cosmological principle.

6 An oscillating model requires that gravity be strong enough to halt the expansion of the Universe. This amount of gravity can be produced by a certain minimum density of matter, but the matter we can detect is far too thinly spread out to produce the required gravity. This leads some astronomers to look for the "missing matter" needed to make the Universe oscillate. If there is such missing matter, what form do you expect it to have?

7 If a galaxy is 1 billion light years away, why do we have to wait for another billion years to find out what's going on there now? Why can't we, at least in

principle, make a long-distance phone call to the galaxy and ask someone?

8 Suppose the Universe exploded from a very small size 15 billion years ago. How far away would the most distant objects be today?

REFERENCES

General references include *The Structure of the Universe* by E. L. Schatzman (McGraw-Hill, New York, 1968); *Galaxies and Cosmology* by P. W. Hodge (McGraw-Hill, New York, 1966); and "Big-Bang Cosmology" by G. B. Field in *The Emerging Universe*, edited by W. C. Saslaw and K. C. Jacobs (University of Virginia Press, Charlottesville, 1972). Two excellent books at a somewhat more advanced level but mostly understandable to the layman are: *Physical Cosmology* by P. J. E. Peebles (Princeton University Press, Princeton, 1971); and *Cosmology*, 2nd ed., by H. Bondi (Cambridge University Press, Cambridge, England, 1961).

A number of *Scientific American* articles related to cosmology are reprinted in *New Frontiers in Astronomy*, edited by O. Gingerich (Freeman, San Francisco, 1975). Some of the more speculative aspects of the subject are emphasized by W. J. Kaufmann in *Relativity and Cosmology*, 2nd ed. (Harper & Row, New York, 1977) and by N. Calder in *Violent Universe* (Viking, New York, 1969).

Discussions of the relevant observations are given by A. R. Sandage in "Cosmology" [*Physics Today* 23 (February 1970), 34] and by J. R. Gott, J. E. Gunn, D. N. Schramm, and B. M. Tinsley in "Will the Universe Expand Forever?" [*Scientific American* 236 (March 1976), 62].

Some aspects of the ways in which different models of the Universe affect chemical abundances are discussed by J. M. Pasachoff and W. A. Fowler in "Deuterium in the Universe" [*Scientific American* 230 (May 1974), 108]. G. R. Burbidge gives reasons for suspecting that the Universe might not have undergone a big bang in "Was There Really a Big Bang?" in *The Emerging Universe*, edited by W. C. Saslaw and K. C. Jacobs (University of Virginia Press, Charlottesville, 1972,) pp. 165–180.

APPENDIX ONE
GENERAL REFERENCES

There are several texts that cover the field of astronomy in a much more complete fashion than is intended in the present book. These are essentially encyclopedias of astronomy and include

Abell, G. O. *Exploration of the Universe*, 3rd ed. Holt, New York, 1975.

Fredrick, L. W., and R. H. Baker. *Astronomy*, 10th ed. Van Nostrand, New York, 1976.

Wyatt, S. P. *Principles of Astronomy*, 2nd ed. Allyn and Bacon, Boston, 1971.

A very good elementary book on the structure and evolution of the stars is

Aller, L. H. *Atoms, Stars, and Nebulae*, 2nd ed. Harvard University Press, Cambridge, 1971.

Many of the important problems that astronomers were concerned with during the first half of the twentieth century are discussed in an interesting way by

Struve, O., and V. Zebergs. *Astronomy of the Twentieth Century*. Macmillan, New York, 1962.

Passages from the works of astronomers from Copernicus's day to 1950 are collected in

Source Book in Astronomy, edited by H. Shapley and H. E. Howarth. McGraw-Hill, New York, 1929.

Source Book in Astronomy 1900–1950, edited by H. Shapley. Harvard University Press, Cambridge, 1960.

There are many good books and articles on space programs. These include

Apollo Expeditions to the Moon, edited by E. M. Cortright. NASA SP-350. U.S. Government Printing Office, Washington, D.C., 1975.

"Apollo 17." *National Geographic* 144 (September 1973), 289.

Exploring Space with a Camera, edited by E. M. Cortright. NASA SP-168. U.S. Government Printing Office, Washington, D.C., 1968.

Fimmel, R. O., W. Swindell, and E. Burgess. *Pioneer Odyssey*. NASA SP-349. U.S. Government Printing Office, Washington, D.C., 1974.

"Skylab." *National Geographic* 146 (October 1974), 441.

There are many books written at a more advanced level that might interest a person with a modest background in mathematics and science. Some of these are

Astronomy: A Handbook, edited by G. D. Roth. Springer-Verlag, New York, 1975.

Blanco, V. M., and S. W. McCuskey. *Basic Physics of the Solar System*. Addison-Wesley, Reading, Massachusetts, 1961.

Brandt, J. C., and P. Hodge. *Solar System Astrophysics*. McGraw-Hill, New York, 1964.

Danby, J. M. *Fundamentals of Celestial Mechanics*. Macmillan, New York, 1962.

Haymes, R. C. *Introduction to Space Science*. Wiley, New York, 1971.

Kaula, W. M. *An Introduction to Planetary Physics*. Wiley, New York, 1968.

Mihalas, D., and P. M. Routly. *Galactic Astronomy*. Freeman, San Francisco, 1968.

Peebles, P. J. E. *Physical Cosmology*. Princeton University Press, Princeton, 1971.

Schwarzschild, M. *Structure and Evolution of the Stars*. Princeton University Press, Princeton, 1958.

Smart, W. M. *Spherical Astronomy*, 4th ed. Cambridge University Press, Cambridge, England, 1956.

Smith, E. v. P., and K. C. Jacobs. *Introductory Astronomy and Astrophysics*. Saunders, Philadelphia, 1973.

Swihart, T. L. *Astrophysics and Stellar Astronomy*. Wiley, New York, 1968.

Unsöld, A. *The New Cosmos*. Springer-Verlag, New York, 1968.

A tremendous amount of astronomical data is contained in

Allen, C. W. *Astrophysical Quantities*, 3rd ed. Athlone Press, London, 1973.

Less technical and less complete is

Robison, J. H. *Astronomy Data Book*. Wiley, New York, 1972.

The following journals are published each year and consist of a series of review articles that give the current state of research in a number of areas. They are intended for the scientific nonspecialist, but most of the articles can be easily understood by an interested member of the general public

Annual Review of Astronomy and Astrophysics. Annual Reviews, Inc., Palo Alto, California.

Annual Review of Earth and Planetary Sciences. Annual Reviews, Inc., Palo Alto, California.

Useful sky maps and guides to the stars and constellations include:

Howard, N. E. *The Telescope Handbook and Star Atlas*. Crowell, New York, 1967.

Levitt, I. M., and R. K. Marshall. *Star Maps for Beginners*. Simon & Shuster, New York, 1964.

Menzel, D. H. *A Field Guide to the Stars and Planets*. Houghton Mifflin, Boston, 1964.

Norton, A. P. *Norton's Star Atlas and Reference Handbook*, 16th ed. Sky Publishing, Cambridge, 1973.

Ottewell, Guy. *Astronomical Calendar*. Department of Physics, Furman University, Greenville, South Carolina 29613. Annual.

Sky Calendar. Abrams Planetarium, Michigan State University, East Lansing, Michigan 48823. Monthly.

The following journals are intended for the interested public:

Astronomy. Astromedia Corp., 411 E. Mason Street, Milwaukee, Wisconsin 53202.

The Astronomy Quarterly, Pachart Publishing House, P.O. Box 6721, Tucson, Arizona 85733.

Griffith Observer. Griffith Observatory, P.O. Box 27787, Los Angeles, California 90027.

Mercury. Astronomical Society of the Pacific, 1244 Noriega Street, San Francisco, California 94122.

Scientific American. 415 Madison Avenue, New York, New York 10017.

Sky and Telescope. Sky Publishing Corp., 49 Bay State Road, Cambridge, Massachusetts 02138.

APPENDIX TWO
VERY LARGE AND VERY SMALL NUMBERS

Scientists often work with numbers that are very much larger or smaller than those commonly used in everyday living. Let's consider first the names and the relations between some of these unusual numbers:

$$1 \text{ million} = 1{,}000{,}000 = 1000 \times 1000$$

$$1 \text{ millionth} = \frac{1}{1 \text{ million}} = 0.000001 = \frac{1}{1000} \times \frac{1}{1000}$$

$$1 \text{ billion} = 1{,}000{,}000{,}000 = 1000 \times 1 \text{ million}$$

$$1 \text{ billionth} = \frac{1}{1 \text{ billion}} = 0.000{,}000{,}001 = \frac{1}{1000} \times \frac{1}{1 \text{ million}}$$

$$1 \text{ trillion} = 1{,}000{,}000{,}000{,}000 = 1000 \times 1 \text{ billion}$$

$$1 \text{ trillionth} = \frac{1}{1 \text{ trillion}} = 0.000{,}000{,}000{,}001 = \frac{1}{1000} \times \frac{1}{1 \text{ billion}}$$

$$1 \text{ quadrillion} = 1{,}000{,}000{,}000{,}000{,}000 = 1000 \times 1 \text{ trillion}$$

$$1 \text{ quadrillionth} = \frac{1}{1 \text{ quadrillion}} = 0.000{,}000{,}000{,}000{,}001 = \frac{1}{1000} \times \frac{1}{1 \text{ trillion}}$$

Beyond quadrillion we have quintillion, sextillion, septillion, octillion, nonillion, decillion, and so on. Eventually we will run out of names, but the numbers go on forever.

In writing very large or very small numbers, we find it convenient to use a shorthand notation, often called scientific, or exponential, notation. Note that any large number consisting of the digit 1 followed by some zeros is equal to the number 10 multiplied by itself a certain number of times; the number of zeros in the large number is equal to the number of times 10 is used as a factor. Thus, if we wish to use 10 as a factor four times, we can express the product as 10,000; we can also write the same number as 10^4, where the number 4, called an exponent, tells how many times 10 is used as a factor. In general, we have

$$100 = 10 \times 10 = 10^2$$

$$1000 = 10 \times 10 \times 10 = 10^3$$

$$10000 = 10 \times 10 \times 10 \times 10 = 10^4$$

and so on. The numbers 1 million, 1 billion, 1 trillion, and so forth, then can be written as 10^6, 10^9, and 10^{12}. Very small numbers are written with negative exponents:

$$1 \text{ millionth} = \frac{1}{10^6} = 10^{-6}$$

$$1 \text{ billionth} = \frac{1}{10^9} = 10^{-9}$$

and so on.

The number 41 is only slightly larger than 40, so

we are apt to think that 10^{41} is only slightly larger than 10^{40}. But what 10^{41} actually does is use 10 as a factor 41 times as opposed to 40 times, which means that it is 10 times larger. When written out in full, 10^{41} is the digit 1 followed by 41 zeros, a tremendously large number. This illustrates the power (!) of the exponential notation.

APPENDIX THREE
VALUES OF CONSTANTS

ASTRONOMICAL CONSTANTS

Diameter of Earth through equator: 7926 miles = 12,756 kilometers

Diameter of Earth through poles: 7899 miles = 12,712 kilometers

Mass of Earth: 6.6×10^{21} tons = 6.0×10^{27} grams

Distance to Moon: 238,000 miles = 384,400 kilometers

Distance to Sun: 92,960,000 miles = 149,600,000 kilometers

Diameter of Sun: 865,000 miles = 1,392,000 kilometers

Mass of Sun: 2.2×10^{27} tons = 2.0×10^{33} grams

1 light year = 5.88×10^{12} miles = 9.46×10^{12} kilometers

1 year = 365.2422 days = 31,557,000 seconds

ENGLISH-METRIC CONVERSION

1 mile = 1.6093 kilometers = 1.6093×10^{5} centimeters

1 inch = 2.54 centimeters

1 centimeter = 0.3937 inch

1 kilometer = 0.6214 mile

1 pound = 453.6 grams = 0.4536 kilograms

1 kilogram = 1000 grams = 2.2046 pounds

TEMPERATURE

degrees Celsius (centigrade) = $\frac{5}{9}$ (degrees Fahrenheit − 32)

degrees F = $\frac{9}{5}$ (degrees C) + 32

degrees Kelvin (absolute) = degrees C + 273

APPENDIX FOUR
DATA ON PLANETS (plus Moon and Ceres)

Planet	Average distance from Sun (millions of miles)	Average distance from Sun (Earth = 1.0)	Period around Sun (years)	Orbit eccentricity (circle = 0)	Satellites
Mercury	36.0	0.387	0.241	0.206	0
Venus	67.2	0.723	0.615	0.007	0
Earth	92.9	1.000	1.000	0.017	1
(Moon)	—	—	—	0.055	0
Mars	141.6	1.524	1.881	0.093	2
(Ceres)	257	2.766	4.602	0.079	0
Jupiter	484	5.203	11.86	0.048	13
Saturn	887	9.539	29.46	0.056	10
Uranus	1780	19.18	84.01	0.047	5
Neptune	2790	30.06	164.8	0.009	2
Pluto	3670	39.44	247.7	0.250	0

Planet	Diameter (miles)	Mass (Earth = 1.0)	Average density (water = 1.0)	Period of rotation (days, hours, minutes)	Surface gravity (Earth = 1.0)
Mercury	3030	0.055	5.4	59^d	0.37
Venus	7520	0.815	5.2	243^d	0.91
Earth	7930	1.000	5.5	23^h56^m	1.00
(Moon)	2160	0.012	3.3	$27\frac{1}{3}^d$	0.17
Mars	4220	0.108	3.9	24^h37^m	0.38
(Ceres)	593	0.00017	2.2	9^h05^m	0.03
Jupiter	88,700	318	1.3	9^h50^m	2.54
Saturn	74,600	95.2	0.7	10^h14^m	1.08
Uranus	32,200	14.6	1.2	10^h49^m	0.89
Neptune	30,800	17.2	1.7	15^h48^m	1.14
Pluto	3700(?)	0.1(?)	?	6^d	?

APPENDIX FIVE
STELLAR DATA

The two tables in this appendix list some of the interesting properties of the brightest and the nearest stars, respectively. In both tables, the apparent brightness is given in *magnitudes*, a measure used by astronomers. Magnitudes are a backward scale in the sense that the brighter a star appears, the smaller is its magnitude. The stars that appear brightest in the night sky have magnitudes between 0 and 1. The faintest stars we can see on a dark night without a telescope are about magnitude 6. The Sun appears so bright that it has a magnitude of about −27. Note that most of the nearest stars to us have magnitudes greater than 6, meaning that they are too faint to be seen without a telescope.

It is surprising to find that the nearest stars are not those that appear brightest to us. Only four of the 21 nearest stars are among the 20 brightest: the Sun, Sirius A, alpha Centauri, and Procyon. The nearest stars are rather close to a random sample of stars. Note that most of these stars are smaller, cooler, and much less luminous than the Sun. The brightest stars, on the other hand, are much more luminous than average, which is the main reason that they appear so bright in the sky.

Most of the brightest stars are at rather large distances from us. As a result, the distances, luminosities, and sizes of such stars are not known with high accuracy. Most of the nearest stars are very faint in apparent brightness and have very low surface temperatures. This also causes inaccuracies in the determined properties of the stars. As a result, most of the numbers in these tables are subject to change as more accurate data become available.

There are four binary-star systems and one triple-star system in the table of the nearest stars. The different members of these systems are indicated by the letters A, B, and C. Several of the nearest stars also have unseen companions, objects too faint to be seen whose presence can be detected only through their gravitational effects.

Star and constellation	Apparent brightness (magnitudes)	Distance (light years)	Surface temperature (degrees Fahrenheit)	Luminosity (Sun = 1.0)	Diameter (Sun = 1.0)
Sun	−26.7	0.000016	10,000	1.0	1.0
Sirius (serious) Canis Major	−1.5	8.6	18,000	26	1.7
Canopus (ka-NO-pus) Carina	−0.7	200	13,000	5500	45
α Centauri (sen-TAU-ri) Centaurus	0[a]	4.3	10,000	1.5	1.2
Arcturus (arc-TUR-us) Bootes	−0.1	35	7200	180	25
Vega (VEE-ga) Lyra	0	27	17,000	65	2.9
Capella (ka-PEL-la) Auriga	0	45	9200	180	16
Rigel (RYE-jel) Orion	0.1	850	20,000	90,000	70
Procyon (PRO-see-on) Canis Minor	0.4	11.4	11,000	7	2.1
Betelgeuse (BEE-tl-jooz) Orion	0.4[b]	500	4800	70,000	1000
Achernar (A-ker-nar) Eridanus	0.5	140	26,000	3300	9
β Centauri (sen-TAU-ri) Centaurus	0.6	400	40,000	75,000	18
Altair (AL-tare) Aquila	0.8	17	14,000	10	1.7
α Crucis (KROO-sis) Crux	1.4[a]	400	43,000	40,000	12
Aldebaran (al-DEB-a-ran) Taurus	0.9	68	5800	470	60
Spica (SPY-ka) Virgo	0.9[b]	240	43,000	18,000	8
Antares (an-TARE-eez) Scorpius	0.9[b]	450	5000	30,000	600
Pollux (PO-lux) Gemini	1.2	35	8000	45	10
Fomalhaut (FO-mal-hot) Piscis Austrinus	1.2	23	15,000	14	1.6
Deneb (DEN-eb) Cygnus	1.3	1500	16,000	70,000	110

The nearest stars

Star	*Distance (light years)*	*Apparent brightness (magnitudes)*	*Surface temperature (degrees Fahrenheit)*	*Luminosity (Sun = 1.0)*	*Diameter (Sun = 1.0)*
Sun	0.000016	−26.7	10,000	1.0	1.0
α Centauri A	4.3	−0.1	10,000	1.5	1.2
α Centauri B	4.3	1.3	9000	0.5	0.8
α Centauri C	4.3	11.1	4400	0.002	0.2
Barnard's Star	5.9	9.5	5100	0.004	0.2
Wolf 359	7.6	13.5	4000	0.0011	0.2
BD^{+}36°2147	8.1	7.5	5800	0.022	0.4
Sirius A	8.6	−1.5	18,000	26	1.7
Sirius B	8.6	8.4	57,000	0.06	0.008
Luyten 726-8 A	8.9	12.4	4400	0.002	0.2
Luyten 726-8 B	8.9	12.9	4300	0.0012	0.17
Ross 154	9.5	11.0	4900	0.004	0.2
Ross 248	10.3	12.3	4600	0.002	0.2
ε Eridani	10.8	3.7	8200	0.36	0.9
Luyten 789-6	10.8	12.2	4500	0.004	0.26
Ross 128	10.8	11.1	5100	0.004	0.2
61 Cygni A	11.1	5.2	7300	0.14	0.7
61 Cygni B	11.1	6.0	6600	0.09	0.7
ε Indi	11.2	4.7	7500	0.24	0.8
Procyon A	11.4	0.4	12,000	7	2.1
Procyon B	11.4	10.7	14,000	0.0006	0.01

APPENDIX SIX
GLOSSARY

Absolute magnitude A measure used by astronomers to describe the luminosity of a star. It is equal to what the apparent magnitude would be if the distance to the star were 10 parsecs or 32.6 light years.

Absolute zero The temperature at which an object has no available heat energy left, and the motions of its atoms and molecules have stopped. Absolute zero occurs at about −460°F or −273°C.

Absorption A process in which a photon of radiation disappears and its energy is added to the energy of a nearby atom or other particle.

Absorption line A particular frequency of a continuous spectrum in which much of the radiation energy has been absorbed. Each absorption line is caused by the absorption of a special kind of atom or molecule.

Absorption spectrum A continuous spectrum with absorption lines superposed.

Acceleration A change of velocity; speeding up, slowing down, or changing direction.

Albedo (al-BEE-do) The fraction of light incident on a body that the body reflects.

Aldebaran (al-DEB-a-ran) A bright red giant star in Taurus.

Alpha Centauri A bright star in the southern skies. It is the nearest star to the solar system and it is nearly identical in properties to the Sun.

Alpha particle The nucleus of the helium atom.

Altair (AL-tare) A bright star in Aquila.

Altitude The angular distance of an object above the horizon.

Andromeda (an-DROM-e-da) A constellation containing the nearest spiral galaxy. It is named after a mythological character.

Angstrom A unit of length equal to 10^{-8} centimeters or about 4×10^{-9} inches. It is commonly used to describe the wavelengths of radiation.

Angular momentum A measure of the amount of sideways or rotational motion a body has around a given axis. It is found by multiplying the sideways speed by the mass by the distance of the body from the axis. Angular momentum can be transferred from one body to another, but it cannot be created or destroyed.

Antarctic circle A circle on the Earth's surface that is everywhere 66½ degrees south of the equator. For points south of this circle, the Sun is above the horizon 24 hours a day on and around December 22, and it is below the horizon 24 hours a day on and around June 22.

Antares (an-TARE-eez) A bright red star in Scorpius. It is a cool supergiant of very large size.

Aphelion For a body revolving around the Sun, the point on its orbit when it is farthest from the Sun.

Apogee The point in the orbit of an Earth satellite in which it is at the greatest distance from the Earth.

Apparent magnitude A measure used by astronomers to describe how bright objects appear from the Earth. The brightest stars of the night sky are roughly of the first magnitude, while the faintest stars that can be seen with the naked eye are of the sixth magnitude. The Sun has an apparent magnitude of about −27.

Aquarius A constellation of the zodiac. Its name means "a person who carries water."

Aquila A constellation containing the bright star Altair. Its name means "eagle."

Arctic circle A circle on the Earth's surface that is 66½ degrees north of the equator. For points north of this circle, the Sun is above the horizon 24 hours a day on and around June 22, and it is below the horizon 24 hours a day on and around December 22.

Arcturus (ark-TUR-us) A bright orange giant star in the constellation Bootes.

Aries A constellation of the zodiac. Its name means "ram."

Aristotle (AIR-is-totl) 384–322 B.C. A Greek philosopher and scientist who had a great influence on thought in the latter part of the Middle Ages.

Asteroid One of the small or minor planets. Most asteroids have orbits that lie between the orbits of Mars and Jupiter.

Astrology The superstition that astronomical bodies have a mysterious influence on human affairs.

Astrometry The branch of astronomy that covers the measurement of positions and motions of bodies.

Astronomical unit The average distance between the Earth and the Sun, equal to about 93 million miles or 150 million kilometers.

Astrophysics The physics of astronomy; specifically, the study of the structure and physical properties of stars, galaxies, and interstellar matter.

Atmosphere The gases that are above the solid or liquid surface of a planet. Also, the outer layers of a gaseous object, such as a star.

Atom A tiny particle consisting of a nucleus surrounded by electrons. Atoms are the basic particles of the chemical elements.

Auriga (au-RI-ja) A constellation whose name means "chariot driver." It contains the bright star Capella.

Aurora A glow from the upper atmosphere of the Earth. The air atoms and molecules are made to emit light because of collisions with particles ejected from the Sun.

Autumnal equinox One of two points in the sky where the Sun's path crosses the celestial equator. The Sun is at the autumnal equinox on about September 22 of each year.

Axis A line about which a body rotates.

Azimuth Angular distance measured along the horizon from the north through the east. This can be determined approximately by a magnetic compass.

Balmer lines Absorption or emission lines that are prominent in the visible light part of the spectra of most stars. They are due to the element hydrogen.

Barred spiral A spiral galaxy that shows a conspicuous bar passing through the nucleus.

Betelgeuse (BEET-l-jooz) A bright red star in the

constellation Orion. It is a cool supergiant of very large size.

Big bang The great explosion which the Universe appears to have experienced about 15 billion years ago and which started the expansion of the Universe which is observed today.

Binary star Two stars that move in orbits around each other and are kept close to each other by the force of gravity.

Black dwarf A white dwarf that has cooled so much it no longer emits a significant amount of light.

Black hole A possible end state for a star in which contraction has increased the surface gravity so much that nothing can escape the star, including light.

Blueshift A change to higher frequencies and shorter wavelengths. A blueshift in the radiation of a star or galaxy usually indicates that the object is moving toward the Earth.

Bode-Titius law A sequence of numbers that follows very closely the distances of the planets from the Sun.

Bootes (bō-Ō-tez) A constellation whose name means "herdsman" or "plowman." It contains the bright star Arcturus.

Cancer A constellation of the zodiac whose name means "crab."

Canis Major A constellation whose name means "large dog." It contains Sirius, the brightest star in the night sky.

Canis Minor A constellation whose name means "small dog." It contains the bright star Procyon.

Canopus (kan-Ō-pus) A bright star in the southern skies.

Capella A bright giant star in the constellation Auriga.

Capricornus A constellation of the zodiac whose name means "goat."

Carbon cycle (*Also* carbon-nitrogen cycle.) A series of nuclear reactions that have the effect of converting hydrogen into helium. The carbon cycle is important in the more massive main sequence stars.

Cassegrain telescope A type of mirror telescope in which the light is brought to a focus through a small hole in the mirror.

Cassiopeia (KAS-i-o-PE-a) A northern constellation shaped like a W. It is named after a mythological character.

Castor A bright star in Gemini.

Celestial equator The circle that goes around the sky directly over the Earth's equator.

Celestial mechanics The theoretical study of the motions of astronomical bodies under the influence of gravity and, sometimes, other forces.

Celestial poles The two points on the sky that are directly over the poles of the Earth.

Celestial sphere The sky imagined as a huge sphere with the Earth at its center.

Centaurus A southern constellation containing the nearest star, Alpha Centauri. Its name means "centaur."

Cepheid variable A certain type of pulsating star.

Cepheus (SĒ-fē-us) A northern constellation. It is named for a mythological character.

Ceres (SEER-eez) The first asteroid discovered and the largest of the asteroids.

Cetus (SĒ-tus) A constellation whose name means "whale."

Chromosphere The thin layer of gases above the surface of the Sun and below the corona.

Cluster A physical group of stars or galaxies.

Color (*Also* color index.) A measure of the relative amounts of energy in two specified frequency regions of radiation.

Color excess The distortion of the color of a star due to reddening produced by interstellar dust between the star and us.

Comet A small solar-system object that contains dust and material that is easily vaporized. The dust and gases released when the comet approaches the Sun often form a long tail, giving the comet its characteristic appearance.

Conjunction A planet is in conjunction with another body when it appears very close to it in the sky.

Constellation 1. A group of stars that forms a recognized pattern on the sky. 2. An area of the sky that contains such a pattern of stars.

Continuous spectrum A spectrum in which the radiation energy varies smoothly with frequency and does not contain emission or absorption lines.

Convection Carrying heat energy from one place to another by currents of gas or liquid.

Copernicus, Nicholas (ko-PER-ni-kus) 1473–1543. A Polish astronomer. He supported the view that the Earth is not at rest at the center of the Universe but travels in orbit around the Sun.

Core (1) The central region of any object. (2) One of the two regions near the center of the Earth that are thought to be mostly composed of iron. The outer core is liquid, the inner core is probably solid.

Corona The very hot outermost part of the atmosphere of the Sun.

Corona Borealis A constellation whose name means "northern crown."

Cosmic rays Extremely high energy particles that are observed to exist in space.

Cosmological principle The statement that on a large enough scale, each part of the Universe is identical to all other parts.

Cosmology The study of the large-scale properties of the Universe.

Crater A circular feature on the surface of the Moon or a planet caused, in most cases, by meteoric impact in the past. Some craters are of volcanic origin.

Crescent phase A phase of the Moon or a planet in which only a small part of the surface is visible.

Crust The outermost solid layers of the Earth or other body.

Cygnus (SIG-nus) A constellation whose brighter stars form a cross. It contains the supergiant star Deneb, and its name means "swan."

Dark nebula A thick cloud of interstellar dust that blocks off the light of distant stars.

Daylight saving time Time that is 1 hour later than standard time.

Declination The angular distance of an object in the sky north or south of the celestial equator.

Degeneracy The state of material that has been compacted to the maximum. The higher the temperature, the greater the density needed for degeneracy.

Deneb A bright supergiant star in Cygnus.

Density Mass divided by volume.

Deuterium An isotope of hydrogen in which the nucleus has one proton and one neutron.

Diffraction The bending of rays of light at the edge of an object.

Disk star A star formed in the plane of a spiral galaxy that usually moves in a nearly circular orbit about the center.

Diurnal circle The apparent path of a celestial object across the sky due to the rotation of the Earth.

Doppler effect The change in the observed frequency and wavelength of a wave due to the motion of the source toward or away from the observer.

Draco (DRĀ-ko) A constellation whose name means "the dragon."

Dwarf Generally, it is a star of low luminosity; specifically, a main sequence star.

Earth The planet on which we live, the third one from the Sun.

Earthquake A sudden shift in the rocks below the surface of the Earth.

Earthquake wave A disturbance that travels through the Earth caused by an earthquake.

East The direction of the Earth's rotation.

Eccentricity A measure of the shape of an ellipse and a hyperbola.

Eclipse An object being in a position to block off a light source to another object.

Eclipsing binary A binary system in which the stars periodically eclipse each other. This causes them to appear to fade in brightness during the eclipse.

Ecliptic The apparent path of the Sun in the sky as a reflection of the Earth's annual motion around the Sun.

Electric charge A property possessed by any material that produces an electric field and responds to the electric force.

Electric field A region of space capable of exerting a force on an electric charge.

Electric force The force that electric charges exert on each other. It can be either attractive or repulsive, depending on the types of charges.

Electromagnetic radiation The waves, including radio, infrared, light, ultraviolet, x ray, and gamma ray, that are the source of practically all our information concerning the astronomical Universe.

Electron One of the basic particles of which all matter is composed. Electrons have a negative electric charge and normally occur around the nucleus of an atom.

Element (*Also* chemical element.) A substance consisting of a number of atoms of one particular kind.

Elements The quantities needed to specify the orbit of a celestial body.

Ellipse An oval-shape curve. Most astronomical bodies follow paths that are either elliptical or hyperbolic with a high degree of accuracy.

Elliptical galaxy A galaxy that appears elliptical in shape.

Elongation The angular distance between two objects in the sky. One of them is usually the Sun.

Emission A process in which a photon of radiation is created and its energy is supplied by a nearby atom or other particle.

Emission line The radiation energy emitted at or near a particular frequency by a thin, emitting gas. Each emission line is caused by the emission from a special kind of atom or molecule.

Emission nebula An interstellar gas cloud that is heated by a nearby hot star so that it emits radiation.

Emission spectrum A spectrum consisting of a number of emission lines. This type of spectrum is emitted by a hot gas that is not too thick.

Energy A property of matter and radiation that measures their ability to do work, in other words, to push things around and to emit radiation.

Ephemeris A list of the positions of an astronomical body at various times.

Epicycle A small circle whose center moves on a larger circle. This was used in ancient and medieval times to model the observed motions of the planets.

Equation of time The difference in east-west position, usually expressed as a time interval, between the actual Sun in the sky and the mean Sun.

Equator The great circle on the Earth that lies in the east-west direction and is halfway between the two poles.

Equinox Either of the two points on the sky where the Sun's path crosses the celestial equator.

Escape velocity The least velocity needed to escape from the gravitational influence of a body.

Evolution The process of changing or aging as time goes on.

Extinction Dimming of starlight by the Earth's atmosphere or by interstellar dust.

Fireball An unusually bright meteor seen flashing across the night sky.

Fission A nuclear reaction in which a nucleus is broken into smaller pieces.

Flare An explosion that suddenly emits high-energy particles and radiation from a small region on the surface of the Sun.

Flare star A star that flares up in brightness at irregular intervals.

Fluorescence The emission of lower energy photons by atoms or molecules following the absorption of higher energy photons.

Focal length The distance from a lens or mirror to the point where the light from a distant source converges to a focus.

Focal ratio The ratio of the focal length of a lens or mirror to its diameter.

Focus (1) The point at which a lens or mirror, or combination of the two, forms an image. (2) Either of two points that have special mathematical properties relating to ellipses and hyperbolas.

Forbidden line A spectral line that has a very low probability of being absorbed or emitted.

Force A push or pull that accelerates an object.

Fraunhofer line An absorption line.

Frequency The number of vibrations per unit of time. For a wave, it is the number of waves passing a point per unit of time.

Full phase A phase of the Moon or a planet in which the entire disk is visible.

Fusion A nuclear reaction in which heavier nuclei are built up out of lighter ones.

Galactic cluster The same as open cluster.

Galactic equator The great circle where the plane of the Milky Way intersects the sky.

Galactic latitude Angular distance from the galactic equator.

Galactic longitude Angular distance measured along the galactic equator from the direction to the center of the Milky Way.

Galaxy A huge group of stars. Galaxies typically contain billions of stars. The Galaxy with capital G means the Milky Way, the galaxy we are located in.

Galileo (gal-i-LAY-o) 1564–1642. An Italian astronomer who made the first important astronomical observations with a telescope.

Gamma rays Photons of extemely high energy.

Gemini A constellation of the zodiac containing the bright stars Castor and Pollux. Its name means "twins."

Geocentric Centered on the Earth.

Geology, geophysics The scientific study of the Earth.

Giant A star of relatively high luminosity that has used up all of its core hydrogen.

Giant planet Jupiter, Saturn, Uranus, or Neptune.

Gibbous phase A phase of the Moon or a planet in which more than one-half of the disk is visible.

Globular cluster A large star cluster containing very old stars and often occurring in the halo of a galaxy.

Globule A small, dense dust cloud.

Gnomon (NO-mon) A stick in the ground whose shadow can be used to determine the motions of the Sun.

Granulation The small light and dark areas that show on the surface of the Sun.

Grating A surface with tiny parallel lines ruled on it, used to separate radiation of different frequencies.

Gravitational energy A form of potential energy in which the basic force is the force of gravity.

Gravity One of the basic forces between material particles. Gravity is important only if a rather large amount of mass is present.

Great circle A circle on the surface of a sphere in which the circle and sphere have the same center.

Greenhouse effect An effect in which warming radiation is allowed to enter an enclosure but an atmosphere or envelope impedes the escape of radiation emitted by the enclosure. The result is a heating of the enclosure.

Greenwich (GREN-ich) A suburb of London, England, that has been chosen to have a longitude of zero.

Gregorian calendar The calendar in general use today, introduced by Pope Gregory XIII in 1582.

H I region A cool region of interstellar space in which the hydrogen is not ionized.

H II region A hot region of interstellar space in which the hydrogen is ionized.

Half-life The length of time it takes one-half of the atoms of a radioactive substance to decay.

Halo Those very old stars and star clusters in a galaxy that extend to great distances in all directions from the center.

Harvest Moon The full Moon that occurs nearest the time of the autumn equinox.

Heat A form of energy consisting of the random kinetic energies of all of the atoms and molecules in an object.

Heavy elements A term that in astronomy usually means all elements except hydrogen and helium.

Heliocentric Centered on the Sun.

Helium The second lightest and second most abundant element in the Universe. Helium nuclei have two protons.

Helium flash The explosive beginning of helium burning in the nearly degenerate core of a red giant.

Hercules A constellation named after a mythological character. It contains the globular cluster M 13.

High-velocity star A star with a large motion with respect to the Sun. Such stars are not moving with nearly circular orbits around the center of the Milky Way.

Hipparchus (hip-PAR-kus) ca 130 B.C. A very accurate observer of ancient Greece.

Horizon The great circle 90° from directly overhead (zenith).

Hour angle The angular distance of an object in the sky east or west of the meridian. It is often measured in time units.

H-R diagram (Hertzsprung-Russell diagram) A plot of luminosity versus surface temperature for stars.

Hubble law The relation between the distance of a galaxy and its speed away from us.

Hydrogen The lightest and most abundant element in the Universe. Hydrogen nuclei always have one proton.

Hyperbola A type of open curve that extends to infinity in two directions. Most astronomical bodies follow paths that are either hyperbolic or elliptical to a good approximation.

Image A picture of an object caused by light from that object being brought to a focus by a lens or mirror and displayed in some fashion.

Inclination The angle between a plane and a reference plane.

Inferior conjunction Mercury or Venus being essentially between the Earth and the Sun.

Inferior planet Mercury or Venus.

Infrared Radiation whose frequency and energy are somewhat less than those of visible light.

Inner core The small central part of the Earth that is thought to be solid, compressed iron.

Interferometer A specialized instrument for measuring very small angular separations of sources of radiation.

Intergalactic space Space between galaxies.

International date line A line on the Earth extending from the North Pole to the South Pole at approximately 180° longitude. Time west of the line is 24 hours later than time just east of the line.

Interplanetary space Space between the planets in the solar system.

Interstellar dust The tiny grains of matter that occur in interstellar space.

Interstellar gas The very low density gas, mostly hydrogen, that occurs in interstellar space.

Interstellar lines Absorption lines observed in the spectra of some stars due to the interstellar gas between the Earth and the stars.

Interstellar matter The very low density gas and dust that occur in interstellar space.

Interstellar space The space within a galaxy between the stars.

Ion A particle having an electric charge, usually referring to an atom that has lost some of its electrons.

Ionization A violent process in which one or more electrons are knocked free from an atom.

Ionization potential The energy needed to ionize an atom.

Ionosphere The upper part of the Earth's atmosphere in which many atoms are ionized.

Irregular galaxy A galaxy that shows no obvious symmetry, being neither spiral nor elliptical.

Isotopes Forms of a given chemical element that have a different number of neutrons in the nucleus.

Jovian planet One of the giant planets; Jupiter, Saturn, Uranus, or Neptune.

Julian calendar The calendar introduced by Julius Caesar in 46 B.C. It is only slightly different from the Gregorian calendar used today.

Julian day The number of days since January 1, 4713 B.C., used by certain astronomers in observational work.

Jupiter The largest planet and the fifth one from the Sun.

Kepler, Johannes 1571–1630. A German astronomer who discovered the famous three laws of planetary motion.

Kiloparsec One thousand parsecs or about 3260 light years.

Kinetic energy Energy of motion.

Latitude Angular distance north or south of the Earth's equator.

Leap year A year with 366 days instead of the usual 365. Leap years are used in order to keep the calendar year as close as possible to the year according to the seasons.

Leo A constellation of the zodiac whose name means "lion." It contains the bright stars Regulus and Denebula.

Libra A constellation of the zodiac whose name means "scales" or "balance."

Light Radiation that is visible to the eye.

Light curve A graph that shows the apparent brightness of a body at different times.

Light year The distance light travels in 1 year. It is about 6 trillion miles or 9.5×10^{12} kilometers.

Limb darkening The edge of the disk of a celestial body appearing less bright than the center of the disk.

Line broadening Effects that cause a spectral line to become broadened. That is, to cover a wider range of frequencies.

Line profile The energy of radiation plotted against frequency or wavelength for a spectral line.

Local group The small cluster of galaxies of which the Milky Way is a member.

Local time Any time-measuring system that depends on the precise longitude of the observer on the Earth. This contrasts with zone or standard time, which is the same for everyone within the same standard time zone.

Longitude Angular distance east or west of Greenwich, England, on the surface of the Earth.

Long-period variable A type of cool, pulsating, and variable star.

Luminosity The radiation energy emitted into space each second by a star or other body.

Lunar eclipse An eclipse of the Moon, occurring when the Moon passes into the shadow of the Earth.

Lyman lines Very strong absorption or emission

lines occurring in the ultraviolet region of the spectra of most stars. They are due to the element hydrogen.

Lyra A northern constellation containing the bright star Vega. Its name means "lyre."

Magellanic clouds Two irregular galaxies visible from southern latitudes on the Earth. They are the galaxies nearest to the Milky Way.

Magnetic field A region of space that produces a characteristic force on magnetic materials and on moving electric charges.

Magnetosphere The region around a planet in which its magnetic field is appreciable.

Magnitude Any of a number of systems used by astronomers to describe the brightness of a source of radiation. These systems can be either intrinsic, called absolute magnitudes, measuring the luminosity, or they can be apparent, measuring the energy that arrives at the Earth.

Main sequence Those stars that, in the H-R diagram, lie along a diagonal line running from the upper left (hot, luminous) to the lower right (cool, faint). All stars begin on the main sequence and stay there until they have exhausted essentially all of their core hydrogen.

Major axis The long diameter of an ellipse.

Mantle The rocky part of the Earth's interior that lies above the cores and below the crust.

Mare (*plural* maria). The large dark areas on the Moon that are now known to be old lava flows.

Mars The fourth planet from the Sun.

Mass A measure of the amount of material in an object.

Mean Sun An imaginary object invented for accurate timekeeping. The actual motions of the Sun in the sky are somewhat irregular, while the mean Sun, by definition, moves uniformly.

Megaparsec One million parsecs or about 3.26 million light years.

Mercury The smallest planet and the one closest to the Sun.

Meridian (1) The great circle in the sky that passes directly overhead in the north-south direction. (2) Any great circle on the surface of the Earth that passes through two poles.

Messier catalog (mes-YEA) A catalog of galaxies, star clusters, and nebulae compiled about 200 years ago. An object known as M 31, for example, is number 31 in the Messier catalog.

Meteor A flash of light seen in the night sky, often called a shooting star. Also, the particle that causes the flash, usually a tiny particle that burns up on entering the Earth's atmosphere at high speed.

Meteorite The particle that causes a meteor, especially a large one that survives to strike the ground.

Meteoroid A meteor particle.

Meteor shower The occurrence of more meteors than usual, caused by the Earth's intersecting a stream of meteors.

Microwave The highest frequency radio waves.

Milky Way (1) The name of the galaxy in which we live. (2) The faint, white band of light along the plane of our galaxy.

Minor planet An asteroid.

Mira (1) A cool supergiant star in the constellation Cetus. It is a pulsating star. (2) A type of pulsating and variable star of which Mira itself is an example; a long-period variable.

Model A theoretical calculation of the properties of a system that, it is hoped, will be in good agreement with the real system; a model atmosphere, model star, model Universe, and so on.

Molecule Two or more atoms held together by the electric forces and acting in many respects like a single particle; the basic particle of a chemical compound.

Momentum The product of the mass of an object multiplied by its velocity. The total amount of momentum is conserved in any interaction.

Monochromatic Radiation having a very small range of frequency or color.

Moon (1) The name of the natural satellite of the Earth. (2) A satellite of any planet.

Nadir The point straight down toward the center of the Earth.

Neap tide The smallest tide. It occurs at the quarter phase of the Moon.

Nebula A cloud of interstellar matter.

Neptune The eighth planet from the Sun, discovered in 1846.

Neutrino One of the fundamental particles of physics, having no mass or charge. Neutrinos are produced in many nuclear reactions.

Neutron A particle having no charge and about the same mass as the proton. It occurs in atomic nuclei.

Neutron star A highly compacted degenerate star whose particles are mostly neutrons. This is a possible final state for a star.

New phase The phase of the Moon or a planet when it is nearly between the Earth and Sun, making it almost invisible.

Newton, Isaac 1642–1727. An English physicist who formulated the laws of motion and the law of gravity.

N galaxy A type of galaxy having an unusually bright, small nucleus.

NGC The initials of New General Catalog which lists numerous nebulae, clusters, and galaxies. NGC 4188, for example, is number 4188 in this catalog.

Nonthermal radiation Radiation emitted in which the energy comes from a source other than heat energy. Synchrotron radiation is a common type.

North On the Earth, directed toward the North Pole; on the sky, directed toward the north celestial pole.

North celestial pole The point in the sky directly over the North Pole of the Earth. The star Polaris in the Little Dipper is very nearly at the north celestial pole.

North Pole One of two points where the rotation axis intersects the Earth's surface. One faces the North Pole by keeping east to one's right; on looking down on the North Pole, the Earth rotates in a counterclockwise direction.

Nova A star that undergoes an explosion, temporarily increasing very much in luminosity.

Nuclear force The very strong but short-range force exerted by protons and neutrons on each other which holds atomic nuclei together.

Nuclear reaction A reaction in which old nuclei are changed into new ones.

Nucleosynthesis The building up of different nuclei through nuclear reactions.

Nucleus (1) The most massive part of an atom, consisting of neutrons and protons about which the electrons move. (2) The central part of an object such as a comet or galaxy.

Objective The main lens or mirror of a telescope.

Occultation The Moon or a planet moving in front of a star or planet.

Opacity A measure of the ability of matter to absorb radiation.

Open cluster A cluster of disk stars, generally much smaller and younger than globular clusters; a galactic cluster.

Ophiuchus (ŌF-i-OO-kus) A constellation whose name means "man holding a snake."

Opposition Being opposite the Sun in the sky.

Orbit The path in space of a celestial body.

Orion (o-RĪ-on) A constellation on the celestial equator that contains the Orion Nebula plus many bright stars, including the red supergiant Betelgeuse and the hot supergiant Rigel. It is named for a mythological character.

Outer core The region of the Earth between the inner core and the mantle, thought to consist mainly of liquid iron.

Parabola (1) A special hyperbola whose legs become parallel to each other. (2) The orbit of a body having the escape velocity.

Parallax The change in the direction of an object when the observer changes position. In astronomy, it is used specifically to mean the angle subtended by the radius of the Earth's orbit at a star.

Parsec The distance of a star having a parallax of 1 second of arc. It is equal to 3.26 light years or 19.2 trillion miles.

Pegasus A constellation named for a famous winged horse of mythology.

Penumbra A partial shadow in which some of the light source is cut off.

Perigee The point in the orbit of an Earth satellite when it is closest to the Earth.

Perihelion For a body revolving around the Sun, the point in its orbit when it is nearest the Sun.

Period The time it takes an object to repeat its motion.

Phase The appearance of a dark object, such as the Moon or a planet, due to a particular direction of illumination.

Photometry The measurement of apparent brightnesses.

Photon A particle of radiation having a specific frequency, wavelength, and energy.

Photosphere The atmosphere or surface of a star.

Pisces A constellation of the zodiac whose name means "fish."

Planet (1) Any of the nine major bodies moving around the Sun. (2) A similar body orbiting another star.

Planetarium A dome-shaped theater in which lights are projected on the ceiling to simulate the appearance of the night sky.

Planetary nebula A shell of gas ejected from a star in the late stages of its evolution.

Plate One of the large blocks of material covering the surface of the Earth. Motions of the plates cause mountain building, earthquakes, continental drift, and other near-surface activity.

Pluto The outermost planet of the solar system, discovered in 1930.

Polaris A rather bright star in the Little Dipper that is very close to the north celestial pole; the North Star.

Polarization The condition of radiation and certain other waves when one orientation of the wave is more common than others.

Pollux A bright giant star in Gemini.

Population I Stars with moderate amounts of heavy elements and any age from very young to very old. Population I corresponds to disk stars in spiral galaxies.

Population II Very old stars having very few heavy elements in their original compositions; halo stars.

Positron A fundamental particle of physics having the mass of an electron but a positive charge.

Potential energy Energy that an object has because of its position in a force field.

Precession The slow wobbling motion of the axis of a top or other rotating body. The axis of the Earth has a precession period of 26,000 years, and this causes the celestial poles to slowly change position in the sky.

Prism A triangular-shaped piece of glass that produces a spectrum or rainbow when white light passes through it.

Procyon (PRO-see-on) A bright star in Canis Minor.

Prominence A mass of bright gas observed beyond the edge of the Sun.

Proper motion The very slow change in direction to a star due to its space motion.

Proto- A prefix meaning an object in the making; for example, protoplanet, protostar, protogalaxy.

Proton One of the fundamental particles of physics. It has nearly 2000 times the mass of an electron and a positive charge and occurs with neutrons in atomic nuclei.

Proton-proton reaction A series of nuclear reactions that convert hydrogen into helium. This reaction is important in the centers of low-mass main sequence stars.

Ptolemy (TOL-e-mee) ca A.D. 150. The last of the important Greek astronomers, he perfected a geocentric model to describe planetary motions.

Pulsar A neutron star whose rapid rotation and unusual nonthermal radiation cause it to be observed in a series of pulses.

Pulsating star A star that periodically expands and contracts because it is unable to find exact internal balance. A pulsating star is also a variable star.

Quadrature A planet or the Moon being 90° from the Sun.

Quarter phase The phase of the Moon or a planet when exactly one-half of its disk is visible.

Quasar An object that appears very small in size whose radiation shows an extremely large redshift. The consensus of opinion holds that quasars are the most remote and most luminous objects known.

QSO (quasistellar object) See quasar.

Radar A method of measuring accurate distances by sending radio signals and timing the reflected wave.

Radial velocity The component of velocity that is along the line of sight; motion directly toward or away from the observer.

Radio Radiation of the smallest frequency and energy and of the longest wavelength.

Radioactivity The process of the spontaneous breaking apart of certain unstable atomic nuclei.

Radio galaxy A galaxy that emits much more radio energy than most.

Radio telescope A telescope designed to collect and measure radiowaves.

Reddening The distortion of the color of starlight by interstellar dust that makes the light more red in color.

Red giant A large, cool, luminous star that has evolved past the main sequence stage.

Redshift A change to lower frequencies and larger wavelengths. A redshift in the radiation from a star or galaxy usually indicates that the object is moving away from the Earth.

Red spot A semipermanent feature observed on the planet Jupiter.

Reflecting telescope A telescope that uses a mirror to bring the light to a focus.

Reflection nebula A cloud of interstellar dust that is illuminated by nearby stars.

Refracting telescope A telescope that uses a lens to bring the light to a focus.

Refraction The change in direction of radiation as it passes from one medium to another.

Relativity The branch of physics that concerns measurements made of high speeds or large gravitational fields.

Resolution A measure of how much detail can be made out in an image.

Revolution The motion of one body about another.

Rigel (RĪ-jel) A bright star in Orion; it is a hot supergiant and among the most luminous stars known.

Right ascension The east-west position of an object in the sky measured from the vernal equinox.

Roche limit (RŌSH) The closest a satellite can get to a planet without being torn apart by tidal forces.

Rotation Turning around on an axis passing through the body itself.

RR Lyrae A type of pulsating star that has a short period.

Russell-Vogt theorem The statement that mass and chemical composition determine all other properties of a normal, stable star.

Sagittarius A constellation of the zodiac that contains the direction to the center of the Milky Way galaxy. Its name means "archer."

Satellite An object moving in orbit around a much larger one; especially a moon revolving around a planet.

Saturn The ringed planet; it is the sixth one from the Sun.

Scorpius A constellation of the zodiac that contains the red supergiant star Antares. Its name means "scorpion."

Seeing The unsteadiness of the Earth's atmosphere that blurs the images of celestial objects and causes the twinkling of stars.

Seismic Having to do with earthquakes.

Seyfert galaxy A type of spiral galaxy whose nucleus shows broad emission lines and nonthermal radiation.

Sidereal period The true rotation or revolution period of a body.

Sidereal time Time according to the stars. Specifically, it is the time interval in appropriate units since the vernal equinox crossed the meridian.

Sirius The brightest star in the night sky; it is a rather hot main sequence star.

Sirius B A binary companion to Sirius; it is a white dwarf star.

Solar system The Sun plus all of the bodies, such as planets, satellites, asteroids, comets, and meteors, that are held by its gravity.

Solar time Time based on the Sun.

Solar wind The outflow of particles from the Sun in all directions.

Solstice Either of two points in the sky where the Sun has its greatest distance north or south of the celestial equator. The Sun is at the summer solstice on about June 22 and at the winter solstice on about December 21.

South On the Earth, directed toward the South Pole; in the sky, directed toward the south celestial pole.

South celestial pole The point in the sky directly over the South Pole of the Earth.

South Pole One of two points where the axis or rotation intersects the surface of the Earth. One faces the South Pole by keeping east, the direction of the Earth's rotation, to one's left.

Spectral class or type A division of the stars according to the appearance of their spectra.

Spectral line An absorption or emission line.

Spectrograph An instrument for producing or displaying the spectrum of an object.

Spectroscopic Relating to spectra.

Spectroscopic binary A binary star whose orbital motion can be detected from its spectrum through the Doppler effect.

Spectrum (*plural* spectra) A display of the radiation energy from an object at different frequencies. A rainbow is the visible-light part of the spectrum of the Sun.

Spica (SPY-ka) A bright, hot star in Virgo.

Spiral arm A chain of young stars and interstellar matter that takes a spiral shape in some galaxies.

Spiral galaxy A galaxy that contains spiral arms.

Sporadic meteor A meteor that does not belong to a shower.

Spring tide A tide that occurs during new or full Moon. At these times, the Sun and Moon are pulling together, and the tides are greatest.

Standard time Timekeeping according to which all points within a given zone on the surface of the Earth have the same time. These time zones are long, narrow longitudinal strips that are about 15° wide.

Star A large, hot, luminous body. Astronomers usually define a star as a luminous object that is now, or was in the past, on the main sequence.

Star cluster A physical group of stars held together by gravity.

Stellar evolution The study of the formation and aging of stars.

Subdwarf A type of low-luminosity star that lies below the ordinary main sequence in the H-R diagram.

Subgiant A star whose luminosity is between normal giants and main sequence stars of the same surface temperature.

Summer solstice The point in the sky where the ecliptic reaches its northernmost position. The Sun is at the summer solstice on about June 22.

Sun (1) The main body in the solar system. (2) Any star in general (not capitalized).

Sundial An instrument for keeping time by the position of the shadow of a pointer.

Sunspot A dark area on the surface of the Sun. A sunspot is cooler than the surrounding area.

Sunspot cycle The 11-year period with which the number of sunspots varies.

Supergiant An extremely luminous star that has evolved past the main sequence stage.

Superior conjunction A planet being on the other side of the Sun from the Earth.

Superior planet A planet farther from the Sun than the Earth.

Supernova A stellar explosion of extremely great magnitude. A supernova will be billions of times more luminous than ordinary stars for a brief time.

Synchrotron radiation Radiation emitted by very high-energy-charged particles moving in a magnetic field.

Synodic period The time interval from opposition to opposition. For the Moon and the inferior planets, the time from a phase back to the same phase.

Taurus A constellation of the zodiac that contains two star clusters, the Pleiades and Hyades, plus the bright red giant star Aldebaran. Its name means "bull."

Tektite A glassy type of meteorite.

Telescope An instrument used by astronomers primarily to collect as much radiation as possible for analysis or recording.

Temperature A measure of the amount of chaotic motion in the atoms and molecules in an object.

Terrestrial Of the Earth. The terrestrial planets are Mercury, Venus, Earth, and Mars, those that are most Earth-like.

Thermal energy Heat energy.

Thermal radiation Radiation emitted by an object whose energy is provided by the heat energy of the object.

Thermonuclear energy Energy released by nuclear reactions brought on by the high temperature of the material.

Tidal force A stretching force that one object exerts on another. It arises from the fact that the gravity of the first object is stronger on the near side than the far side of the second object.

Tide Deformation of a body due to a tidal force. Tides on Earth are caused primarily by the Moon and, to a smaller extent, by the Sun.

Transit (1) An instrument for measuring when a star crosses the meridian. (2) The appearance of a smaller body crossing in front of the disk of a larger body.

Triple-alpha process The nuclear reactions that result in three helium nuclei forming one carbon nucleus. This process is important in red giants.

Tropical year The year according to the seasons; the time from the vernal equinox to the next vernal equinox.

Tropic of Cancer The circle on the Earth of latitude 23½° north. This is the northernmost part of the Earth from which the Sun passes through the zenith, and it is the northern boundary of the tropics.

Tropic of Capricorn The circle of 23½° south latitude on the Earth. It is the southern boundary of the tropics.

T Tauri stars A type of variable star that is thought to occur just prior to the main sequence stage.

Tycho Brahe (TIE-ko BRAH-he, also TEE-ko) 1546–1601. Danish astronomer who was a very accurate observer.

Ultraviolet Radiation having somewhat higher frequency and energy and shorter wavelength than visible light.

Umbra Total shadow in which all of the light source is cut off.

Universal time Standard time at Greenwich, England.

Universe The collection of all matter and radiation that we can, at least in principle, detect.

Uranus The seventh planet from the Sun, discovered in 1781.

Ursa Major The famous northern constellation containing the seven stars known as the Big Dipper. Its name means "large bear."

Ursa Minor The northern constellation containing the seven stars known as the Little Dipper. The brightest star is Polaris, which is very close to the north celestial pole. Ursa Minor means "small bear."

Variable star A star whose brightness changes with time. Usually this is due to a changing luminosity, but it also includes eclipsing binary stars.

Vega (VEE-ga, also VAY-ga) A bright star in the constellation Lyra; it is a rather hot main sequence star.

Velocity The motion of an object, including both its speed and its direction; more loosely, speed.

Venus The second planet from the Sun.

Vernal equinox One of two points on the sky where the ecliptic crosses the celestial equator. The Sun is at this point on about March 21 of each year.

Virgo A constellation of the zodiac that contains the bright star Spica. Its name means "virgin."

Visual binary A binary in which the two stars can be seen separately.

W Ursae Majoris A type of eclipsing binary in which the two stars are essentially in contact.

Wavelength The distance between peaks in a wave.

Weight The strength of the force of gravity on an object, especially an object on the surface of the Earth.

West The direction opposite the rotation of the Earth.

White dwarf A star whose electrons have become degenerate. It is a possible final state for a star.

Winter solstice The point in the sky where the ecliptic reaches its southernmost position. The Sun is at the winter solstice on about December 22 each year.

X ray A very high frequency and high energy type of radiation.

Year The period of the Earth's orbit around the Sun.

Zenith The point in the sky directly overhead, opposite the direction of gravity.

Zero-age main sequence The line in the H-R diagram representing stars of different masses that have just arrived on the main sequence. Their distinguishing characteristic is that they have not yet burned up a significant amount of their core hydrogen.

Zodiac The 12 constellations covering the ecliptic.

Zodiacal light A glow along the ecliptic, best seen near sunrise and sunset, due to the scattering of sunlight by interplanetary dust.

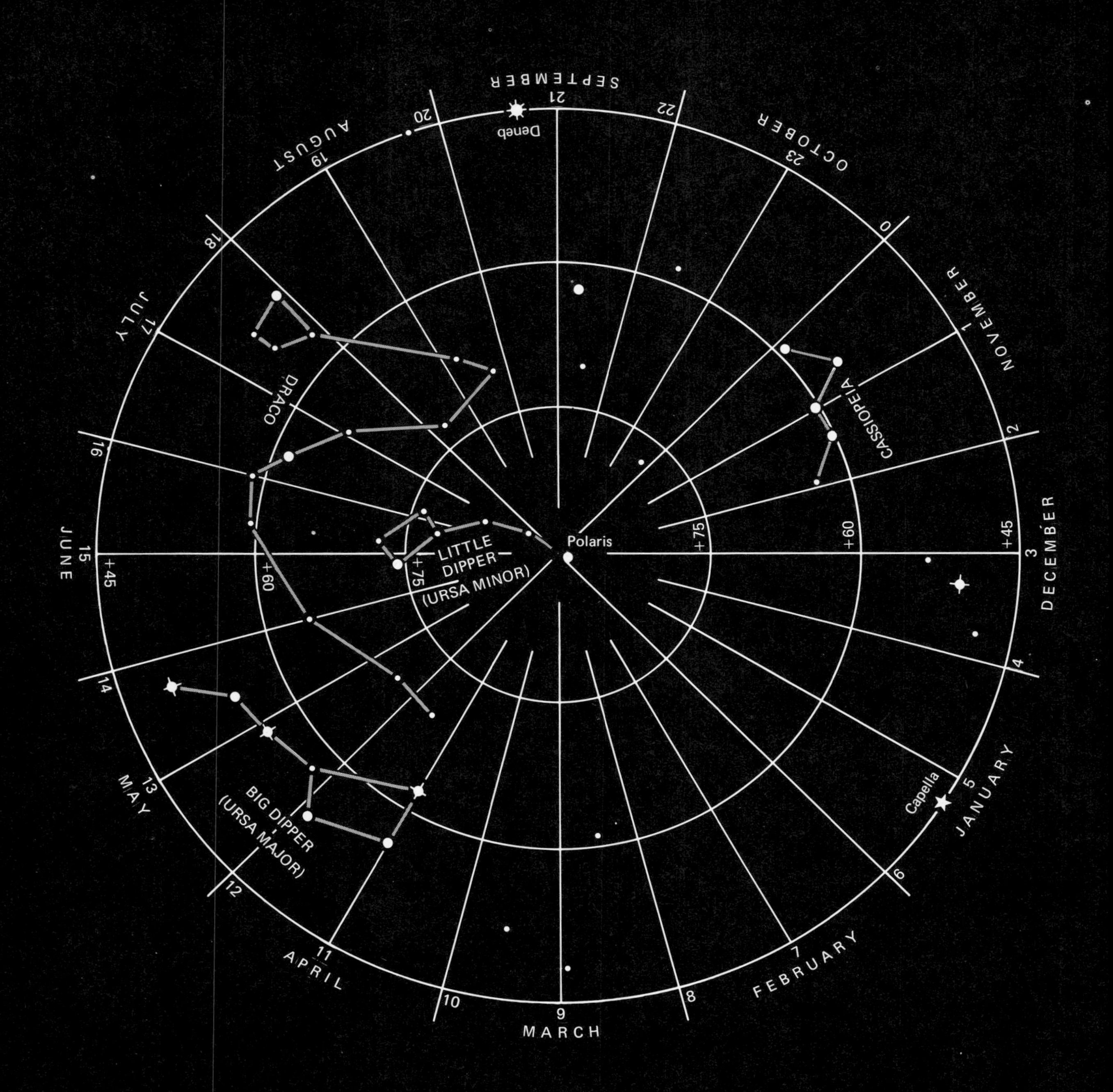
SEPTEMBER
21
22
20
Deneb
AUGUST
19
OCTOBER
23
0
18
JULY
17
NOVEMBER
1
DRACO
CASSIOPEIA
2
16
LITTLE
DIPPER
(URSA MINOR)
Polaris
+75
+60
+45
JUNE
15
+45
+60
+75
DECEMBER
3
4
14
13
MAY
BIG DIPPER
(URSA MAJOR)
Capella
5
JANUARY
6
12
11
APRIL
7
FEBRUARY
10
9
MARCH
8

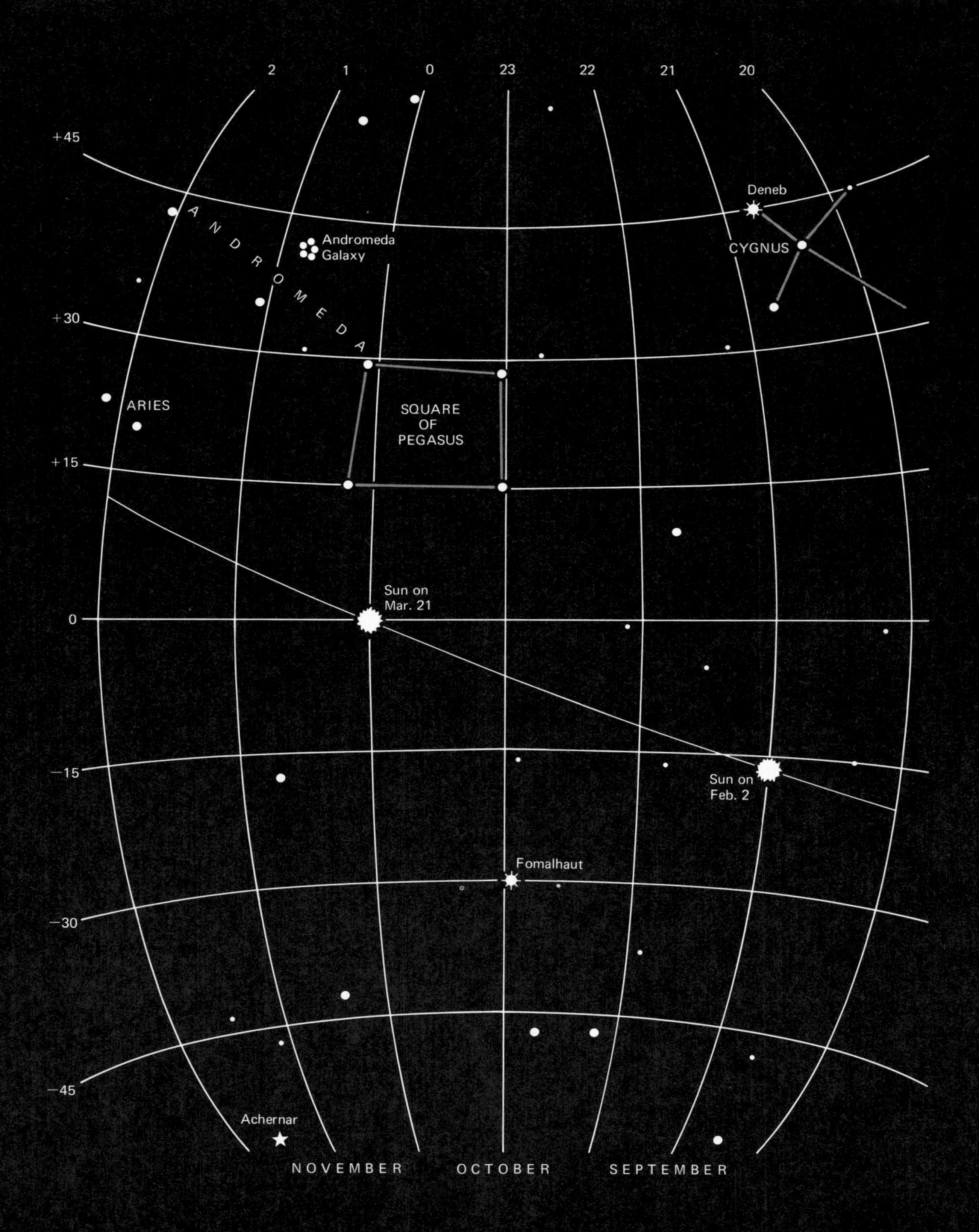
2
1
0
23
22
21
20
+45
+30
+15
0
−15
−30
−45
ANDROMEDA
Andromeda
Galaxy
Deneb
CYGNUS
ARIES
SQUARE
OF
PEGASUS
Sun on
Mar. 21
Sun on
Feb. 2
Fomalhaut
Achernar
NOVEMBER
OCTOBER
SEPTEMBER

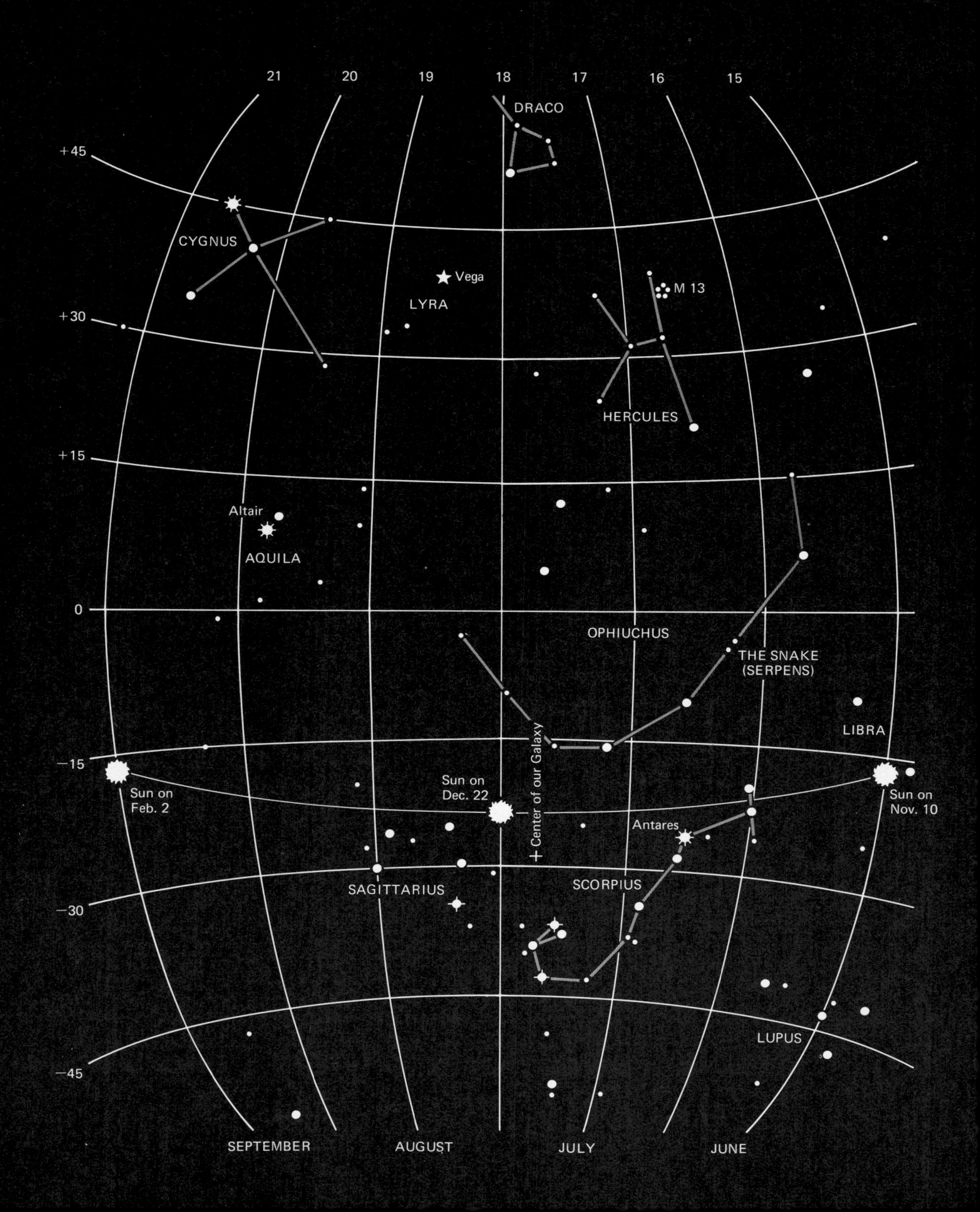

21
20
19
18
17
16
15
DRACO
+45
CYGNUS
Vega
LYRA
M 13
+30
HERCULES
+15
Altair
AQUILA
0
OPHIUCHUS
THE SNAKE
(SERPENS)
LIBRA
Center of our Galaxy
−15
Sun on
Feb. 2
Sun on
Dec. 22
Sun on
Nov. 10
Antares
SAGITTARIUS
SCORPIUS
−30
LUPUS
−45
SEPTEMBER
AUGUST
JULY
JUNE

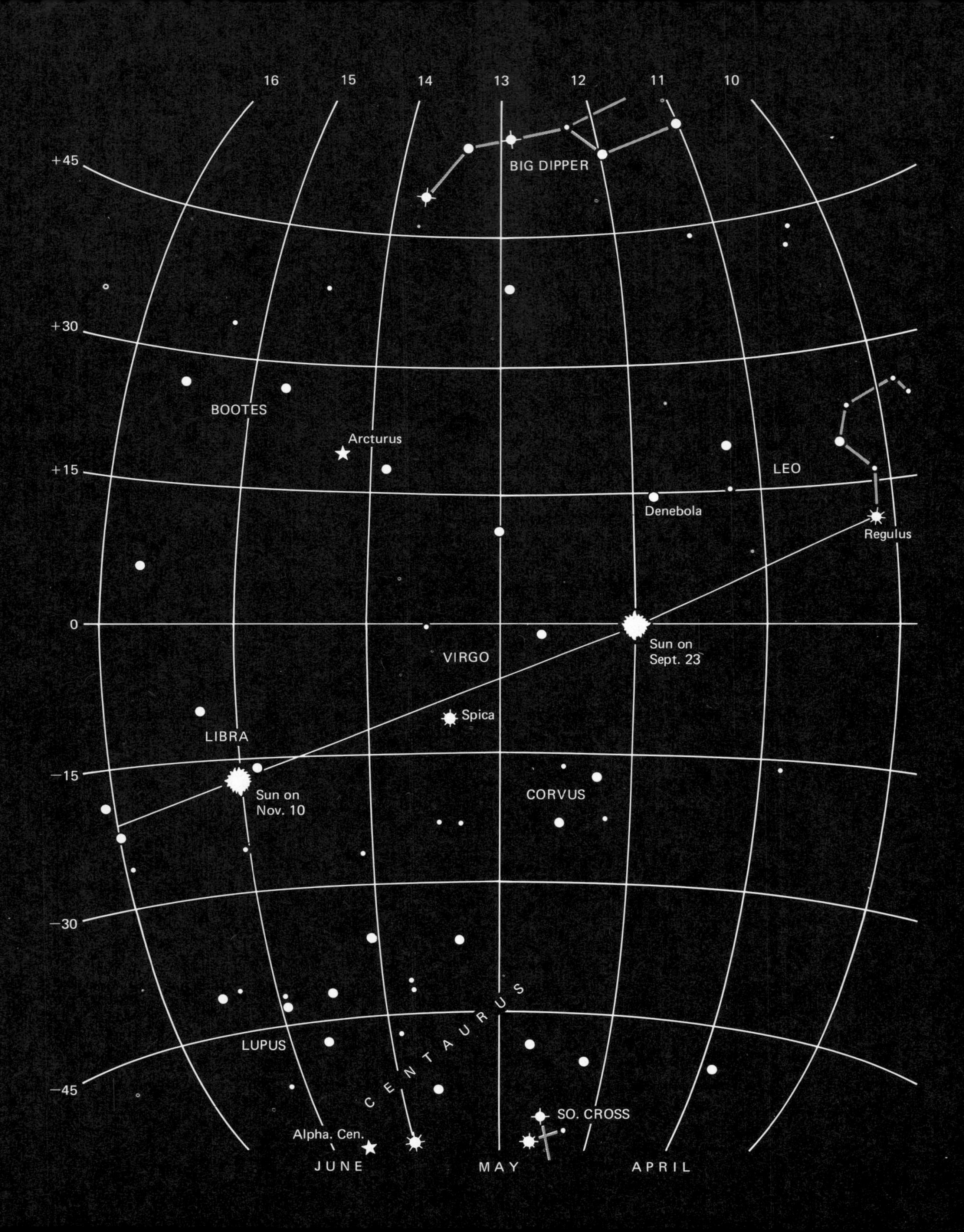
16
15
14
13
12
11
10
+45
+30
+15
0
−15
−30
−45
BIG DIPPER
BOOTES
Arcturus
LEO
Denebola
Regulus
VIRGO
Sun on
Sept. 23
Spica
LIBRA
Sun on
Nov. 10
CORVUS
CENTAURUS
LUPUS
SO. CROSS
Alpha. Cen.
JUNE
MAY
APRIL

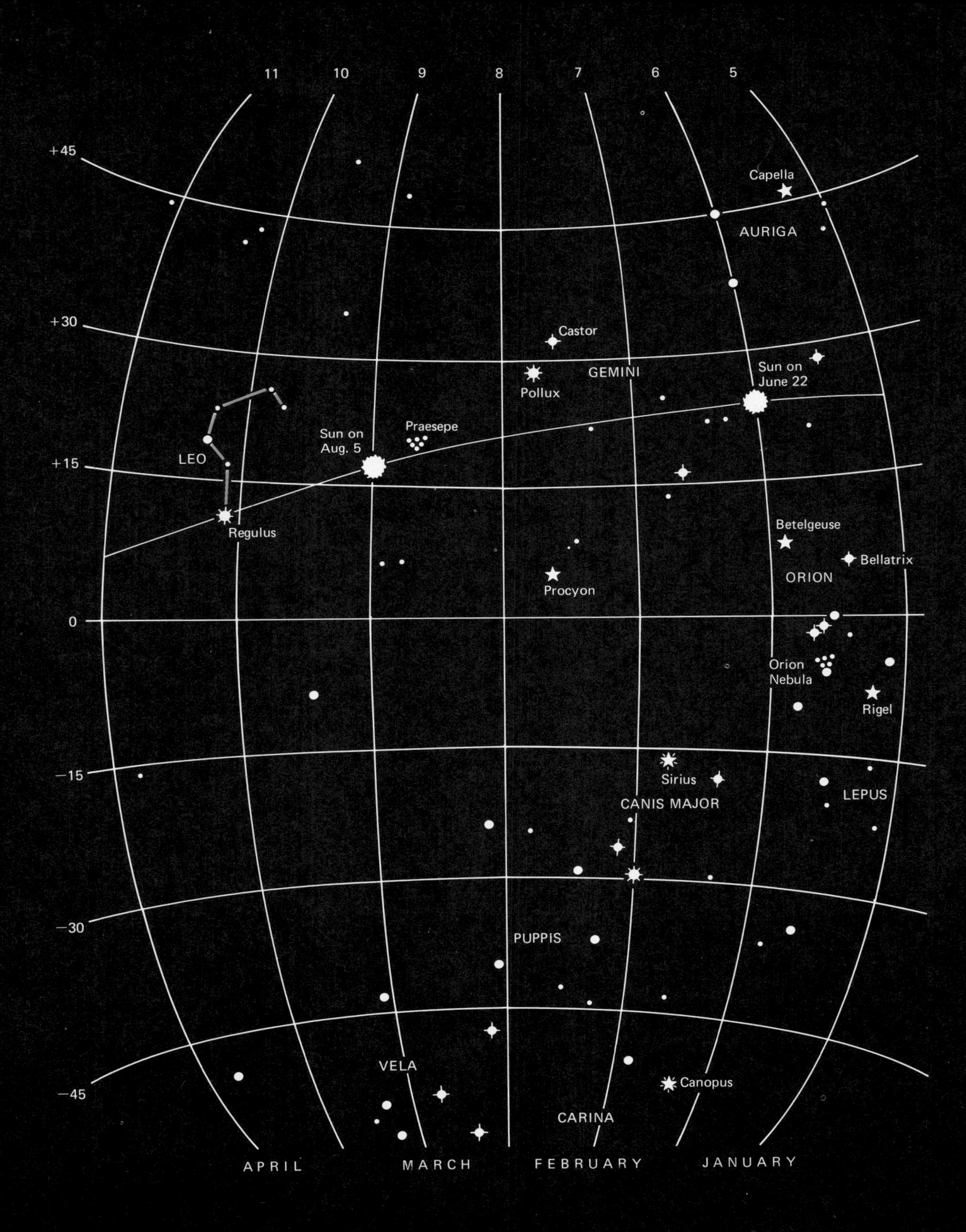

11
10
9
8
7
6
5
+45
+30
+15
0
−15
−30
−45
Capella
AURIGA
Castor
GEMINI
Pollux
Sun on
June 22
LEO
Praesepe
Sun on
Aug. 5
Regulus
Betelgeuse
Bellatrix
ORION
Procyon
Orion
Nebula
Rigel
Sirius
CANIS MAJOR
LEPUS
PUPPIS
VELA
Canopus
CARINA
APRIL
MARCH
FEBRUARY
JANUARY

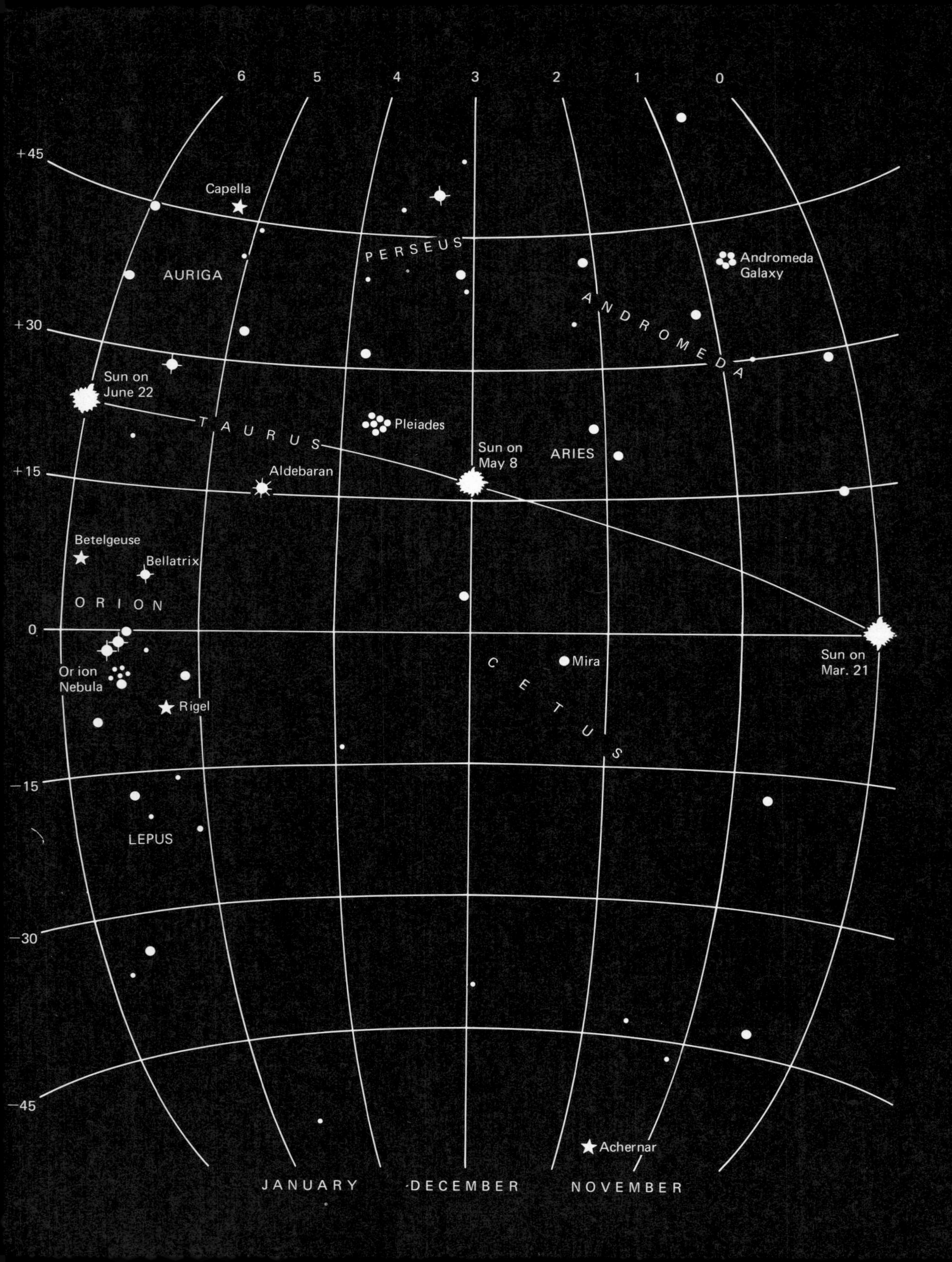

6
5
4
3
2
1
0
+45
+30
+15
0
−15
−30
−45
Capella
PERSEUS
AURIGA
Andromeda
Galaxy
ANDROMEDA
Sun on
June 22
TAURUS
Pleiades
Sun on
May 8
ARIES
Aldebaran
Betelgeuse
Bellatrix
ORION
Sun on
Mar. 21
Mira
Or ion
Nebula
CETUS
Rigel
LEPUS
Achernar
JANUARY
DECEMBER
NOVEMBER

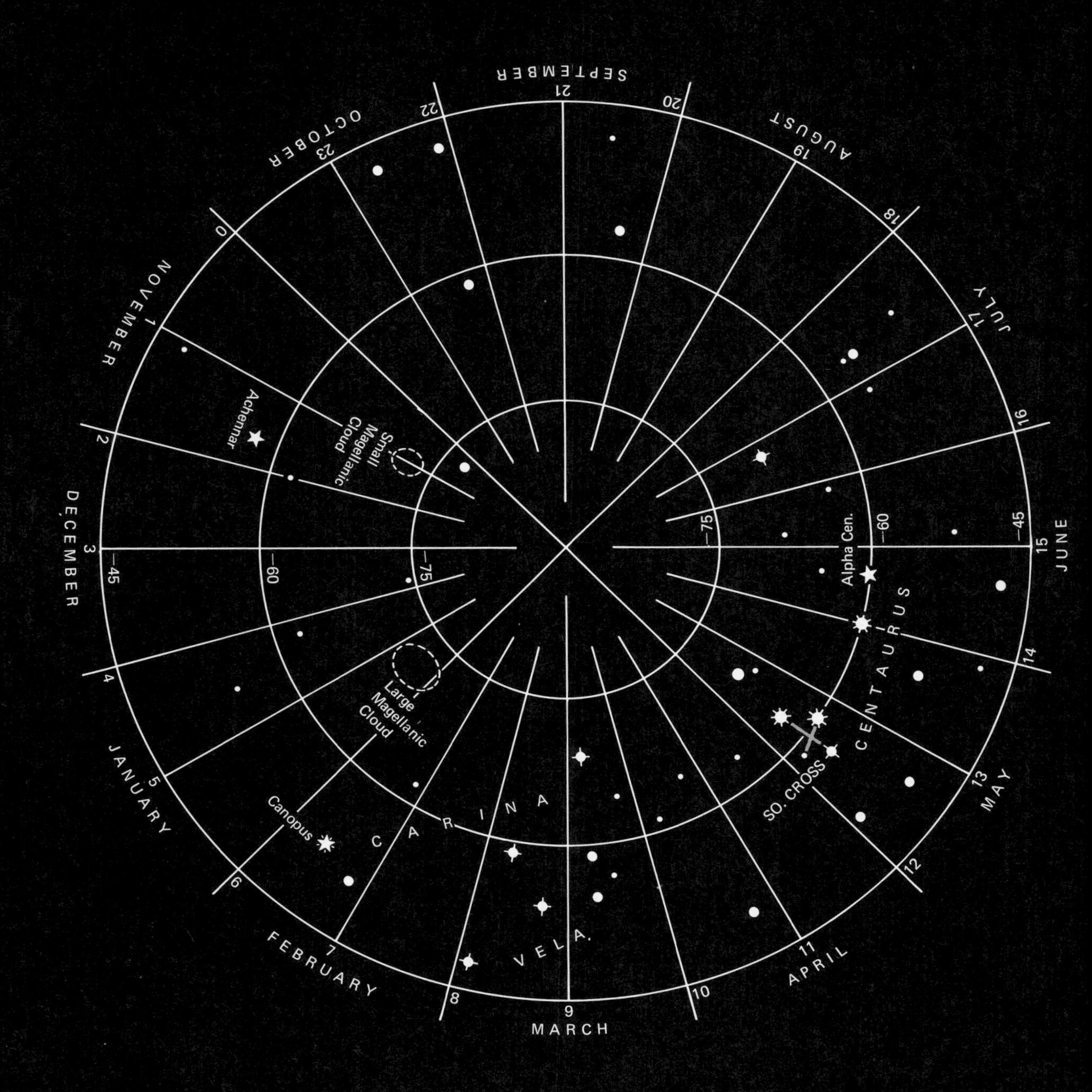

SEPTEMBER
AUGUST
JULY
JUNE
MAY
APRIL
MARCH
FEBRUARY
JANUARY
DECEMBER
NOVEMBER
OCTOBER
Achennar
Small Magellanic Cloud
Large Magellanic Cloud
Canopus
CARINA
VELA
SO. CROSS
CENTAURUS
Alpha Cen.

INDEX

ABCDEFGHIJ–VB–798